Waves in Gradient Metamaterials

Waves in Gradient Metamaterials

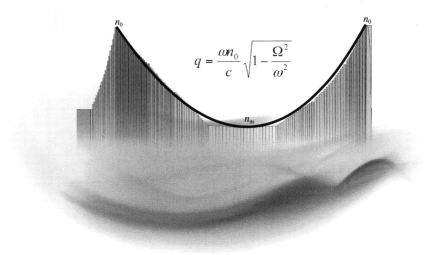

Alexander B Shvartsburg
Russian Academy of Sciences, Russia

Alexei A Maradudin
The University of California, Irvine, USA

NEW JERSEY · LONDON · SINGAPORE · BEIJING · SHANGHAI · HONG KONG · TAIPEI · CHENNAI

Published by

World Scientific Publishing Co. Pte. Ltd.
5 Toh Tuck Link, Singapore 596224
USA office: 27 Warren Street, Suite 401-402, Hackensack, NJ 07601
UK office: 57 Shelton Street, Covent Garden, London WC2H 9HE

British Library Cataloguing-in-Publication Data
A catalogue record for this book is available from the British Library.

WAVES IN GRADIENT METAMATERIALS
Copyright © 2013 by World Scientific Publishing Co. Pte. Ltd.

All rights reserved. This book, or parts thereof, may not be reproduced in any form or by any means, electronic or mechanical, including photocopying, recording or any information storage and retrieval system now known or to be invented, without written permission from the Publisher.

For photocopying of material in this volume, please pay a copying fee through the Copyright Clearance Center, Inc., 222 Rosewood Drive, Danvers, MA 01923, USA. In this case permission to photocopy is not required from the publisher.

ISBN 978-981-4436-95-3

In-house Editor: Song Yu

Typeset by Stallion Press
Email: enquiries@stallionpress.com

Printed in Singapore by World Scientific Printers.

*To the memory of a dear friend and colleague,
Tamara Aleksandrovna Leskova*

CONTENTS

1. Introduction … 1
   Bibliography … 12
2. Non-local Dispersion of Heterogeneous Dielectrics … 15
   2.1. Giant Heterogeneity-Induced Dispersion of Gradient Photonic Barriers … 18
   2.2. Reflectance and Transmittance of Subwavelength Gradient Photonic Barriers: Generalized Fresnel Formulae … 21
   2.3. Non-Fresnel Reflectance of Unharmonic Periodic Gradient Structures … 28
   Comments and Conclusions to Chapter 2 … 38
   Bibliography … 41
3. Gradient Photonic Barriers: Generalizations of the Fundamental Model … 43
   3.1. Effects of the Steepness of the Refractive Index Profile near the Barrier Boundaries on Reflectance Spectra … 44
   3.2. Asymmetric Photonic Barriers … 47
   3.3. Inverse Functions and Parametric Presentations — New Ways to Model the Photonic Barriers … 55
   Comments and Conclusions to Chapter 3 … 63
   Bibliography … 65

4. Resonant Tunneling of Light Through Gradient
   Dielectric Nanobarriers   67

   4.1. Transparency Windows for Evanescent
        Modes: Amplitude — Phase Spectra of
        Transmitted Waves   73
   4.2. Energy Transfer in Gradient Media
        by Evanescent Waves   78
   4.3. Weakly Attenuated Tunneling of Radiation
        Through a Subwavelength Slit, Confined
        by Curvilinear Surfaces   82
   Comments and Conclusions to Chapter 4   91
   Bibliography 92

5. Interaction of Electromagnetic Waves with
   Continuously Structured Dielectrics   93

   5.1. Reflectance/Transmittance Spectra of Lossy
        Gradient Nanostructures   97
   5.2. Interplay of Natural and Artificial Dispersion
        in Gradient Coatings   100
   5.3. EM Radiation in Gradient Superlattices   109
   Comments and Conclusions to Chapter 5   117
   Bibliography 119

6. Polarization Phenomena in Gradient Nanophotonics   121

   6.1. Wideangle Broadband Antireflection Coatings . . .   124
   6.2. Polarization-Dependent Tunneling of Light
        in Gradient Optics   133
   6.3. Reflectionless Tunneling and Goos–Hänchen
        Effect in Gradient Metamaterials   140
   Comments and Conclusions to Chapter 6   147
   Bibliography 150

7. Gradient Optics of Guided and Surface
   Electromagnetic Waves ... 151

   7.1. Narrow-Banded Spectra of S-polarized Guided
        Electromagnetic Waves on the Surface of a Gradient
        Medium: Heterogeneity-Induced Dispersion 153
        7.1.1. $0 < \omega < \Omega_c$ ... 159
        7.1.2. $\omega > \Omega_c$ ... 164
   7.2. Surface Electromagnetic Waves on a Curvilinear
        Interface: Geometrical Dispersion ... 165
   7.3. Surface Electromagnetic Waves on Rough Surfaces:
        Roughness-Induced Dispersion ... 177
        7.3.1. Periodically corrugated surfaces ... 182
        7.3.2. A randomly rough surface ... 187
   Comments and Conclusions to Chapter 7 ... 197
   Bibliography ... 199

8. Non-local Acoustic Dispersion of Gradient Solid Layers ... 203

   8.1. Gradient Acoustic Barrier with Variable Density:
        Reflectance/Transmittance Spectra of Longitudinal
        Sound Waves ... 207
   8.2. Heterogeneous Elastic Layers: "Auxiliary Barrier"
        Method ... 210
   8.3. Double Acoustic Barriers: Combined Effects
        of Gradient Elasticity and Density ... 217
   Comments and Conclusions to Chapter 8 ... 224
   Bibliography ... 225

9. Shear Acoustic Waves in Gradient Elastic Solids ... 227

   9.1. Strings with Variable Density ... 229
   9.2. Torsional Oscillations of a Graded Elastic Rod ... 234
   9.3. Tunneling of Acoustic Waves Through a Gradient
        Solid Layer ... 243

Comments and Conclusions to Chapter 9	246
Bibliography	248

10. Shear Horizontal Surface Acoustic Waves on Graded Index Media — 251

10.1. Surface Acoustic Waves on the Surface of a Gradient Elastic Medium	252
10.2. Surface Acoustic Waves on Curved Surfaces	262
10.2.1. Surface acoustic waves on a cylindrical surface	266
10.2.2. A variable radius of curvature	276
10.3. Surface Acoustic Waves on Rough Surfaces	284
10.3.1. A periodic surface	288
10.3.2. A randomly rough surface	292
Comments and Conclusions to Chapter 10	298
Bibliography	300

Appendix: Fabrication of Graded-Index Films — 303

A.1. Co-Evaporation	305
A.2. Physical and Chemical Vapor Deposition	306
A.3. Plasma-Enhanced Chemical Vapor Deposition (PECVD)	308
A.4. Pulsed Laser Deposition (PLD)	311
A.5. Graded Porosity	312
A.6. Ion-Assisted Deposition (IAD)	315
A.7. Sputtering	317
Comments and Conclusions to the Appendix	319
Bibliography	319

Index — 323

CHAPTER 1

INTRODUCTION

This book presents the first self-contained introduction to a newly shaping branch of applied physics on the frontier between modern photonics, electromagnetics, acoustics of heterogeneous media, and design of heterogeneous metamaterials and nanostructures with special properties unattainable in nature. This branch opens up the new avenues in the creation of miniaturized subwavelength systems, governing wave flows by means of gradient metamaterials characterized by technologically controlled smooth spatial distributions of dielectric or elastic parameters. Some results, obtained in these fields on a case-by-case basis, were dispersed hitherto in a number of journals devoted to quantum electronics, optics, radiophysics and material science; this work revealed the appearance of overlapping problems. However, the analysis of each such problem resembled sometimes a kind of art, and no standardized approach to these problems was elaborated. In contrast, the generality of physical fundamentals and the mathematical basis for wave phenomena in electromagnetics and acoustics of gradient media pervades this entire book. The current interest in these topics is threefold. First, the progress in fabrication of nanogradient structures attracts growing attention due to the ability of such structures to control the propagation of electromagnetic waves on and below the wavelength scale. Second, the possibility to replace metallic dispersive elements in traditional plasmonics-based photonic crystals by gradient glass and polymer films without free carriers indicates a new way for the creation of low cost components for optoelectronic circuitry. Third, the one-to-one correspondence between a series of electromagnetic and acoustic wave phenomena in gradient structures promotes parallel researches, making the corresponding problems more tractable and accelerating the design of innovative devices for science, technology, and defence.

Never before published information about new trends in directed energy and information transfer by electromagnetic and acoustic waves in heterogeneous media with giant artificial dispersion forms the "skeleton" of this book. These trends are represented in the framework of a unified consideration of physical fundamentals and mathematical basis of wave phenomena in different subfields of electromagnetics and acoustics, namely, in gradient nanophotonics for visible and infrared spectral ranges and acoustics of gradient solids. The scientific dividend, earned in this way, is provided by a series of flexible models of metamaterial structures, containing several free parameters, which are applicable for both of these subfields simultaneously.

This book is intended to bridge the gaps between the novel physical concepts of wave phenomena in gradient metamaterials and the use of these phenomena for innovative engineering purposes. The following are the main goals of this book:

1. To indoctrinate the concept of gradient wave barriers of finite thickness as the perspective dispersive elements for photonic and phononic crystals;
2. To highlight the effects of reflectionless tunneling of EM and acoustical waves, habitual namely to gradient metamaterial structures, as a powerful tool for governing radiation flows in wave circuitries;
3. To elaborate the standardized mathematical basis for optimization of parameters of gradient barriers, providing the amplitude-phase reflectance/transmittance spectra for any barrier and any needed spectral range.

The following key concepts, inspired by consideration of the above-mentioned goals, run throughout this book:

1. Strong heterogeneity-induced non-local dispersion of gradient dielectric layers, both normal and anomalous, which can be formed

in an arbitrary spectral range by means of an appropriate geometry of refractive index profile, the host material being given.
2. Flexibility of reflectance—transmittance spectra of heterogeneous dielectric nanofilms, controlled by the gradient and curvature of the refractive index profiles of photonic barriers, and the appearance of a cut-off frequency in barriers fabricated from dispersiveless host materials (generalized Fresnel formulae).
3. Effective energy transfer by evanescent EM modes, tunneling without attenuation through the "window of transparency" in gradient dielectric photonic barriers with concave spatial profiles of their dielectric permittivity. Manifestation of similar effects in gradient acoustic barriers, formed by solid layers with coordinate-dependent density and/or elasticity.
4. Formation of plasma-like dispersion in gradient dielectrics in arbitrary spectral ranges, imitating some dispersive properties of solid plasmas.
5. Scalability of results, obtained by means of exactly solvable models of gradient photonic barriers, between different spectral ranges and different thicknesses of barriers.

This book presents an example of how an appropriate mathematical language causes the creation of new physical insights, such as, e.g. non-local dispersion or reflectionless tunneling. Due to a coordinate-dependent velocity of wave propagation in these media the leading and trailing parts of waveforms travel with different velocities; this difference results in distortions of waveforms. In particular, when waves with harmonic envelopes are incident on the surface of a stratified medium, the spatial shapes of these envelopes inside the medium become non-sinusoidal. In this case the exact analytical solutions of the wave equations for stratified media, revealing features of the structure of the wave fields in such media, acquire fundamental importance. Introduction of "phase" coordinates, caused by the spatial distributions of refractive index, is shown to enable the solution of problems of propagation of waves of different physical nature in a similar fashion. The mathematical fundamentals of wave theory in gradient media, derived here "from first principles", are based on the exact analytical solutions of wave equations for media with continuous spatial variations of dielectric or elastic parameters. These solutions, obtained beyond of the scope of any truncations, perturbations, or other WKB-like assumptions about the smallness or slowness of variations of wave fields or media, are needed for analysis of wave phenomena in gradient metamaterial structures with subwavelength spatial scales, for which these simplifying assumptions become invalid. The wide classes of new simple exact analytical solutions of the Maxwell equations, as well as of the acoustic wave equations in solids, expressed sometimes via elementary functions, are used in standardized algorithms, represented in this book for solving the milestone wave problems for gradient metamaterials. If they are guided by these solutions, considered to be the etalon ones, researchers can decrease the risk of losing a great deal of physical insight in the processes of numerical simulations.

Theoretical fundamentals of wave physics for heterogeneous media have a time-honored history. While discussing the stationary propagation of a plane wave, several models of a coordinate-dependent velocity, allowing exact analytical solutions of wave equation, can be mentioned. Maxwell was among the first to consider inhomogeneous media in optics, when as long ago as in 1854, he described a lens, called "fish-eye". One of the first exactly solvable models was pioneered by Rayleigh as long ago as in 1880 in a solution of the wave equation describing sound propagation in a stratified atmosphere with a monotonic inverse square dependence of the sound velocity on the altitude $v(z)$ [1.1]. Later on these and more complicated models, containing, e.g. combinations of several exponents [1.2, 1.3], attracted much attention in fields as different as acoustics [1.4], plasma electromagnetics [1.5] and magnetic hydrodynamics [1.6]. Treatment of a series of such problems in the framework of the WKB — approximation was summarized in [1.7]. Unlike these models, describing natural media, the advent of lasers stimulated a burst of interest to man-made heterogeneous media, such as, e.g. thin transparent layers and multilayer systems, used as optical filters, polarizers, and antireflection coatings; during the past two decades, the engineered dielectric properties of thin films became a well-developed field of microelectronics and nanotechnology [1.8]–[1.9]. Modeling of continuous distributions of refractive index across the transparent film by means of step-like piece-wise profiles, developed in [1.10], was complemented by analysis of reflectance of films with sandwich-like metal-dielectric structures [1.11]. Moreover, the interest in optics of thin films is strengthened nowadays by the overall attention to the tunneling of photons through nanostructured metal films [1.12]. The enhanced optical transmittance of these periodic structures, supported by surface plasmon polariton modes in metal nanofilms, was shown to be an effective mechanism for resonant transfer of EM energy in optoelectronic devices [1.13]. These problems, covered by detailed reviews [1.14] and monographs [1.15], lie outside the scope of this book.

In contrast, the architecture of this book is determined by a step-by-step in-depth development of the concept of giant controllable dispersion of gradient metamaterials as a dominant paradigm, applied to nanooptics (Chs. 2–7) and acoustics of heterogeneous solids (Chs. 8–10). Harnessing of dielectrics with strong artificial heterogeneity-induced dispersion opens new horizons for the synthesis of optoelectronic systems. The physics of gradient dielectric photonic barriers and coatings, whose optical parameters are not connected with free carriers, constitutes the subjects of Chs. 2–7 of this book.

Chapter 2 is intended to provide the first acquaintance with the cornerstone concepts of gradient electromagnetics, such as an exactly solvable multiparameter flexible model of a gradient wave barrier, its characteristic frequencies and heterogeneity-induced dispersion, both normal and anomalous. These types of dispersion are inherent to concave and convex spatial distributions of the dielectric permittivity inside the barrier $\varepsilon(z)$; the characteristic frequencies are determined by the first and second derivatives of the profile $\varepsilon(z)$. Generalized Fresnel formulae for normal incidence, illustrating the decisive influence of the gradient and curvature of the refractive index distribution across the barrier on its reflectance/transmittance spectra, are derived "from first principles". To help the reader to assimilate this new approach, a simple "key model" visualizing the optical properties of the barrier by means of the simple elementary functions is suggested, and detailed examples of analytical calculations of these spectra for the suggested model are given; this model is used in several subsequent chapters. The classical Fresnel formulae for homogeneous films are shown to be limiting cases of the generalized expressions, obtained, related to the special case of vanishing heterogeneity. The list of exactly solvable models for gradient media is broadened in Ch. 3 by examples of both symmetric and asymmetric, as well as direct $(n = n(z))$, inverse $(z = z(n))$ and parametric dependencies of the refractive index n on the coordinate z inside the barrier.

The intriguing effect of reflectionless, or resonant, tunneling of EM waves through a transparent gradient dielectric barrier is considered in Ch. 4. Tunneling is a basic wave phenomenon, opening many tempting scientific and engineering perspectives. Pioneered by Gamow [1.16], this phenomenon gave rise to its numerous applications in optoelectronics, quantum mechanics, and solid state physics, as well as to the long term debates concerning the "superluminality" of tunneling processes [1.17, 1.18]. The concept of the nonlocality of matter-wave interactions linked this assumption with the inability to localize photons in space [1.19]. However, the exponentially small transfer of tunneling radiation through opaque barriers constricts the effectiveness of this transfer and impedes the observations of the tunneling effects for thick barriers. In contrast, the reflectionless tunneling of light in gradient media, visualized in Ch. 4, can provide a powerful tool for governing radiation flows in wave circuitry [1.20]. A new channel for energy and information transfer is shown to exist in media with some definite types of heterogeneity. Having nothing in common with the widely discussed surface plasmon polariton-assisted mechanism of tunneling of light through structured metal films [1.13], this effect is linked with the interference of evanescent and antievanescent waves, reflected from all the parts of gradient layer; here the reflection coefficient can vanish at some frequency, providing complete transparency for this frequency, and almost complete transparency in a finite spectral range surrounding this frequency. Amplitude-phase spectra of transmitted monochromatic CW flows, illustrating the location of these peaks of transparency and large phase shifts of tunneled waves, are presented in Ch. 4; the velocity of tunneling-assisted wave energy transfer through the gradient barrier in question is found to be subluminal. Perspectives for the design of miniaturized spectral filters and phase shifters, based on these results, are illustrated. Another promising effect, involving the effective energy transfer by evanescent waves, is connected with the reflectionless tunneling of a guided mode through a smoothly shaped narrowing in the waveguide, with the width of the slit formed by this narrowing being 2.5–3 times smaller than the wavelength.

Gradient antireflection coatings, based on the interplay of absorption, and natural and heterogeneity-induced dispersion, are considered in Ch. 5. Attention is given here to the interaction of light with superlattices, formed by gradient nanolayers. Propagation of waves through these structures is examined in the framework of a new exactly solvable model, generalizing the classical Kronig–Penny model in solid state physics [1.21]. The same approach is used for the calculation of reflectance spectra of superlattices containing metamaterials with a negative refractive index $n < 0$ [1.22–1.24] Unlike Chs. 2–5, focused on the normal incidence of waves on the barrier, Ch. 6 is devoted to the oblique incidence and, respectively, to the difference of reflectance/transmittance spectra for S- and P-polarized waves in gradient nanophotonics. Polarization-dependent tunneling of these waves is shown to provide the potential for new types of large angle polarizers and wide angle frequency-selective interfaces. The lateral displacement of rays in the traditional bi-prism configuration with an air-filled slit (Goos–Hänchen effect [1.18]) is reconsidered for the configuration where the slit is filled by a gradient dielectric multilayer structure; the complete tunneling-assisted transmission of radiation through this system is shown to promote the observation of the Goos–Hänchen effect.

By generalizing the previous analytical approach we examine in a straightforward way the influence of different dispersive parameters, characterizing the subsurface layer of a dielectric without free carriers, on the spectra of surface electromagnetic waves (Ch. 7). In contrast to Chs. 2–5 and 6, which treat problems of the normal and oblique incidence of radiation on the surface of a gradient layer, the propagation of waves along this surface is considered. Thus, the heterogeneity-induced dispersion of a dielectric is shown to support the new branches of S-polarized guided waves in narrow banded spectral intervals in the visible and infrared frequency ranges, if their frequencies are smaller, than a cut-off frequency, defined by the spatial distribution of $\varepsilon(z)$, decreasing in the depth of subsurface layer; this cut-off frequency is found by means of an exactly solvable model of the subsurface layer. The influence of dispersion, caused by periodically corrugated surfaces, as well as of roughness-induced dispersion, on the spectra of surface waves are investigated as well.

Emphasizing the paramount role of gradient nanophotonics, which dominates currently the frontline research in this field, one could be tempted to implement the corresponding ideas in other branches of cross-disciplinary physics; thus, the current penetration of concepts of heterogeneity-induced dispersion into the acoustics of gradient solids is described in Chs. 8–10. This penetration signifies the formation of a timely new topic-gradient acoustics of solid metamaterials. The main goals of this topic are connected with the creation of acoustic dispersion in the spectral range in need, forming the controlled reflectance and transmittance spectra of acoustical barriers as well as the resonant tunneling of sound through these barriers. Thus, Ch. 8 is centered on the reflection and transmission spectra of solid layers with coordinate-dependent distributions of density and/or elasticity, exemplified, e.g. by composite metamaterials, metallic glasses or alloys with graded concentration of components. Naming these layers by analogy with optics as "gradient acoustic barriers", and using again the "key model" of heterogeneity from Ch. 2, one can visualize the effects of non-local acoustical dispersion in arbitrary spectral ranges [1.25]. These effects form the physical basis for the elaboration of dispersive acoustical reflectors, frequency filters, antireflection coatings, phase shifters. As compared with gradient nanooptics, which deals with spatial variations of only one parameter-the refractive index, the problems of gradient acoustics of solids, operating in a general case with spatial distributions of two parameters-density and elastic Young's modulus — are more complicated [1.26]. Considering initially for simplicity elastic layers with variations of either density or Young's modulus, we find the spectra of longitudinal and shear sound waves, reflecting from these layers; the special algorithm, the "auxiliary barrier" method, is elaborated in Ch. 8 for the exact analytical solution of these problems [1.25]. Being armed by this knowledge, we examine the reflectance of the abovementioned gradient solid layer, characterized by independent distributions of both density and elasticity.

Some trends of the penetration of the key concepts of gradient electromagnetics to neighbouring fields of wave physics are shown in the Ch. 9, devoted to the calculation of the eigenfrequencies of acoustic oscillations for heterogeneous strings, layers, and rods. To illustrate the applicability of the methods of gradient optics to these problems, the ancient acoustic problem-calculation of eigenfrequencies of an elastic string with a heterogeneously distributed density — is investigated in Ch. 9. This problem was treated initially by Lagrange and Lord Rayleigh in the framework of perturbation theory. Their results were included in Rayleigh's classical book "Theory of Sound" [1.27]. However, borrowing the model of density distribution from the nanooptical problem in Ch. 2, one can obtain the eigenfrequencies of a heterogeneous string rigorously, without any assumptions concerning the smallness of its density or cross-section variations. The same model proves to be useful for the analysis of the eigenfrequencies of a gradient elastic layer [1.28]. The discrete spectrum of torsional oscillations, inherent to the chain of elastic rods with different lengths, is shown to represent the acoustic analogy of the Wannier–Stark ladder, found first in the quantum mechanics [1.29]. Since the image of an elastic string is widely used in analysis of the eigenoscillations of distributed systems in mechanics, instrumentation, and electrical engineering, the spectra obtained may become interesting for several subfields.

The far reaching analogies between electromagnetic and acoustic waves in graded media are continued in Ch. 10. Comparison of Secs. 7.1 and 10.1 illustrates the effects of heterogeneity-induced dispersion on the surface waves in optics and acoustics. Comparisons of

Secs. 7.2 and 10.2 as well as Secs. 7.3 and 10.3 visualize the one-to-one correspondence between the dispersive effects in spectra of light and sound surface waves, originated by curved and rough surfaces, respectively.

A detailed overview of each chapter is included to the relevant chapter's introduction; moreover, each chapter is completed by a list of references and a short conclusion marking the next steps in the development of the topics under discussion. A brief discussion of several methods for fabricating gradient nanofilms is given in the Appendix.

This book may be of interest to different groups of readers:

For scientists, interested in basic physics, this book points out a series of new perspectives, which remained hitherto unexplored. The concept of heterogeneity-induced dispersion, highlighted in this book, is shown to form the cornerstone of several subfields of optics and acoustics, which are now entering a new era. The remarkable similarity in dynamics of wave fields of different physical nature in gradient metamaterials promotes the development of a valid "wave intuition" in the solution of corresponding problems in cross-disciplinary physics. New intriguing horizons in the aforesaid subfields, based on peculiar reflectionless tunneling of electromagnetic and acoustic waves in gradient metamaterials, are expected to provide a key to the creation of new miniaturized dispersive elements for photonic and phononic crystals, bridging the gap between the current achievements in 1D and 2D structures and the future challenges in the development of 3D photonic devices. A new analytical approach to these problems stimulates the introduction of novel physical concepts and images to modern optoelectronics and acoustics.

For designers and users of communication systems this book can provide new trends in the miniaturization of wave circuitry elements up to subwavelength scales, accompanied by the related decrease of losses, which are needed in optical communication nets. Systematic use of the heterogeneity-induced dispersion concept is shown to enable the design of a series of key elements of such circuitry, such as, e.g. filters, phase shifters, frequency-selective interfaces, large angle polarizers, and lossless antireflection coatings of subwavelength spatial scales. Replacement of metallic foils, widely used in plasmonics-based photonic crystals, by thin gradient glass and polymer layers with heterogeneity-induced cut-off frequencies, mimicking the dispersive properties of solid plasmas, can decrease the cost of such crystals and broaden the list of materials used in optoelectronics. Scaling between the different spectral ranges of electromagnetic waves and, moreover, one-to-one mapping of similar effects of gradient electromagnetics to gradient acoustics and vice versa promotes parallel researches in these fields and optimization of the parameters of such wave systems.

For lecturers and students the logical scheme of the book and arrangement of information within each Chapter is suitable for didactic goals and instructive analysis of wave phenomena, giving a simple tool for the design of the next generation of wave-assisted devices, based on novel physical fundamentals. The detailed mathematical approach, using the recently discovered exact analytical solutions of wave equations in complex media, yields standardized algorithms for the calculation of wave field parameters in such media. Bringing computations to masses, these standardized algorithms may be useful for seminar discussions and self-study of wave phenomena in heterogeneous media. Presentation of classical results in optics of homogeneous media as limiting cases of effects of gradient electromagnetics, widely used in the book, promotes the elaboration of a fresh insight on the traditional optical concepts. Acquaintance with the rapid, explosion-like development of this field of science can convince newcomers that, despite the almost bicentennial history of discoveries, even the linear branch of wave physics is not exhausted yet, while the non-linear effects in the metamaterials are coming into play only now [1.30].

The Chs. 1–6, 8, the Secs. 9.1 and 9.3 are written by A.B. Shvartsburg, the Chs. 7, 10, Appendix and Sec. 9.2 are written by A.A. Maradudin. During the last years many people have contributed to the authors' understanding of this field either knowingly in scientific collaboration and discussions or unwittingly through their support and encouragement. Among those, who have contributed scientifically, we thank our colleagues N. Erokhin, R. Fitzgerald, V. Kuzmiak, G. Petite, J. Polanco, and M. Zuev. The discussions with Professors T. Arecchi, T. Brabec, P. Corcum, S. Haroche, V. Konotop, A. Migus, L. Vazquez, V. Veselago, and E. Wolf are highly appreciated. It is our pleasure to thank Professors V. Fortov, A. Kuz'michev, V. Vorob'ev, and L. Zelenii for their immutable interest to this work. Authors express the deep gratitude to Prof. E. Sheftel and Prof. O. Rudenko for their invisible influence on this work. We are much obliged to Prof. M.D. Malinkovich for providing the density profile of gradient photonic barrier, presented on the cover of this book. Special thanks go to Dr. O.D. Volpian and all colleagues from the R&D Company "Fotron—Auto Ltd", for carrying out the first experiments with dispersive optical nanofilms. Authors are indebted to Dr. E. Voroshilova and Dr. S. Lokshtanov for providing Figs. 2.5–2.7, 3.4–3.5, 4.2, 5.1 and 6.2–6.3, respectively, to Prof. M. Fitzgerald and Mr. J. Polanco for providing Figs. 7.1–7.6 and 7.8–7.10, to Dr. S. Chakrabarti and Dr. E. Chaikina for providing Figs. 7.11–7.14 and Figs. 7.6, 7.7, 9.2, 10.4–10.6, respectively.

Authors apologize to all, whose work in this rapidly developing field has not been assessed appropriately.

Authors.

Bibliography

[1.1] J. W. S. Rayleigh, *Proc. Lond. Math. Soc.* **11**, 51 (1980).
[1.2] V. L. Ginzburg, *Propagation of Electromagnetic Waves in a Plasma* (Pergamon Press, Oxford, 1968).
[1.3] E. W. Marchland, *Gradient Index Optics* (Academic Press, NY, 1978).
[1.4] L. M. Brekhovskikh and O. A. Godin, *Acoustics of Layered Media, I, II, Berlin* (Springer–Verlag, 1990).
[1.5] A. B. Mikhailovskii, "Electromagnetic Instabilities in an Inhomogeneous Plasma", A. Hilger, (1992).
[1.6] Z. E. Musielac, J. M. Fontenla and R. L. Moore, *Fhys. Fluids. B* **4**, 13 (1992).
[1.7] Yu. A. Kravtsov, *Geometric Optics in Engineering Physics* (Alpha Science International, Harrow, UK, 2005).
[1.8] P. Yeh, *Optical Waves in Layered Media* (NY, Wiley, 1988).
[1.9] C. Gomez-Reino, M. Peres and C. Bao, *Gradient Index Optics. Fundamentals and Applications* (Springer–Verlag, 2002).
[1.10] F. Abeles, *Progr. Opt.* (Ed. by E. Wolf), **2**, 249 (1963).

[1.11] S. A. Maier, *Plasmonics: Fundamentals and Applications* (Springer, NY, 2007).
[1.12] D. Sarid and W. Challener, *Modern Introduction to Surface Plasmons: Theory, Matematical Modelling and Applications* (Cambridge University Press, NY, 2010).
[1.13] E. Ozbay, *Science* **311**, 189 (2006).
[1.14] A. V. Zayatz, I. I. Smolyaninov and A. A. Maradudin, *Physics Reports* **408**, 131–314 (2005).
[1.15] Y. Toyozawa, *Optical Processes in Solids* (Cambridge University Press, 2003).
[1.16] G. A. Gamow, *Z. Phys.* **51**, 204–212 (1928).
[1.17] T. E. Hartman, *Appl. Phys.* **33**, 3427 (1962).
[1.18] V. S. Olkhovsky, E. Recami and J. Jakiel, *Physics Reports* **398**, 133–178 (2004).
[1.19] O. Keller, *JOSA B* **18**(2), 206–217 (2001).
[1.20] A. B. Shvartsburg and G. Petite, *Opt. Lett.* **31**, 1127–1130 (2006).
[1.21] A. B. Shvartsburg, V. Kuzmiak and G. Petite, *Physics Reports* **452**, 33–88 (2007).
[1.22] V. G. Veselago, *Soviet Phys. — Uspekhi*, **10**, 509–514 (1968).
[1.23] J. B. Pendry, D. Schurig and D. R. Smith, *Science* **312**, 1777 (2006).
[1.24] C. M. Soukolis and M. Wegener, *Nat. Photon.* **5**, 523–530 (2011).
[1.25] A. B. Shvartsburg and N. S. Erokhin, *Physics — Uspechi*, **54**, 627–646 (1911).
[1.26] A. V. Granato, *J. Phys. — Chem. Solids* **55**(10), 931–939 (1994).
[1.27] J. W. S. Rayleigh, *The Theory of Sound* (Macmillan & Co, London, 1937).
[1.28] L. D. Landau and E. M. Lifshitz, *The Theory of Elasticity* (Pergamon Press, Oxford, 1986).
[1.29] J. L. Mateos and G. Monsivias, *Physics A* **207**, 445–451 (1994).
[1.30] K. Bush, G. von Freumann, S. Linden, S. F. Mingaleev, L. Tkeshelashvili and M. Wegener, *Physics Reports* **444**, 101–202 (2007).

CHAPTER 2

NON-LOCAL DISPERSION OF HETEROGENEOUS DIELECTRICS

The salient features of the reflection and transmission of electromagnetic waves through gradient dielectric wave barriers are investigated here analytically in the framework of a simple one-dimensional problem. Let us consider propagation of a plane wave in a non-uniform non-magnetic dielectric, whose dielectric permittivity ε depends upon the z-coordinate. To stress the effects associated with the non-uniformity of ε, we assume initially, that the wave absorption and material dispersion are insignificant in the range of frequencies ω under consideration. In this case the dependence of $\varepsilon(z)$ in the transparency region $\varepsilon \geq 0$ can be represented as

$$\varepsilon(z) = n_0^2 U^2(z); \quad U|_{z=0} = 1. \tag{2.1}$$

Here n_0 is the refractive index of the medium at the boundary $z = 0$, and the dimensionless function $U^2(z)$ describes the spatial distribution of the permittivity.

The Maxwell equations for a linearly polarized wave with components E_x and H_y, traveling in the z-direction through the medium (2.1), are of the form

$$\frac{\partial E_x}{\partial z} = -\frac{1}{c}\frac{\partial H_y}{\partial t}; \tag{2.2}$$

$$-\frac{\partial H_y}{\partial z} = \frac{n_0^2 U^2(z)}{c}\frac{\partial E_x}{\partial t}. \tag{2.3}$$

The function $U^2(z)$ still remains unknown.

While discussing one-dimensional (1D) stationary propagation of an EM wave, several models of a coordinate-dependent dielectric permittivity $\varepsilon(z)$, allowing exact analytical solutions of Maxwell equations, can be outlined. One of the first such profiles was pioneered by Rayleigh in a solution of the wave equation for the acoustic problem of sound propagation with a coordinate-dependent velocity $v(z)$ [2.1]. Application of this result to the wave equation, governing the EM wave propagation, brought later the widely used model of $\varepsilon(z)$, expressed via the function $U_R^2(z)$ [2.2]

$$U_R^2(z) = \left(1 \pm \frac{z}{L}\right)^{-2}. \tag{2.4}$$

Almost half a century ago the development of optics and microwave physics stimulated the using of several exactly solvable models for U_1^2 [2.3], U_2^2 [2.4] and U_3^2 [2.5],

$$U_1^2 = 1 + \frac{z}{L}; \quad U_2^2 = \left(1 + \frac{z}{L}\right)^{-1}; \quad U_3^2(z) = \exp\left(-\frac{2z}{L}\right); \tag{2.5}$$

Here the characteristic length L is a unique free parameter.

It is remarkable, that the models (2.5), which at first appear are different, may be viewed as particular cases of one generalized distribution, containing two free parameters — the characteristic scale L and some real number m:

$$U^2(z) = \left(1 - \frac{1}{m-2}\frac{z}{L}\right)^{-m}; \quad m \neq 2. \tag{2.6}$$

The values m = 1 and m = -1 in (2.6) relate to the profiles $U_1^2(z)$ and $U_2^2(z)$ in (2.5).

Moreover, by using the classical formula

$$\lim\left(1 + \frac{1}{x}\right)^x \bigg|_{x \to \infty} = e, \tag{2.7}$$

one can show, that in the limit $m \to \infty$ the distribution (2.6) tends to the exponential profile U_3^2 (2.5).

Models (2.4)–(2.6) describe only monotonic variations of the refractive index. A model of a non-monotonic barrier, built from broken straight lines ("trapezoidal barrier" [2.6]) contains several unphysical angle points, formed by the crossing of these lines. The restricted flexibility of models (2.4) and (2.5), containing only one free parameter, hampers the optimization of regimes of wave propagation through the realistic gradient photonic barriers. To visualize the physically meaningful parameters, important for such optimization, one has to use more flexible models of photonic barriers.

Unlike the exactly solvable models of $U^2(z)$, given in (2.4) and (2.5), the series of more flexible exactly solvable models of gradient photonic barriers, containing several free parameters, will be obtained below by means of a special transformation of Maxwell equations (2.2) and (2.3) from physical space to a phase space. The corresponding analytical solutions of Maxwell equations (2.2) and (2.3) and reflectance/transmittance spectra for these barriers will be found on this way. The spectra obtained display the decisive role of heterogeneity-induced dispersion, depending upon the shape and spatial scales of the profile $U^2(z)$, in wave processes inside the gradient photonic barriers. Expressing the E_x and H_y components of the electromagnetic wave field in terms of some generating function Ψ permits reducing the system of two first-order equations (2.2) and (2.3) to one second-order equation for the Ψ function. This transform can be accomplished by two different methods:

1. The generating function Ψ is chosen so, that Eq. (2.2) becomes an identity, while the function Ψ is determined from Eq. (2.3).
2. The function Ψ that makes Eq. (2.3) an identity is determined from Eq. (2.2).

It is appropriate to consider separately the solutions and wave propagation regimes, obtained by these methods; this Section is focused on method 1. Section 2.1 is devoted to the artificial positive and negative dispersion for the simple model of gradient wave barriers. Generalized Fresnel formulae for these barriers are derived in Sec. 2.2. Multilayer systems, formed by combination of simple barriers, examined in Sec. 2.1, are considered in Sec. 2.3.

2.1. Giant Heterogeneity-Induced Dispersion of Gradient Photonic Barriers

To operate in the framework of the abovementioned method 1, let us express the field components in terms of the vector potential $\vec{A}$

$$\vec{E} = -\frac{1}{c}\frac{\partial \vec{A}}{\partial t}; \quad \vec{H} = \operatorname{rot}\vec{A}. \tag{2.8}$$

In the geometry of the problem under study the vector potential $\vec{A}$ has only one component $A_x (A_y = A_z = 0)$. Presenting the A_x component as a product of some normalization constant A_0 and dimensionless generating function Ψ permits Eq. (2.3), which determines the function Ψ, to be written as

$$\frac{\partial^2 \Psi}{\partial z^2} - \frac{n_0^2 U^2(z)}{c^2}\frac{\partial^2 \Psi}{\partial t^2} = 0. \tag{2.9}$$

One can see from Eq. (2.9), that an unknown function Ψ obeys a wave equation with a coordinate-dependent speed of wave propagation.

Equation (2.9) can be solved by introducing new functions F and Q and a new variable η [2.7]:

$$\Psi = \frac{F}{\sqrt{U(z)}}; \quad U(z) = \frac{1}{Q(z)}; \quad \eta = \int_0^z U(z_1)dz_1. \tag{2.10}$$

In this case Eq. (2.9) takes the form

$$\frac{\partial^2 F}{\partial \eta^2} - \frac{n_0^2}{c^2}\frac{\partial^2 F}{\partial t^2} = F\left[\frac{1}{2}Q\frac{d^2Q}{dz^2} - \frac{1}{4}\left(\frac{dQ}{dz}\right)^2\right]. \tag{2.11}$$

The coordinate-dependent coefficient is eliminated from the left side of Eq. (2.11), but the function $Q(z)$ still remains unknown.

Consider, for instance, a simple particular solution of Eq. (2.11), which corresponds to the function $Q(z)$ defined by the conditions

$$\frac{1}{2}Q\frac{d^2Q}{dz^2} - \frac{1}{4}\left(\frac{dQ}{dz}\right)^2 = p^2; \tag{2.12}$$

Here p^2 is some constant, which will be defined below. Assuming, that the time dependence of the field F is harmonic, Eq. (2.11) can be rewritten in a form

$$\frac{\partial^2 F}{\partial \eta^2} + F\left(\frac{n_0^2 \omega^2}{c^2} - p^2\right) = 0; \qquad (2.13)$$

Thus, owing to the transformations (2.10), Eq. (2.9) is reduced to a standard equation with constant coefficients [2.7].

Introducing the quantities, linked with the constant p^2,

$$q = \frac{\omega n_0}{c} N; \quad N = \sqrt{1 - \frac{\Omega^2}{\omega^2}}; \quad \Omega^2 = \frac{c^2 p^2}{n_0^2}; \qquad (2.14)$$

one can write the solution of Eq. (2.13) in the form of a harmonic wave, traveling along the η-direction: $F \approx \exp[i(q\eta - \omega t)]$; the quantities q and N in (2.14) play the role of wavenumber and refractive index respectively in η-space. Substitution of this solution into (2.10) yields the expression for the generating function Ψ:

$$\Psi = \frac{\exp[i(q\eta - \omega t)]}{\sqrt{U(z)}}. \qquad (2.15)$$

Till now the profile $U(z)$ remains unknown. This profile, expressed via the function $Q(z)$ (2.10), can be found from the solution of Eq. (2.12) in the form

$$U(z) = \left(1 + \frac{s_1 z}{L_1} + \frac{s_2 z^2}{L_2^2}\right)^{-1}; \quad s_1 = 0, \pm 1; \quad s_2 = 0, \pm 1. \qquad (2.16)$$

containing two arbitrary spatial scales L_1 and L_2. Here the case $s_1 = -1$, $s_2 = +1$ corresponds to a convex profile, while the case $s_1 = +1$, $s_2 = -1$ describes a concave one (Fig. 2.1). Let us stress that, unlike profiles (2.4) and (2.5), profile (2.16) relates to a non-monotonic distribution of the dielectric permittivity inside the gradient barrier. In the case of opposite signs of s_1 and s_2 profile (2.16) has either a maximum ($s_1 = -1$, $s_2 = +1$) or a minimum ($s_1 = +1$, $s_2 = -1$) with a value U_m. The scales L_1 and L_2 are linked in these cases with the layer's thickness d and the extremal value U_m via the

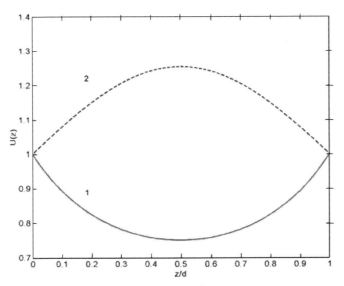

Fig. 2.1. Profiles of dielectric permittivity $U^2(z)$ vs the normalized thickness in the gradient barrier (2.16).

gradient parameter y:

$$U_m = (1 + s_1 y^2)^{-1} \quad y = L_2/2L_1 \quad L_2 = d(2y)^{-1}; \quad L_1 = d(4y^2)^{-1}. \tag{2.17}$$

Substitution of $U(z)$ (2.16) into (2.12) yields the value of the constant p^2:

$$p^2 = \frac{s_1^2}{4L_1^2} - \frac{s_2}{L_2^2}; \tag{2.18}$$

Thus, the model (2.16) has four free parameters — the layer's thickness d, the extremal value U_m and the signs $s_{1,2}$.

Subject to the shape of the profile $U(z)$ (Fig. 2.1), the sign s_2 in (2.18) may be positive, negative or equal to zero. These possibilities relate to different types of non-local dispersion, determined by the parameter N in (2.14), which may be viewed as the refractive index in dispersive η-space:

a. concave profile ($s_2 = -1$); using the quantities y and barrier width d (2.17), one can write the "refractive index" N and characteristic frequency Ω in the forms:

$$N = \sqrt{1-u^2}; \quad u = \frac{\Omega_1}{\omega}; \quad \Omega_1 = \frac{2cy\sqrt{1+y^2}}{n_0 d}. \tag{2.19}$$

Expression (2.19) for N resembles the refractive index for a plasma, where the cut-off frequency Ω_1 is analogous to the plasma frequency. The quantity N increases with the increase of the frequency ω; here the condition $\omega \geq \Omega_1$ is assumed to be fulfilled. This condition is known to determine the negative (normal) dispersion. The opposite case, $\omega < \Omega_1$, will be discussed in Ch. 4.

b. convex profile ($s_2 = +1$); in this case one can obtain by analogy with (2.20):

$$N = \sqrt{1+u^2}; \quad u = \frac{\Omega_2}{\omega}; \quad \Omega_2 = \frac{2cy\sqrt{1-y^2}}{n_0 d}; \quad y^2 \leq 1. \tag{2.20}$$

Unlike the monotonic dependence $\Omega_1(y)$, related to a concave profile $U(z)$, the function $\Omega_2(y)$ has a maximum at $y^2 = 0.5$ (2.19). The expression (2.20), describing the increase of N due to a decrease of the frequency ω, relates to the case of positive (anomalous) dispersion. These effects of artificial heterogeneity-induced dispersion are shown below to play the fundamental role in all the complex of wave phenomena in gradient barriers.

Thus, the EM fields in gradient wave barriers, described by different modifications of the $\varepsilon(z)$ profile (2.16), can be represented via the amplitude-modulated harmonic waves (2.15) in dispersive η-space. One can now use this representation for the calculation of the complex reflectance/transmittance coefficients characterizing these barriers.

2.2. Reflectance and Transmittance of Subwavelength Gradient Photonic Barriers: Generalized Fresnel Formulae

The standard way to examine the reflectance/transmittance properties of a plane layer contains the consideration of the complete field inside the layer, formed by the interference of forward and backward waves, and the use of the sing of continuity conditions for electric and magnetic field components at the boundaries of this layer and the surrounding media. Making use of the solution (2.15), one can write the generating function Ψ inside the barrier (2.16) in a form of superposition of forward and backward waves:

$$\Psi = \frac{A}{\sqrt{U(z)}}[\exp(iq\eta) + Q\exp(-iq\eta)]\exp(-i\omega t). \qquad (2.21)$$

Here A is some normalization constant, the dimensionless quantity Q describes the reflectivity of the far boundary $z = d$. Let us suppose, that the wave $E = E_0 \exp[i(kz - \omega t)]$ is incident on the barrier interface $z = 0$ from the air ($z < 0$). To find the complex reflection coefficient R, one has to use the continuity conditions on the interfaces $z = 0$ and $z = d$. Substituting (2.21) to (2.8) and omitting for simplicity the exponential factor $\exp(-i\omega t)$, one can calculate the electric E_x and magnetic H_y components of the EM field inside the barrier. The continuity condition for E_x on the plane $z = 0$ is

$$E_0(1 + R) = E_1(1 + Q). \qquad (2.22)$$

Use of the derivatives of profile (2.16),

$$\frac{1}{U^2}\frac{dU}{dz}\bigg|_{z=0} = -\frac{s_1}{L_1}, \quad \frac{1}{U^2}\frac{dU}{dz}\bigg|_{z=d} = \frac{s_1}{L_1}, \qquad (2.23)$$

brings the continuity condition for H_y into the form

$$ikE_0(1 - R) = \frac{i\omega}{c}E_1\left[-\frac{i\gamma s_1}{2}(1 + Q) + n_e(1 - Q)\right]; \qquad (2.24)$$

$$k = \frac{\omega}{c}; \quad \gamma = \frac{c}{\omega L_1}; \quad n_e = n_0 N. \qquad (2.25)$$

It is noteworthy that the parameter n_e (2.25) can be viewed as the effective refractive index of the gradient material, describing its heterogeneity-induced dispersion. Division of (2.22) by (2.24) yields the expression for the reflection coefficient R:

$$R = \frac{1 + \frac{i\gamma s_1}{2} - n_e(1-Q)(1+Q)^{-1}}{1 - \frac{i\gamma s_1}{2} + n_e(1-Q)(1+Q)^{-1}}. \quad (2.26)$$

The unknown parameter Q in (2.26) can be found from the boundary conditions on the far side of the barrier, $z = d$. Assuming the barrier to be located on the surface of a half-space, formed by a homogeneous lossless dispersiveless dielectric with refractive index n, one can write the electric component of the EM field in this half-space in the form $E = E_2 \exp[i(k_2 z - \omega t)]$. The continuity conditions on the plane $z = d$ are:

$$E_1[\exp(iq\eta_0) + Q\exp(-iq\eta_0)] = E_2; \quad (2.27)$$

$$ikE_1 \left\{ \frac{i\gamma s_1}{2}[\exp(iq\eta_0) + Q\exp(-iq\eta_0)] \right.$$

$$\left. + n_e[\exp(iq\eta_0) - Q\exp(-iq\eta_0)] \right\} = ik_2 E_2; \quad (2.28)$$

$$k_2 = \frac{\omega}{c} n; \quad \eta_0 = \int_0^d U(z)dz. \quad (2.29)$$

Division of (2.27) by (2.28) leads to the value of the dimensionless parameter Q:

$$Q = \exp(2iq\eta_0) \frac{n_e + \frac{i\gamma s_1}{2} - n}{n_e - \frac{i\gamma s_1}{2} + n}. \quad (2.30)$$

Finally, substitution of (2.30) to (2.26) yields the complex reflection coefficient of the gradient photonic barrier [2.7]:

$$R = \frac{\sigma_1 + i\sigma_2}{\chi_1 + i\chi_2} = |R|\exp(i\phi_r);$$

$$\sigma_1 = t\left(n + \frac{\gamma^2}{4} - n_e^2\right) - n_e \gamma s_1; \quad \sigma_2 = -(n-1)\xi; \quad t = \mathrm{tg}(q\eta_0);$$

$$\chi_1 = t\left(n - \frac{\gamma^2}{4} + n_e^2\right) + n_e \gamma s_1; \quad \chi_2 = (n+1)\xi; \quad \xi = n_e - \frac{\gamma s_1 t}{2}.$$

$$(2.31)$$

Equation (2.31) presents the generalized Fresnel formula for the reflection coefficient of a single layer for both concave ($s_1 = 1$) and convex ($s_1 = -1$) profiles $U(z)$, shown in Fig. 2.1. The reflectance spectrum, described by (2.31), is characterized by non-local dispersion, determined by the dependence of the parameters n_e and γ (2.25) on the normalized frequency u (2.17). The phase path length η_0, calculated from (2.29), as well as the parameter y (2.17), are different for concave and convex profiles. The explicit expressions for the quantities n_e, u, y, η_0 and $q\eta_0$, determined for concave and convex profiles $U(z)$, are listed below:

Concave Profile $U(z)$.

a. $n_e = n_0\sqrt{1 - u^2}$;

b. $u = \dfrac{2cy\sqrt{1+y^2}}{n_0 d\omega}$;

c. $y = \sqrt{\dfrac{1}{U_{\min}} - 1}$;

d. $\gamma = \dfrac{2un_0 y}{\sqrt{1+y^2}}$;

e. $\eta = \dfrac{L_2}{2\sqrt{1+y^2}}\ln\left(\dfrac{1+y_+ z/L_2}{1-y_- z/L_2}\right)$;

f. $\eta_0 = \dfrac{d}{2y\sqrt{1+y^2}}\ln\left(\dfrac{y_+}{y_-}\right)$;

g. $q\eta_0 = \dfrac{\sqrt{1-u^2}}{u}\ln\left(\dfrac{y_+}{y_-}\right)$;

Convex Profile $U(z)$.

a. $n_e = n_0\sqrt{1 + u^2}$;

b. $u = \dfrac{2cy\sqrt{1-y^2}}{n_0 d\omega}$;

c. $y = \sqrt{1 - \dfrac{1}{U_{\max}}}$;

d. $\gamma = \dfrac{2un_0 y}{\sqrt{1-y^2}}$;

e. $\eta = \dfrac{L_2}{\sqrt{1-y^2}}\operatorname{arctg}\left(\dfrac{z/L_2\sqrt{1-y^2}}{1-yz/L_2}\right)$;

f. $\eta_0 = \dfrac{d}{y\sqrt{1-y^2}}\operatorname{arctg}\left(\dfrac{y}{\sqrt{1-y^2}}\right)$;

g. $q\eta_0 = \dfrac{2\sqrt{1+u^2}}{u}\operatorname{arctg}\left(\dfrac{y}{\sqrt{1-y^2}}\right)$.

(2.32)

The dimensionless parameters $y_\pm$ are

$$y_\pm = \sqrt{1+y^2} \pm y; \quad y_+ y_- = 1. \qquad (2.33)$$

Using the expressions for the factor Q (2.30) and the reflection coefficient R, (2.31), one can calculate the transmission coefficients with respect to the electric T_E and magnetic T_H components of the EM field for the lossless barrier:

$$T_E = \frac{2in_e}{\cos(q\eta_0)(\chi_1 + i\chi_2)} = |T_E|\exp(i\phi_t); \quad T_H = nT_E. \quad (2.34)$$

Substitution of expression for R (2.31) into (2.34) brings the formula for the transmission coefficient with respect to energy $|T|^2 = T_E T_H^*$ into an explicit form:

$$|T|^2 = \frac{4n_e^2 n(1+t^2)}{\left|t\left(n - \frac{\gamma^2}{4} + n_e^2\right) + n_e \gamma s_1\right|^2 + (n+1)^2 \left(n_e - \frac{\gamma s_1 t}{2}\right)^2}. \quad (2.35)$$

All the quantities in (2.35) as well as in (2.31) have to be chosen for concave and convex profiles of $U(z)$ according to the definitions (2.32). The reflection and transmission coefficients for stratified media are known to be unique [2.8].

Examples of spectra of reflectance and transmittance of gradient photonic barriers with convex and concave profiles of the refractive index, related to different values of the gradient parameter y and the substrate refractive index n, are shown in Figs. 2.2 and 2.3. These spectra as well as other reflectance and transmittance spectra for the gradient structures discussed in this book, are presented for some given values of the refractive indices n_0 and n and gradient parameter y as functions of the normalized frequency of the incident wave u; the values of $|R|^2$ and $|T|^2$ from these spectra relate to the frequency ω, determined by expressions, following from the definitions of the normalized and characteristic frequencies u and $\Omega_{1,2}$ (2.19)–(2.20):

$$\frac{\omega d}{c} = \frac{2y\sqrt{1+y^2}}{n_0 u}; \quad (2.36)$$

This universal nature of the spectra in Figs. 2.2 and 2.3 allows to use each value of $|R(u)|^2$ and $|T(u)|^2$ for analyses of the propagation of different wavelengths through barriers with a given profile

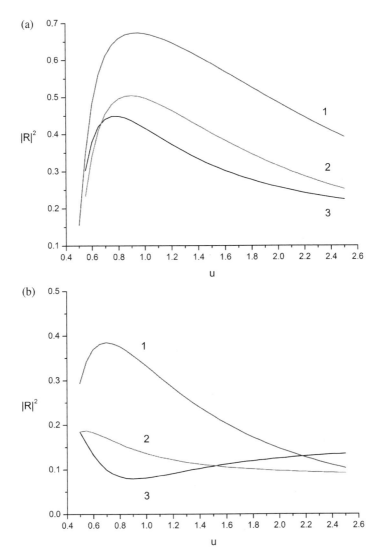

Fig. 2.2. Reflection coefficient $|R|^2$ for a convex profile, $n_0 = 1.47$. (a): $y = 0.75$. (b): $y = 0.577$. Curves 1, 2 and 3 correspond to the values of $n = 1$, 1.8 and 2.3 respectively.

$U(z)$ but different thicknesses. Thus, e.g. fixing the value $u = 0.5$ we find for graph 1, depicted in Fig. 2.3(b), the value of the transmission coefficient $|T|^2 = 0.835$. This value remains valid for all the barriers (2.16) with $n_0 = 2.3$ and depth of refractive index

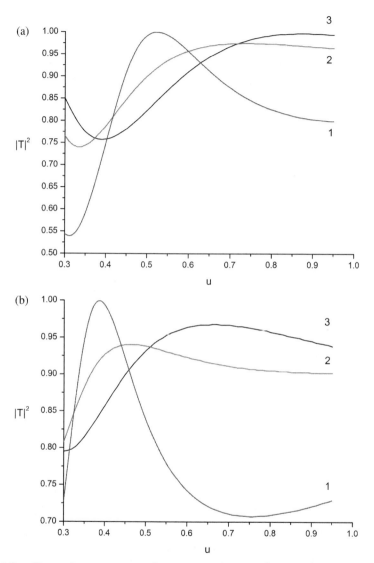

Fig. 2.3. Transmittance spectra for a concave barrier ($n_0 = 2.3$) vs the normalized frequency y for different values of the substrate refractive index n; curves 1, 2 and 3 correspond to $n = 1$; 1.8 and 2.3 respectively. (a) and (b) correspond to the values of the gradient parameter $y = 0.75$ and $y = 0.577$, respectively.

modulation $n/n_0 = (1 + y^2)^{-1} = 0.64$, meanwhile the barrier's width d and frequency ω may be distinguished, but are linked by the relation following from (2.36): $\omega d/c = 0.58$. According to this relation the propagation of waves with wavelengths $\lambda = 800$ (620) nm through the barrier with width $d = 74$ (57) nm is characterized by the equal transmission coefficients $|T|^2 = 0.835$. This similarity proves to be useful for optimization of parameters of gradient optical structures.

2.3. Non-Fresnel Reflectance of Unharmonic Periodic Gradient Structures

Periodic dielectric multilayer nanostructures possess a considerable flexibility of reflection-transmission properties. The traditional multilayer structure consists of alternating homogeneous dielectric layers of two materials with high and low refractive indices n_1 and n_2 and layer thicknesses d_h and d_l respectively [2.9]. In contrast to these structures, gradient barriers can be designed from alternating concave or convex profiles as well as from more complicated configurations, e.g. alternating gradient and homogeneous barriers. The series of dielectric nanostructures, containing adjacent gradient barriers, can form periodic systems with unharmonic profiles of the refractive index and unusual reflectance/transmittance spectra. Side by side with the reflection of waves due to the discontinuity of the refractive index at the boundaries of films, habitual to adjacent homogeneous films as well, the reflectance of waves from a gradient structure is influenced by discontinuities of the gradient and curvature of the profiles $n(z)$ on these boundaries. The interplay of all these phenomena provides a huge diversity of reflectance/transmittance spectra of unharmonic periodic and sandwich structures.

Rigorously speaking, the generalized Fresnel formulae, obtained in Sec. 2.2 for one gradient film, located on a substrate, include the contributions of discontinuities of both refractive index $U(z)$ and its gradient and curvature on the boundaries of the film with homogeneous media — air and substrate. However, to emphasize the importance of effects of both gradient and curvature discontinuities to the layer's reflectance, we will examine these effects separately, considering two configurations of adjacent films, located on a homogeneous substrate with refractive index n:

1. The gradients $n(z)$ on the boundary of adjacent films $z = d$ are unequal, meanwhile the curvatures of profile $n(z)$ on this boundary are equal [2.10].

To illustrate the details of such a generalization let us start from a stack of similar adjacent concave barriers (Fig. 2.4(a)), supported by a thick homogeneous dielectric substrate with refractive index n, located on the far side of system. Considering the normalized profile of refractive index $U(z)$ (2.16), one can see, that the values of $\mathrm{grad}\, U$, expressed in normalized coordinates $x = z/d$, possess a jump on the boundary $x = 1$ from $dU/dx|_{1-0} = 4y^2$ to $dU/dx|_{1+0} = -4y^2$, while the curvatures of both concave profiles on this boundary remain equal: $K_1 = 8y^2(4y^2 + 1)(1 + 16y^4)^{-3/2}$. Attributing the number $m = 1$ to the first layer at the far side of the stack, we will find the parameter Λ_1 describing the interference of forward and backward waves inside the first barrier; this parameter, connected with the value Q (2.30) by the relation $\Lambda_1 = (1 - Q)(1 + Q)^{-1}$, is

$$\Lambda_1 = \frac{n - \frac{i\gamma s_1}{2} - i n_e t_\pm}{n_e - \left(in + \frac{\gamma s_1}{2}\right) t_\pm}. \qquad (2.37)$$

Here $t_\pm = \mathrm{tg}(q\eta_0)$, where the values $q\eta_0$, as well as the quantities γ and n_e, are defined for concave (t_+) and convex (t_-) profiles in (2.32). It is worthwhile to introduce, by analogy with parameter Λ_1, determined by the first layer, the analogous parameter Λ_m (m $\geq$ 1), describing the interference of forward and backward waves in the m-th layer by means of a factor Q_m (2.21):

$$\Lambda_m = \frac{1 - Q_m}{1 + Q_m}; \qquad (2.38)$$

The formula (2.26), defining the reflectance of one gradient barrier, relates to the case $m = 1$. Using the continuity conditions on each boundary between adjacent layers, we'll find a simple recursive

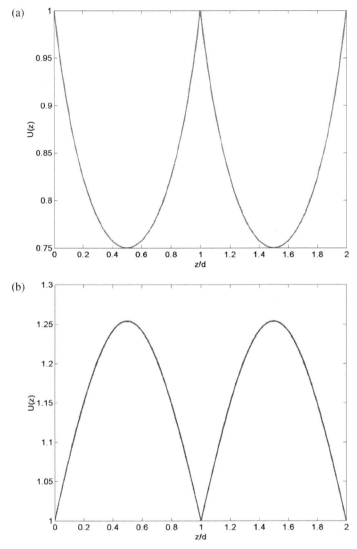

Fig. 2.4. Unharmonic periodical structure, formed from gradient barriers: normalized profiles of the refractive index $U(z)$ are plotted vs the normalized coordinate z/d. Figures 2.4(a) and 2.4(b) show the parts of periodic structures, consisting of similar barriers with normal and anomalous non-local dispersion, respectively.

relation between parameters Λ_m and Λ_{m-1}:

$$\Lambda_m = \frac{n_e(\Lambda_{m-1} - it_\pm) - i\gamma s_1}{n_e(1 - i\Lambda_{m-1}t_\pm) - \gamma t_\pm s_1}; \quad m \geq 2. \tag{2.39}$$

Since the wave is incident from $z < 0$ on the interface of m-th layer, one can find the reflectance of the entire periodic structure, from a formula, that generalizes the corresponding expression for a single gradient barrier (2.26):

$$R_m = \frac{1 + \frac{i\gamma s_1}{2} - n_e \Lambda_m}{1 - \frac{i\gamma s_1}{2} + n_e \Lambda_m}. \tag{2.40}$$

Amplitude-phase spectra of the reflectance of periodic structures, containing several gradient barriers with concave (Fig. 2.4(a)) and convex (Fig. 2.4(b)) profiles $n(z)$, are shown in Figs. 2.5–2.6. It is worthwhile to emphasize some salient features of these spectra:

a. Spectral maxima, as well as spectral minima, of periodic gradient systems are non-equidistant.
b. A narrow peak of total reflectance (spectral filtration) arises for the multilayer nanostructure with normal non-local dispersion near the frequency $u = 0.22$.
c. Reflectance spectra of multilayer nanostructures contain frequency ranges of finite widths between $0.41 < u < 0.48$ (normal dispersion, Fig. 2.5(a)) and $0.48 < u < 0.63$ (anomalous dispersion, Fig. 2.6(a)), characterized by total reflectance ($|R|^2 = 1$).
d. The phase shifts ϕ_r of reflected waves remain positive (Fig. 2.5(b)) and negative (Fig. 2.6(b)) in the aforesaid spectral ranges of total reflectance; these phase shifts increase with the increase of frequency ($\partial \phi_r / \partial \omega > 0$) in both cases, the reflectance being constant.

2. Gradients of $n(z)$ on the boundary of adjacent films are equal, while the curvatures are unequal [2.11].

To display the influence of discontinuities of curvature of a smooth profile $U(z)$ in the gradient layer on its optical properties, let us consider the structure, whose reflectance is governed by these

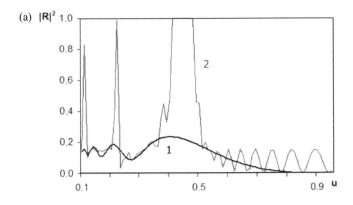

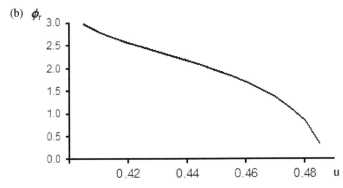

Fig. 2.5. Reflectance spectrum of a periodical structure, containing m = 20 gradient layers with normal dispersion, shown in Fig. 2.4(a) ($n_0 = 2.21875$, $n = 2.3$, $y = 0.75$). (b): variations of the phase of the reflected wave ϕ_r under the conditions, shown in Fig. 2.4(a), are depicted for the spectral range, corresponding to the total reflection: $|R|^2 = 1$.

discontinuities only, the refractive index and its derivative being continuous. The relevant configuration, presenting the "smoothened" transition layer between two media, spaced by distance d, with the refractive indices n_1 and n_2, is depicted on Fig. 2.7(a). The refractive index $n(z)$ is varying along this slit from the value n_1 up to n_2 monotonically and continuously, it's gradient, nullified at the interfaces $z = 0$ and $z = d$, is varying continuously along the slit too, however this layer possesses three discontinuities of curvature — two at the interfaces and one at some point $z = z_0$ inside the layer. Our goal is to find the reflection coefficient R for this "smoothened" sandwich

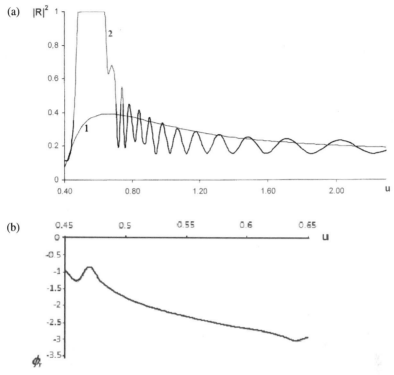

Fig. 2.6. Reflectance spectrum of a periodic structure, containing m = 20 gradient layers with anomalous dispersion, shown in Fig. 2.4(b) ($n_0 = 1.42$, n = 2.3, y = 0.75). (b): variations of the phase of the reflected wave ϕ_r under the conditions, shown in Fig. 2.4(b), are depicted for the spectral range, corresponding to the total reflection: $|R|^2 = 1$.

gradient structure; the parameters n_1, n_2, d, z_0 are supposed to be known.

Let us consider such a structure, containing gradient layers 1 and 2, characterized by different distributions of refractive indices n_- and n_+ respectively:

$$n_- = n_1 U_1; \quad U_1 = \left(1 - \frac{z^2}{l^2}\right)^{-1}. \tag{2.41}$$

$$n_+ = n_0 U_2; \quad U_2 = \left[1 - \frac{z - z_0}{L_1} + \frac{(z - z_0)^2}{L_2^2}\right]^{-1}. \tag{2.42}$$

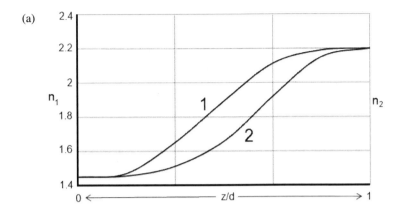

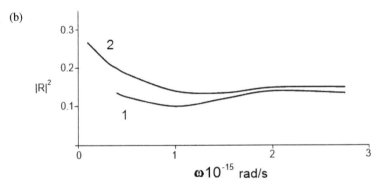

Fig. 2.7. Reflectance of a smooth transition layer due to an internal discontinuity of curvature. (a): gradient transition layer between media with refractive indices n_1 and n_2; distribution of refractive index and its gradient inside the layer is continuous, the discontinuity of curvatures of profiles $U(z)$ arises at the intermediate plane $z = z_0$. (b): Reflectance spectra of the transition layer, shown in (a), in the middle IR range ($n_1 = 1.42$, $n_2 = 2.22$, $d = 150\,\mathrm{nm}$), spectra 1 and 2 correspond to the values $z_0 = 50\,\mathrm{nm}$, $z_0 = 100\,\mathrm{nm}$, respectively.

One can see, that distributions (2.41) and (2.42) are different forms of model (2.16). The profiles n_- and n_+ cross at some point z_0, characterized by the value n_0:

$$n_0 = n_-(z_0) = n_+(z_0); \qquad (2.43)$$

First of all we have to find the geometrical parameters l, L_1 and L_2 in profiles (2.41) and (2.42). These parameters are linked by the condition of a smooth tangent of curves n_- and n_+ at the point $z = z_0$

$$\frac{1}{L_1} = \frac{2z_0}{l^2 - z_0^2}. \tag{2.44}$$

and by the condition that $\operatorname{grad} U_2$ vanishes at the point $z = d$:

$$\frac{1}{L_1} = \frac{2(d - z_0)}{L_2^2}. \tag{2.45}$$

Substitution of l and L_1 from (2.44) and (2.45) into continuity condition $n_+(d) = n_2$ brings the value of L. Assuming, for definiteness, $n_2 > n_1$, we find:

$$L_2^2 = \frac{(d - z_0)\aleph}{n_2 - n_1}; \quad \aleph = n_2 d - z_0(n_2 - n_1); \tag{2.46}$$

Now one can calculate the geometrical parameters l, L_1 and the value n_0 (2.46):

$$L_1 = \frac{\aleph}{2(n_2 - n_1)}; \quad l = \sqrt{\frac{z_0 d n_2}{n_2 - n_1}}; \quad n_0 = \frac{n_1 n_2 d}{\aleph}. \tag{2.47}$$

It is worthwhile to recall, that the "tangent point" z_0 was chosen freely.

With these values of l, L_1 and L_2 in hand we can calculate the reflection coefficient R. The tangent of arcs U_1 and U_2 on the internal boundary z_0 between layers is smooth and, thus, the refractive index and its gradient on this boundary are continuous. It is remarkable here, that the discontinuity of curvatures of the profiles U_1 and U_2 at this boundary makes a contribution to the reflectance of the sandwich. To calculate this reflectance one can use the standardized approach, developed above for a single barrier: the wave fields in layers 1 and 2 are represented by means of wave functions (2.21), and the boundary conditions on the interfaces $z = 0$ and $z = d$ are formulated in (2.22)–(2.24) and (2.27)–(2.28) respectively. A peculiar part of this analysis is connected with the boundary conditions at the internal boundary z_0, where, due to smooth tangent of profiles U_1 and U_2 the values of the parameter L_1 in both profiles is equal.

This condition, linking the quantities Q_1 and Q_2 in the fields $\Psi_{1,2}$ (2.21), reads as:

$$Q_1 = \exp(2iq_1\eta_1)\left[\frac{N_1 - N_2\Lambda_2}{N_1 + N_2\Lambda_2}\right]; \quad \Lambda_2 = \frac{1-Q_2}{1+Q_2}; \quad (2.48)$$

Considering the profile n_- as a half of the concave arc (2.16), one can represent the generating function for the EM field Ψ in the form (2.21) with the values of the "wave number" q_1 and variable η_1, given by formulae

$$q_1 = \frac{\omega}{c}n_1 N_1; \quad N_1 = \sqrt{1 - \frac{\omega_1^2}{\omega^2}};$$

$$\omega_1 = \frac{c}{n_1 l}; \quad \eta_1 = \frac{l}{2}\ln\left(\frac{1+z/l}{1-z/l}\right). \quad (2.49)$$

Let us restrict ourselves here to the high frequency spectral interval $\omega \geq \omega_1$. The expression for R can be derived in this case in a form similar to (2.26) in the limit $\gamma \to 0$, corresponding to the geometry of profile U_1 near the interface $z = 0$:

$$R = \frac{1 - N_1\Lambda_1}{1 + N_1\Lambda_1}; \quad \Lambda_1 = \frac{1-Q_1}{1+Q_1}. \quad (2.50)$$

The generating function Ψ for the arc U_2 (Fig. 2.4(b)), treated as a half of the convex arc (2.16), is written again in the form (2.21) with the values of q_2 and η_2 given by

$$q_2 = \frac{\omega}{c}n_0 N_2; \quad N_2 = \sqrt{1 + \frac{\Omega_2^2}{\omega^2}}. \quad (2.51)$$

Here the characteristic frequency for the convex arc Ω_2 is defined in (2.20), the variable η_2 can be obtained from (2.32e) by the replacement $z \to z - z_0$, and the dimensionless parameter $y = L_2/2L_1$, important for calculation of reflection coefficient, can be found by means of (2.45)–(2.46):

$$y = \sqrt{\frac{d-z_0}{\aleph}}. \quad (2.52)$$

According to expression (2.47), the reflection coefficient R depends on the parameter Λ_1, which, in its turn, depends upon the factor Q_1. The continuity conditions at the internal boundary $z = z_0$ yield the link of Q_1 with the analogous factor Q_2, determined for the arc U_2; this link was found above in (2.48). Finally, the factor Q_2, found from the boundary conditions at the interface $z = d$, is

$$Q_2 = \exp(2iq_2\eta_2) \left[\frac{n_0 N_2 - n_2}{n_0 N_2 + n_2}\right]; \qquad (2.53)$$

Going back along this chain of calculations and substituting Q_2 from (2.53) into (2.48), we find Q_1. Then, substitution of Q_1 into (2.53) yields the complex reflection coefficient R:

$$R = \frac{K_1 - i\Gamma_1}{K_2 - i\Gamma_2}; \qquad (2.54)$$

$$\begin{aligned}
K_1 &= N_1 N_2 (n_0 - n_2) + n_2 t_1 t_2 (N_1^2 - n_0 N_2); \\
K_2 &= N_1 N_2 (n_0 + n_2) - n_2 t_1 t_2 (N_1^2 + n_0 N_2); \\
\Gamma_1 &= n_2 (N_1 t_2 + n_2 t_1) - n_0 N_1 N_2 (N_1 t_1 + N_2 t_2); \\
\Gamma_2 &= n_2 (N_1 t_2 + n_2 t_1) + n_0 N_1 N_2 (N_1 t_1 + N_2 t_2).
\end{aligned} \qquad (2.55)$$

Here the values N_1 and N_2 are defined in (2.49) and (2.51) respectively, while the factors t_1 and t_2 are:

$$t_1 = \operatorname{tg}\left[\sqrt{u_1^{-2} - 1}\operatorname{arctg}\left(\frac{z_0}{l}\right)\right]; \quad u_1 = \frac{\omega_1}{\omega}. \qquad (2.56)$$

$$t_2 = \operatorname{tg}\left[\sqrt{u_2^{-2} + 1}\operatorname{arctg}\left(\frac{y}{\sqrt{1-y^2}}\right)\right]; \quad u_2 = \frac{\Omega_2}{\omega}. \qquad (2.57)$$

The characteristic frequencies ω_1, Ω_2 and the factor y are determined in (2.46), (2.20) and (2.52), respectively.

Formulae (2.55)–(2.57) present the complex reflection coefficient of gradient transition layer, defined only by discontinuities of the curvature of a smooth profile of the refractive index inside the sandwich structure (Fig. 2.7(b)). Variations of the location of the internal boundary in this sandwich (point z_0), the total thickness d being fixed, opens the possibilities to optimize the parameters of a transition layer in a fixed spectral range.

Comments and Conclusions to Chapter 2

1. It is instructive to compare and contrast the exact solution of Eq. (2.9) in the form (2.15), which is not restricted by any assumptions about the smallness or slowness of variations of fields and media, with the solution of (2.9), obtained in the framework of the traditional WKB-approximation, when these variations of profile $U(z)$ are presumed to be small and slow; such a WKB-solution may be written as [2.12]

$$\Psi = \frac{\exp\left[i\left(\frac{\omega n_0}{c}\eta - \omega t\right)\right]}{\sqrt{U(z)}}. \tag{2.58}$$

The variable η in (2.58) is defined, as above, in (2.10). The difference between the exact (2.15) and approximate (2.58) solutions is stipulated by the value of factor N in the wavenumber q defined in (2.14): in the exact solution factor N is distinguished from unity due to the characteristic frequencies Ω_1 and Ω_2, which describe the non-local heterogeneity-induced dispersion, while in the WKB-approach these non-local effects are ignored

$$L_1 \to \infty; \quad L_2 \to \infty; \quad \Omega_1 \to 0; \quad \Omega_2 \to 0. \tag{2.59}$$

and the factor N possesses the constant value $N = 1$. These effects of artificial heterogeneity-induced dispersion are shown below to play the fundamental role in all the complex of wave phenomena in gradient barriers.

It is remarkable, that the case $N = 1$, corresponding to the vanishing of heterogeneity-induced dispersion, can arise even in a medium with non-zero values of the spatial scales L_1 and L_2 due to the special relation between them: $L_2 = 2L_1$. Using this condition, one can obtain from (2.16) the profiles of gradient dispersiveless barriers [2.11]:

$$U(z) = \left(1 \pm \frac{z}{L_2}\right)^{-2}. \tag{2.60}$$

Calculation of new the variable η (2.10) by means of (2.60) brings the expression for a wave, traveling through the barriers (2.60), in a form

$$\Psi = \exp\left[\frac{i\omega n_0 z}{c}\left(1 \pm \frac{z}{L_2}\right)^{-1}\right]. \tag{2.61}$$

2. It has to be emphasized, that the classical Fresnel formulae, describing reflectance and transmittance of homogeneous dielectric layers, can be viewed as the limiting cases of more general formulae (2.31) and (2.35) for gradient layers, corresponding to the condition of vanishing of the heterogeneity in the profiles of the refractive index $U(z)$; the conditions (2.59) are supplemented now by condition $U = 1$. Thus, in this limit expression (2.31) is reduced to the well-known Fresnel formula for the reflectance of homogeneous dielectric layer with thickness d and refractive index n_0, located on a half-space with refractive index n:

$$R = \frac{(n - n_0^2)\mathrm{tg}\alpha - in_0(n - 1)}{(n + n_0^2)\mathrm{tg}\alpha + in_0(n + 1)}; \quad \alpha = \frac{\omega n_0 d}{c}; \tag{2.62}$$

Correspondingly, the expression for the complex transmission coefficient T_E of a gradient layer (2.34) is reduced in the same limit to another classical Fresnel formula:

$$T_E(\alpha) = \frac{2in_0}{(n + n_0^2)\sin\alpha + in_0(n + 1)\cos\alpha};$$

$$\frac{\omega n_0 d}{c} = \alpha = \frac{2y\sqrt{1 + y^2}}{u}; \tag{2.63}$$

Figure 2.8 shows the transmittance spectrum $|T(u)|^2$ for a gradient barrier, calculated by means of (2.35), and the spectrum $|T(\alpha)|^2$ for the homogeneous rectangular barrier (2.63) for the same frequency ω; the parameters d and n_0 for both barriers are equal, and the factor α is determined in (2.62). Inspection of both spectra illustrates the drastic changes in reflectance/transmittance spectra, caused by the smooth heterogeneous structure of the transparent barrier.

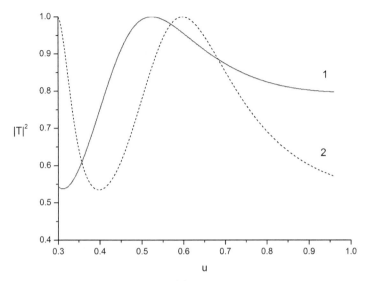

Fig. 2.8. Effect of gradient profile $n(z)$ in transmittance spectra for the heterogeneous barrier (2.16) with normal non-local dispersion ($y = 0.75$, curve 1) and a homogeneous barrier (curve 2); the refractive indices $n_0 = 2.3$, $n = 1$ and thickness d are equal for both barriers.

3. The spectral ranges of total reflectance, shown in Fig. 2.5(a) and Fig. 2.6(a), possess the potential for the design of effective broadband reflectors in the middle IR range.
4. The substantial dispersion of the phase shift of the reflected wave (Figs. 2.5(b) and 2.6(b)) may become useful for fast phase modulation of broadband IR radiation, keeping its amplitude invariant. Thus, considering, e.g. the thickness of gradient films $d = 150\,\text{nm}$, one can find the critical frequencies $\Omega_1 = 1.7 \times 10^{15}\,\text{rad s}^{-1}$ (Fig. 2.5(b)) and $\Omega_2 = 2.1 \times 10^{15}\,\text{rad s}^{-1}$ (Fig. 2.6(b)); the derivatives $\partial \phi_r / \partial \omega$, defined from these graphs, are $3.15 \times 10^{-15}\,\text{s}$ for the normal dispersion and $1.95 \times 10^{-15}\,\text{s}$ for the anomalous one.
5. The exactly solvable model of a gradient barrier $U(z)$, (2.16), possess several salient features:

    a. flexibility, stipulated by interplay of two free parameters (s_1 and s_2), yields the possibility to examine non-monotonic (both convex and concave) profiles of gradient photonic barriers; here the widely used monotonic Rayleigh profile (2.4) proves to be a limiting case of (2.16), related to the value $s_2 = 0$; moreover, as it follows from (2.18), the cut-off frequency $\Omega = c/2n_0L(L = L_1)$ and wavenumber q (2.14) retain the same values for both concave ($s_1 = +1$) and convex ($s_1 = -1$) Rayleigh barriers.

b. mathematical simplicity, providing a standardized analytical calculation of reflectance/transmittance spectra of gradient photonic barriers by means of exact analytical solutions of the Maxwell equations, expressed by elementary functions in a special η-space.

c. scalability of results, which can be applied to waves of different physical nature described by wave equation (2.9); examples of such scalability will be shown below in the analysis of acoustic waves in gradient elastic media [2.13].

Owing to these features the model (2.16) is considered to be a key model for several problems touched in this book. A series of improvements of this model, which bring it closer to the real materials, are discussed in Ch. 3.

Bibliography

[2.1] J. W. S. Rayleigh, *P. Lond. Math. Soc.* **11**, 51–56 (1880).
[2.2] J. R. Wait, *Electromagnetic Waves in Stratified Media* (Pergamon Press, 1970).
[2.3] D. R. Hartree, *P. Roy. Soc. Lond. A Math.* **131**, 428–450 (1931).
[2.4] V. L. Ginzburg, *Propagation of Electromagnetic Waves in Plasma* (Pergamon Press, 1967).
[2.5] O. Rydbeck, *Phil. Mag.* **34**, 342–348 (1943).
[2.6] A. Iwamoto, V. M. Aquino and V. C. Aquilera-Nowarro, *Int. J. Theor. Phys.* **43**, 483–495 (2004).
[2.7] A. B. Shvartsburg and G. Petite, *European Phys. J. D* **36**, 111–118 (2005).
[2.8] A. Berti, *J. Math. Phys.* **47**, 1–14 (2006).
[2.9] W. H. Southwell, *Appl. Opt.* **28**, 5091–5094 (1989).
[2.10] A. B. Shvartsburg, *Physics — Uspekhi* **177**, 43 (2007).
[2.11] A. B. Shvartsburg, V. Kuzmiak and G. Petite, *Phys. Rep.* **452**(2–3), 33–88 (2007).
[2.12] Yu. A. Kravtsov, *Geometric Optics in Engineering Physics, Alpha Science International* (Harrow, UK, 2005).
[2.13] A. B. Shvartsburg and N. S. Erokhin, *Physics — Uspekhi* **181**, 245–264 (2011).

CHAPTER 3

GRADIENT PHOTONIC BARRIERS: GENERALIZATIONS OF THE FUNDAMENTAL MODEL

The simple model of a gradient photonic barrier presented in Ch. 2 shows, nonetheless, several salient features of such structures:

1. A strong technologically managed non-local dispersion, which may be fabricated to be either normal or anomalous;
2. The possibility to design controlled reflectance/transmittance spectra in any visible and IR spectral ranges;
3. Miniaturized, even subwavelength, spatial scales.

The essential dependence of the reflectance/transmittance of gradient barriers on their geometric parameters, illustrated in Ch. 2 in the framework of the key model (2.16), stimulated attempts to broaden the variety of these spectra by means of new exact analytical solutions of Maxwell equations (2.2) and (2.3) for gradient media, characterized, unlike (2.16), by other profiles $U(z)$. Moreover, the analysis of optical properties of gradient photonic barriers (2.16) in Ch. 2 revealed their strong dependence on the gradient and curvature of the profiles $U(z)$ on the boundaries of the barriers. However, while modeling some fabricated barrier by means of model (2.16) and using its technologically controlled parameters — width d and extremal values $U_{\max}$ or $U_{\min}$ — one has to attribute some fixed values to the aforesaid gradient and curvature of $U(z)$ on both sides of the barrier. To make the model (2.16) more practical and flexible, and, still, analytically solvable, the following generalizations are required:

1. The slope of the profile $U(z)$ near to the barrier boundary, e.g. $z = 0$, may be characterized by the dimensionless factor ξ, determined by the derivative

$$\left.\frac{dU^2}{dz}\right|_{z=0} = 2UU_z|_{z=0}; \quad \xi = 2dUU_z|_{z=0}. \quad (3.1)$$

The factor ξ, calculated for profile (2.16), is fixed: using the relation between d and L_1 (2.17), we find a constant value of ξ:

$$\xi = -\frac{2s_1 d}{L_1} = -8s_1 y^2. \quad (3.2)$$

Thus, the dependence of reflectance/transmittance spectra on the slope of the profile $U(z)$ near the barrier boundaries cannot be considered in the framework of model (2.16). To investigate this dependence, a more general model of the barrier, containing an additional free parameter, is considered below in Sec. 3.1.
2. Some models of asymmetric barriers with thickness D, characterized, unlike (2.16), by the property $U(0) \neq U(D)$, can be obtained by a simple truncation of the symmetrical profile (2.16) (Sec. 3.2).
3. It has to be emphasized, that, side by side with the explicit expression for the refractive index profile $U(z)$, like (2.16), there is a multitude of exactly solvable implicit models, presented both by means of inverse dependence $z = z(u)$ and in parametric form $(z = z(\eta), U = U(\eta)$, where η is some parameter); the examples of such presentations, illustrating the algorithm for the formation of implicit models, is discussed in Sec. 3.3.

3.1. Effects of the Steepness of the Refractive Index Profile near the Barrier Boundaries on Reflectance Spectra

To visualize the sensitivity of the reflectance spectra of a barrier $U(z)$ on the variable gradient of its profile, the other barrier parameters being fixed, let us consider the generalized model of barrier $U_1(z)$

$$U_1(z) = \left(1 + \frac{s_1 z}{L_1} + \frac{z^2}{L_2^2}\right)^{-\mu} = [U(z)]^\mu. \quad (3.3)$$

Here the profile $U(z)$ is given by the key model (2.16), where the dimensionless power μ is the new free parameter, determining the slope of profile $U_1(z)$ with respect to the boundaries of the barrier; thus, the model (2.16) may be viewed as the limiting case of (3.3), related to the value $\mu = 1$. The extrema of the symmetric profiles (3.3) are (see (2.17)):

$$U_m = (1 + s_1 y^2)^{-\mu}. \quad (3.4)$$

Other relations between the width d, heterogeneity scales L_1 and L_2 in (2.17) remain valid for the profile (3.3) as well. Substituting the function $U_1(z)$ instead of $U(z)$ into the wave equation for the generating function Ψ (2.9), and introducing, by analogy with (2.10), the new variable η and new functions F and Y

$$F = \Psi\sqrt{U_1(z)}; \quad U(z) = [Y(z)]^v; \quad \eta = \int_0^z [U(z_1)]^\mu dz_1, \quad (3.5)$$

one can rewrite Eq. (2.9) in a generalized form

$$\frac{d^2 F}{d\eta^2} + F\left\{\left(\frac{\omega n_0}{c}\right)^2 - \frac{v\mu}{2}\left[\left(1 + \frac{v\mu}{2}\right)\frac{Y_z^2}{Y^{2v\mu+2}} - \frac{Y_{zz}}{Y^{2v\mu+1}}\right]\right\} = 0. \quad (3.6)$$

Till now no suppositions concerning the numbers v and μ have been made. However, when these numbers obey the condition

$$v\mu = -1, \quad (3.7)$$

one can reduce Eq. (3.6) to the form (2.11). Moreover, assuming the conditions (2.12) and (2.13) to be fulfilled, the generating function Ψ for profile (3.3) can be represented in a form, coinciding with (2.15) after the replacement $U(z) \to U_1(z)$, and the use of the variable η, defined in (3.5). Following further the scheme of analysis, described in Sec. 2.2, we obtain a formula for the reflection coefficient R, valid for an arbitrary value of the parameter μ; this formula, resembling (2.31), obtained for $\mu = 1$, is distinguished from (2.31) due to replacement

$$\gamma \to \gamma\mu, \tag{3.8}$$

and the new expression for the factor $t = \text{tg}(q\eta_0)$, where $\eta_0 = \eta(d)$ has to be calculated, unlike (2.32g), from (3.5) for each value of the parameter μ. Thus, integration in (3.5) for the values $\mu = 0.5$ and $\mu = 1.5$ for, e.g., the concave profiles $U_1(z)$, leads to the expressions:

$$\begin{aligned}
\mu &= 0.5; \quad t = \text{tg}\left[\Theta(y_1)\sqrt{\frac{1}{u^2} - 1}\right]; \\
\Theta(y_1) &= 2\sqrt{1 + y_1^2}\arcsin\left(\frac{y_1}{\sqrt{1 + y_1^2}}\right); \\
\mu &= 1.5; \quad t = \text{tg}\left[\Theta(y_2)\sqrt{\frac{1}{u^2} - 1}\right]; \\
\Theta(y_2) &= y_2 + \frac{1}{2\sqrt{1 + y_2^2}}\ln\left(\frac{\sqrt{1 + y_2^2} + y_2}{\sqrt{1 + y_2^2} - y_2}\right).
\end{aligned} \tag{3.9}$$

The values y_1 and y_2 are defined here for the given U_m from (3.4).

Reflection coefficients R (2.31) with the correctly defined parameters t and γ enable the calculation of the reflectance for any barrier (3.3); here the case $\mu = 1$ relates to the key model (2.16). To emphasize the importance of the slope of profiles (3.3) for the effectiveness of the reflectance/transmittance, it makes sense to compare, e.g. the reflectance spectra for profiles characterized by equal widths d and equal extrema U_m, but different slopes ξ (3.1), depending on the power μ. Thus, the equality of minima $U_{\min}$ (3.4) for two concave profiles with $\mu = 1$ and $\mu \neq 1$ yields the condition, linking the values of $y(\mu = 1)$ and y_1 ($\mu \neq 1$):

$$1 + y^2 = (1 + y_1^2)^\mu; \tag{3.10}$$

Substitution of y_1, calculated from (3.10), into the expression for the slope parameter ξ yields finally for concave profiles (2.16):

$$\xi(\mu) = -8\mu[(1 + y^2)^{\frac{1}{\mu}} - 1]. \tag{3.11}$$

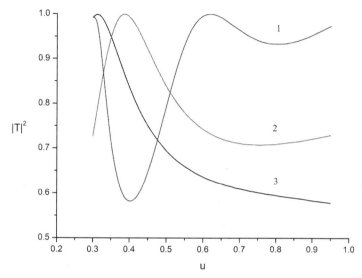

Fig. 3.1. Reflectance spectra for the generalized profile (3.3). Curves 1, 2, 3 relate to concave profiles with the values of the parameter $\mu = 0.5, 1, 1.5$, respectively; $n_0 = 2.3$; $y = 0.577$; $n = 1$.

The difference between barriers, related to these values of the power μ in (3.3), is formed near the film boundaries $z = 0$ and $z = d$; the smaller value of μ, providing the bigger slope (3.11) at the periphery of the profile $[U(z)]^\mu$, results in the flattening of its central part, the profile minimum U_m (3.4) being fixed.

Thus, the profile $[U(z)]^\mu$ for $= 0.5$ is more flattened in its central part, than the profile with $\mu = 1.5$. The reflectance spectra for profile (3.3), calculated for the abovementioned values $\mu = 0.5, 1$ and 1.5, are shown in Fig. 3.1. Determining the slope (3.11) for these profiles, we see the drastic variety of reflectance, caused by the decrease of the slopes (3.12): $\xi(0.5) = -3.104$; $\xi(1) = -2.664$; $\xi(1.5) = -2.536$, other parameters of photonic barriers being the same.

3.2. Asymmetric Photonic Barriers

The barrier (3.3) with thickness d is symmetric with respect to the plane $z = d/2$, containing its extremum $U_m (U(0.5d - z) = U(0.5d + z))$. Violations of this symmetry can provide significant changes of the barrier's reflection spectra. To illustrate some reflection tendencies for asymmetric barriers with thickness D, one can consider two different types of profiles $U(z)$:

a. The barrier is formed due to truncation of concave or convex symmetric profiles (2.16), characterized by thickness d, at same point $D < d$, so, that $U(0) \neq U(D)$;
b. The opposite case of an asymmetric barrier, where $U(0) = U(D)$, but the profile is not symmetric with respect to its extremum U_m.

Considering the first case, it is instructive to examine here a special profile, related to a half of the symmetric barrier (2.16), when the extremum of the profile U_m is located at the far side of the asymmetric barrier $U(z)$ with thickness $D = 0.5d$; in this case one can find from (2.17):

$$U(0) = 1; \quad U(D) = U_m = \frac{1}{1 + s_1 y^2}. \qquad (3.12)$$

In this geometry the gradient of the refractive index $n(z)$ at the interface $z = D$ vanishes $(U_z|_{z=D} = 0)$; however, the curvature of the profile $n(z)$ at this interface will be shown to have an influence on the reflectance of the barrier even in the case where neither the index, nor its gradient possess any discontinuity at the plane $z = D$.

The continuity conditions at the plane $z = 0$, given by (2.24) and (2.25), remain valid, as well as the expression (2.26) for the reflection coefficient R; however, the factor Q has to be redefined from the continuity conditions at the interface $z = D$ $(\eta(D) = \eta_1, \gamma = 0)$:

$$\frac{E_1[\exp(iq\eta_1) + Q\exp(-iq\eta_1)]}{\sqrt{U_m}} = E_2; \qquad (3.13)$$

$$E_1 n_e \sqrt{U_m}[\exp(iq\eta_1) - Q\exp(-iq\eta_1)] = nE_2. \qquad (3.14)$$

Division of (3.14) by (3.13) yields the dimensionless factor Q:

$$Q = \exp(2iq\eta_1)\left(\frac{n_e U_m - n}{n_e U_m + n}\right). \qquad (3.15)$$

On substituting Q (3.15) into the general expression (2.26) and taking into account the relation, valid for the "half" of barrier (2.16)

$\eta_1 = 0.5\eta_0$, we obtain finally an explicit expression for the reflection coefficient R for the asymmetric barrier under discussion:

$$R = \frac{nt - \frac{\gamma n_e U_m s_1}{2} - n_e^2 U_m t - i\left(nn_e - \frac{\gamma n t s_1}{2} - n_e U_m\right)}{nt + \frac{\gamma n_e U_m s_1}{2} + n_e^2 U_m t + i\left(nn_e - \frac{\gamma n t s_1}{2} + n_e U_m\right)};$$

$$t = \operatorname{tg}\left(\frac{q\eta_0}{2}\right). \tag{3.16}$$

Expressions for the product $q\eta_0$ for both concave ($s_1 = 1$) and convex ($s_1 = -1$) profiles were presented in (2.32g).

The examples of transmittance spectra $|T|^2$ of truncated profiles of gradient barriers, calculated due to substitution of the reflection coefficient (3.16) into equation $|T|^2 = 1 - |R|^2$, are depicted on Fig. 3.2 for concave profiles with different values of U_m.

The effect of truncation is seen in a comparison of spectra 1 and 2 of Fig. 3.2 with spectrum 1 in Figs. 2.3(a) and 2.3(b) respectively; the spectra in Figs. 2.3(a) and 2.3(b) relate to the "complete" profiles (2.16), the values of U_m as well as the refractive indices n for each of the pairs compared being equal. The increase of the gradient

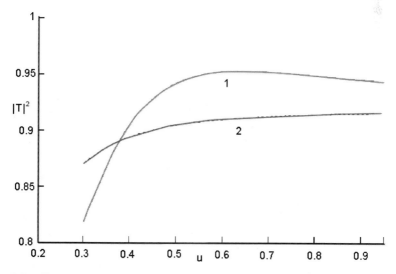

Fig. 3.2. Transmittance spectra for asymmetric barrier, represented by "half" of a symmetric concave barrier, shown by curve 1 (Fig. 2.1). Spectra 1 and 2 relate to the values $y = 0.75$ and $y = 0.577$, respectively, $n_0 = n = 1.8$.

film thickness increases the film's transparency, while the effect of reflectance from the discontinuity of profile's curvature produces the opposite tendency. The interplay of these tendencies permits optimizing the thickness of the film, which is needed to provide, e.g. a better transmittance in some fixed spectral range: thus, in the vicinity of the normalized frequency $u = 0.53$ the transmission coefficient $|T|^2$ of the film with the "complete" profile is $|T(u=0.53)|^2 = 1$ (Fig. 2.3(a), y = 0.75), while the truncated barrier is characterized for the same frequency by the smaller value $|T(u=0.53)|^2 = 0.94$ (Fig. 3.2, curve 1). However, for lower frequencies ($u > 0.61$) the values of $|T|^2$ for the truncated barrier exceed those values for the "complete" one: thus, for u = 0.9 the values $|T|^2$ for the truncated and "complete" barriers are 0.94 and 0.8, respectively. Similar tendencies can be revealed for the spectrum, shown in Fig. 3.2(b).

Recalling the profile (2.16) with thickness d, one can consider another asymmetric barrier, characterized, again, by the aforesaid inequality $U(0) \neq U(d)$, when, however, the signs s_1 and s_2 are the same: $s_1 = s_2 = \pm 1$, so that these profiles $U(z)$ are monotonic. Thus, in the case of a monotonically decreasing profile (2.16), related to the case $s_1 = s_2 = +1$, one can express the scales L_1 and L_2 via the values of the function $U(z)$ at the boundary of barrier $z = d$ and at some point inside the barrier, e.g. at the mid plane $z = d/2$:

$$U_c = U(d) = \left(1 + \frac{d}{L_1} + \frac{d^2}{L_2^2}\right)^{-1};$$

$$U_m = U\left(\frac{d}{2}\right) = \left(1 + \frac{d}{2L_1} + \frac{d^2}{4L_2^2}\right)^{-1}. \quad (3.17)$$

Substitution of these values L_1 and L_2 into Eq. (2.18) yields the values of the factor p^2 and cut-off frequency Ω, determining the nonlocal dispersion of the gradient photonic barrier with monotonically decreasing profile (2.16):

$$p^2 = \frac{1}{4d^2}\left[\left(\frac{4}{U_m} - 1\right)^2 + \frac{1}{U_c}\left(\frac{1}{U_c} - \frac{8}{U_m} - 2\right)\right]; \quad \Omega = \frac{cp}{n_0}. \quad (3.18)$$

The reflectance of these profiles can be examined by means of the familiar solutions (2.15) by analogy with the general approach, presented in Sec. 2.2, although the continuity conditions at $z = d$ are distinguished from the corresponding expressions for the symmetric profile (2.32g).

Let us consider now some other exactly solvable models of asymmetric barriers, belonging to the aforesaid type (b). An example of such a barrier, formed, e.g. by a concave asymmetric profile, is given by the equation

$$U^2(z) = \frac{(1-x)^{\nu-2}}{(1+x)^{\nu+2}}; \quad x = \frac{z}{L}. \tag{3.19}$$

The concave profile (3.19), possessing two free parameters — the spatial scale L and dimensionless index ν-is depicted in Fig. 3.3. The minimum of this profile $U^2_{\min}$ is located at the point $x_m = \nu/2$. The normalized thickness of this barrier, x_0, can be calculated as the non-zero solution of the equation $U(x_0) = 1$; this barrier is asymmetric with respect to its mid plane, $x_m \neq 0.5x_0$.

To solve the wave equation (2.9) with the barrier (3.19) it is worthwhile to represent the function (3.19) as the product of two

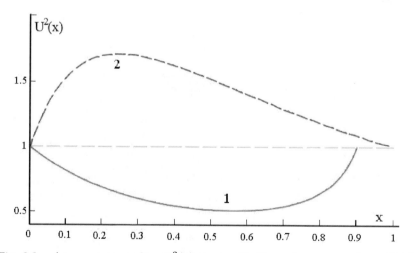

Fig. 3.3. Asymmetric profiles $U^2(x)$, $x = z/L$. Curve 1 corresponds to (3.19) with the value of the parameter $\nu = 1.128$; Curve 2 represents the profile (3.28) with $\nu = 1.5$, $g = 0.2916$.

functions

$$U^2(z) = U_1^2(z)U_2^2(z); \quad U_1^2(z) = \frac{1}{(1-x^2)^2}; \quad U_2^2(z) = \left(\frac{1-x}{1+x}\right)^v. \tag{3.20}$$

Following the scheme of solution developed in Sec. 2.1, let us introduce in Eq. (2.9), according to (2.10), the new variable η and new function F:

$$F = \Psi\sqrt{U_1(z)}; \quad \eta = \int_0^z U_1(z_1)dz_1 = \frac{L}{2}\ln\left(\frac{1+x}{1-x}\right). \tag{3.21}$$

The use of the variable η, (3.21), transforms (2.9) into the form:

$$\frac{d^2F}{d\eta^2} + \left(\frac{\omega^2 n_0^2}{c^2}U_2^2(z) - \frac{1}{L^2}\right)F = 0. \tag{3.22}$$

It is worthwhile to introduce the normalized variable ς and to express the function $U_2^2(z)$ (3.20) via ς

$$\varsigma = \frac{\eta}{L}; \quad U_2^2(z) = \exp(-2v\varsigma). \tag{3.23}$$

Owing to this transformation Eq. (3.21) can be reduced to the standard form of the Bessel equation

$$\frac{d^2F}{d\tau^2} + \frac{1}{\tau}\frac{dF}{d\tau} + \left(q^2 - \frac{s^2}{\tau^2}\right)F = 0; \quad q = \frac{\omega n_0 L}{2cv};$$

$$s = \frac{1}{2v}; \quad \tau = \exp(-v\varsigma). \tag{3.24}$$

Solution of Eq. (3.24) may be written as the superposition of forward and backward waves, given by Hankel functions of the first and second kind $H_s^{(1)}(q\tau)$ and $H_s^{(2)}(q\tau)$ [3.1]. Finally, substitution of this superposition into the definition of the generating function Ψ (3.20) yields the solution of wave equation (2.9) with the asymmetric barrier (3.18),

$$\Psi = A\sqrt{1 - \frac{z^2}{L^2}}[H_s^{(1)}(q\tau) + QH_s^{(2)}(q\tau)]. \tag{3.25}$$

Here the constants A and Q have to be calculated, as usual, from the boundary conditions. The value of product $q\tau$, found from

(3.20), is

$$q^\tau = \frac{1}{4u\nu}\left(\frac{1-x}{1+x}\right)^{\frac{\nu}{2}}; \quad u = \frac{\Omega}{\omega}; \quad \Omega = \frac{c}{2n_0 L}. \qquad (3.26)$$

While calculating these constants, it is convenient to use the expressions for the Hankel functions in terms of the Bessel and Neumann functions J_s and N_s:

$$H_s^{(1)}(q\tau) = J_s(q\tau) + iN_s(q\tau); \quad H_s^{(2)}(q\tau) = J_s(q\tau) - iN_s(q\tau). \quad (3.27)$$

Another type of asymmetric barrier is given by the profile

$$U^2(z) = \left(1+\frac{z}{L}\right)^{-2}\left[1+\frac{1}{g}-\frac{1}{g}\left(1+\frac{z}{L}\right)^{-2\nu}\right]. \qquad (3.28)$$

Unlike profile (3.19), profile (3.28) is characterized by three free parameters-length scale L and dimensionless real parameters g and ν; for definiteness, let us consider the convex asymmetric profile ($g > 0$), shown in Fig. 3.2. The maximum of this profile is located at the point x_m, determined from the equation

$$(1+x_m)^{2\nu} = \frac{1+\nu}{1+g}; \quad x = \frac{z}{L}. \qquad (3.29)$$

Inspection of Fig. 3.2. shows that profile (3.28) is essentially asymmetric, its maximum is strongly shifted towards the plane $z = 0$. Following the algorithm of analysis, developed above, we will represent the function (3.28), by analogy with (3.19), as the product of two functions:

$$U_1^2(z) = (1+x)^{-2}; \quad U_2^2(z) = 1 + \frac{1}{g} - \frac{1}{g}(1+x)^{-2\nu}. \qquad (3.30)$$

Note, that U_1 in Eq. (3.30) is the familiar Rayleigh profile (2.4). Introducing the new variable η and the new function F according to (3.20),

$$\eta = \int_0^z U_1(z_1)dz_1 = L\ln(1+x). \qquad (3.31)$$

we rewrite wave equation (2.9) in a dimensionless form, similar to (3.21):

$$\frac{d^2 F}{d\varsigma^2} + F\left[\left(\frac{\omega n_0 L}{c}\right)^2 \left(1 + \frac{1}{g} - \frac{e^{-2\nu\varsigma}}{g}\right) - \frac{1}{4}\right] = 0; \quad \varsigma = \frac{\eta}{L}. \tag{3.32}$$

Substitution of $\tau = \exp(-\nu\varsigma)$, analogous to (3.23), transforms Eq. (3.32) into the Bessel equation (3.24) with

$$q^2 = -\frac{1}{g}\left(\frac{\omega n_0 L}{c\nu}\right)^2; \quad s^2 = \frac{1}{\nu^2}\left[\frac{1}{4} - \left(1 + \frac{1}{g}\right)\left(\frac{\omega n_0 L}{c}\right)^2\right]. \tag{3.33}$$

Since $q^2 < 0$, solution of Eq. (3.33) in the case $s^2 \geq 0$ is known to be represented by the modified Hankel functions, expressed, in their turn, by means of modified Bessel and Neumann functions $I_s(p\tau)$ and $K_s(p\tau)$ by analogy with (3.27); here $p^2 = -q^2 > 0$. Finally, the generating function Ψ for the wave equation (2.9) with barrier (3.28) is obtained, according to (3.20), in a form resembling (3.25):

$$\Psi = A\sqrt{1+x}\{I_s(\sigma) + iK_s(\sigma) + Q[I_s(\sigma) - iK_s(\sigma)]\}; \tag{3.34}$$

$$\sigma = p\tau = \frac{\omega n_0 L}{c\nu}(1+x)^{-\nu}. \tag{3.35}$$

Representation of the generating function Ψ in the form (3.34) is valid under the condition $s^2 \geq 0$; this means that this solution is valid for low frequencies ω, obeying the condition:

$$\omega \leq \omega_{cr} = \frac{c}{2n_0 L}\sqrt{\frac{g}{g+1}}. \tag{3.36}$$

Thus, for the values of the refractive index $n_0 = 1.8$, spatial scale $L = 100$ nm and $g = 0.2915$ the critical frequency ω_{cr}, found from (3.36), is $\omega_{cr} = 3.9610^{14}$ rad/s, which relates to the infrared spectral range ($\lambda > \lambda_{cr} = 4.65\,\mu$m). The thickness of this barrier d (3.28), given by the non-zero solution of equation $U^2(d/L) = 1$, for, e.g. $\nu = 1.5$, is $d/L = 1$, $d = 100$ nm. Using the standard procedure, i.e. satisfying the continuity conditions on the both sides of this gradient asymmetric barrier ($x = z/L = 0$ and $x = 1$), one can find the reflectance/transmittance spectra of this asymmetric barrier.

Considering asymmetric barriers one can recall the simplest monotonic model of such a barrier, exemplified by the one-parameter Rayleigh profile (2.4); more complicated non-monotonic distributions of dielectric susceptibilities, given by asymmetric profiles (3.19) and (3.28), contain two and three free parameters, respectively. A huge variety of asymmetric profiles $U(z)$, which may be needed for creation of controllable reflectance/transmittance spectra, can be achieved in multilayer nanostructures, containing groups of gradient photonic barriers, spaced by homogeneous layers of unequal thicknesses [3.2–3.4].

3.3. Inverse Functions and Parametric Presentations — New Ways to Model the Photonic Barriers

It was mentioned in Sec. 2, that the system of Maxwell equations (2.2) and (2.3) can be reduced to one equation by introducing another generating function θ [3.5], distinguished from function Ψ (2.8):

$$H_y = \frac{1}{c}\frac{\partial \theta}{\partial t}; \quad E_x = -\frac{1}{n_0^2 U^2}\frac{\partial \theta}{\partial z}. \tag{3.37}$$

Here the function $U(z)$ describes again an unknown barrier, which will be found below. Due to the representation (3.18), Eq. (2.3) becomes an identity, while the generating function θ is governed by the equation, following from (2.2),

$$\frac{d^2\theta}{dz^2} + \left(\frac{\omega n_0}{c}\right)^2 U^2 \theta = \frac{2}{U}\frac{dU}{dz}\frac{d\theta}{dz}. \tag{3.38}$$

To examine the regime of wave propagation through the gradient barrier $U^2(z)$, let us introduce the variable τ in a different way from (2.10)

$$\tau(z) = \int_0^z U^2(z_1)dz_1. \tag{3.39}$$

Using this variable, one can rewrite Eq. (3.38) in a form:

$$\frac{d^2\theta}{d\tau^2} + \frac{\omega^2 n_0^2}{c^2}\frac{\theta}{U^2(z)} = 0. \qquad (3.40)$$

Equation (3.40) contains simultaneously both the function θ, dependent on the variable τ, and the function U, dependent on the variable z. The function $U(z)$ remains unknown, and, to find any solutions of (3.38), one has to assume some link between the function $U(z)$ and the function $U(\tau)$, describing the profile of the barrier via the new variable τ. Since the choice of this link is not restricted by any preliminary conditions, let us consider, e.g. the following link

$$U^2(z)U^2(\tau) = 1, \qquad (3.41)$$

which transforms Eq. (3.38) into a form similar to Eq. (2.9), examined above in Ch. 2, but written in τ-space due to the replacement $\eta \to \tau$:

$$\frac{d^2\theta}{d\tau^2} + \frac{\omega^2 n_0^2}{c^2}U^2(\tau)\theta = 0. \qquad (3.42)$$

Combination of the condition (3.41) with the definition of the variable τ, (3.39), written in differential form $d\tau = U^2(z)dz$, produces the equation governing the function $U(\tau)$:

$$dz = U^2(\tau)d\tau. \qquad (3.43)$$

To solve Eq. (3.42), it makes sense to continue the analogy between (2.9) and (3.42) and to consider the function $U(\tau)$ in a form, coinciding with (2.16) after the replacement $\eta \to \tau$; putting for simplicity, $s_1 = 0$ in (2.16), we obtain:

$$U(\tau) = \left(1 - \frac{\tau^2}{L^2}\right)^{-1}. \qquad (3.44)$$

Differentiating Eq. (3.44), condition (3.41) is taken into account, and substitution of the expression for the differential $d\tau$ into (3.43) leads to the equation [3.5]

$$\frac{dz}{L} = -\frac{dU(z)}{2U^2(z)\sqrt{1-U(z)}}. \qquad (3.45)$$

Gradient Photonic Barriers: Generalizations of the Fundamental Model 57

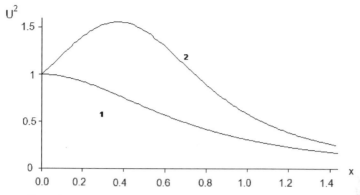

Fig. 3.4. Exactly solvable models of asymmetric gradient photonic barriers $U^2(x)$, represented by implicit functions (3.46) and (3.71)–(3.73), are shown by profiles 1 and 2, respectively, $x = z/L$, the maximum value of $U^2(x)$ for profile 2 is $U_m^2 = 1.5625$.

The integral of Eq. (3.45), written by means of (3.41), represents the normalized profile $U(z)$ in an inverted form:

$$\frac{z}{L} = x = \frac{1}{2}\left[\frac{\sqrt{1-U}}{U} - \ln\left(\frac{1-\sqrt{1-U}}{\sqrt{U}}\right)\right]. \qquad (3.46)$$

Profile (3.46) is shown in Fig. 3.4; one can see from (3.46), that $U(0) = 1$. However, this function $z = z(U)$ cannot be converted into an explicit dependence $U = U(z)$.

Although the profile of this gradient barrier $U(z)$ is represented by means of an inverted dependence, the analytical solution of Maxwell equations (2.2) and (2.3), reduced to Eq. (3.38), can be obtained in an explicit form. Using the formal similarity between Eqs. (2.9) and (3.42) and introducing in (3.42), by analogy with (2.10), the new function θ and new variable ς

$$f = \frac{\theta}{\sqrt{U(\tau)}}; \quad \varsigma = \int_0^\tau U(\tau_1)d\tau_1. \qquad (3.47)$$

We obtain the solution of (3.42) in a form of superposition of forward and backward waves, traveling in ς-space:

$$f = \frac{A[\exp(iq\varsigma) + Q\exp(-iq\varsigma)]}{\sqrt{U(\tau)}}. \qquad (3.48)$$

Here the wavenumber q is given by the familiar formula (2.14), with $p = L^{-1}$ and, thus, the cut-off frequency Ω is defined as

$$\Omega = c/n_0 L. \qquad (3.49)$$

Calculation of reflection/transmission coefficients can be performed now by means of standard continuity conditions on the planes $z = 0 (\tau = \varsigma = 0)$ and $z = d$. To find the value $\varsigma(d)$ one has to find first $\varsigma(\tau)$ from the integral (3.47). Expressing τ via $U(z)$ from (3.44),

$$\frac{\tau}{L} = \sqrt{1 - U(z)}, \qquad (3.50)$$

we'll obtain

$$\frac{\varsigma}{L} = \ln\left[\frac{1 + \sqrt{1 - U(z)}}{1 - \sqrt{1 - U(z)}}\right]. \qquad (3.51)$$

The value $\varsigma(d)$ is determined by $U(d)$ from (3.51), defined from (3.46). Thus, we have obtained an exactly solvable model for an asymmetric gradient photonic barrier, expressed via the inverse dependence $z = z(U)$.

Another exactly solvable model of a photonic barrier $U(z)$ is connected with the representation of the coordinate z and profile U as functions of some parameter while, unlike (3.46), the link between and U cannot be written in an explicit form. Propagation of waves through the gradient barrier $U(z)$, described by Eq. (3.38), can be examined for different barrier profiles, that can not be represented by a direct coordinate dependence of the dielectric susceptibility $\varepsilon = \varepsilon(z)$. Let us use again the variable η (2.10) and introduce the new functions f and W:

$$f = \frac{\theta}{\sqrt{U}}; \quad W^2(z) = \frac{1}{U(z)}. \qquad (3.52)$$

By means of substitutions (3.52), Eq. (3.38) can be rearranged to give [3.6]

$$\frac{\partial^2 f}{\partial \eta^2} - \frac{n_0^2}{c^2}\frac{\partial^2 f}{\partial t^2} = \frac{f}{W(z)}\frac{\partial^2 W(\eta)}{\partial \eta^2}. \qquad (3.53)$$

The right side of Eq. (3.53) contains the functions $W(z)$ and $W(\eta)$, dependent on different arguments; here the variables η and z are linked by the transformation (2.10). Till now these functions in Eq. (3.53) remain unknown, and one can assume any link between them, keeping in mind the link between the variables η and z, given in (2.10). Let us make, for simplicity, two assumptions:

a. $W(z) = -W(\eta)$;
b. the right side of Eq. (3.53) is equal to some real constant p^2.

These assumptions result in the equation, governing the function $W(\eta)$:

$$-\frac{1}{W(\eta)}\frac{d^2 W(\eta)}{d\eta^2} = p^2. \tag{3.54}$$

Substitution of (3.54) into (3.53) permits rewriting the latter for a harmonic time dependence of the EM field in a form, that coincides with (2.13):

$$\frac{\partial^2 f}{\partial \eta^2} + f\left(\frac{n_0^2 \omega^2}{c^2} - p^2\right) = 0. \tag{3.55}$$

Using this analogy, one can represent the solution of (3.55) as a superposition of forward and backward harmonic waves, traveling along the η-direction, ($f \approx \exp[\pm i(q\eta - \omega t)]$). To define the wavenumber q one has to examine the solution of Eq. (3.54), taking into account that these solutions depend on the sign of the constant p^2. Thus, in the case $p^2 > 0$ the function $W(\eta)$ can be written as:

$$W(\eta) = \cos(p\eta) + M\sin(p\eta). \tag{3.56}$$

This solution contains two unknown constants: the scale parameter p and the dimensionless factor M. To express these quantities via the extremal value of the refractive index $n_m = U_m$ and the barrier's thickness d, which are assumed to be known, let us write by means of (3.52) the expression for profile $U(\eta)$ [3.7]:

$$U(\eta) = [\cos(p\eta) + M\sin(p\eta)]^{-2}. \tag{3.57}$$

Calculating the extremal value U_{min} of the function (3.57), we find the constant M, connected with U_{min}:

$$U_{min} = \frac{1}{1+M^2}. \qquad (3.58)$$

Thus, the profile of $U(\eta)$, related to barrier (3.57), is concave: $U_{min} < 1$.

To find the profile $U(z)$ one can follow the method developed above: let us rewrite formula (2.10), determining the variable η, in the differential form $d\eta = U(z)dz$; then, making, according to (3.52), the substitution $U(z) = W^{-2}(\eta)$, where the function $W(\eta)$ is given by (3.56), we obtain the differential equation, defining the coordinate z inside the barrier as a function of the variable η:

$$dz = W^2(\eta)d\eta. \qquad (3.59)$$

Integration of Eq. (3.59) yields the link between z and η

$$pz = \frac{(1+M^2)p\eta}{2} + \frac{(1-M^2)\sin(2p\eta)}{4} + M\sin^2(p\eta). \qquad (3.60)$$

To express the unknown parameter p via the barrier thickness d, let us emphasize, that the condition $U(z) = 1$ is satisfied at the boundaries $z = 0$ and $z = d$ of the barrier, corresponding on the η axis to the points η_1 and η_2, which are the roots of the equation $W(\eta) = 1$:

$$\eta_1 = 0; \quad \eta_2 = \frac{2\mathrm{arctg}M}{p}. \qquad (3.61)$$

Substitution of the value η_2 from (3.61) into (3.60) yields the scale factor p:

$$p = \frac{1}{d}[M + (1+M^2)\mathrm{arctg}M]. \qquad (3.62)$$

Finally, manipulations with (3.52) and (3.56) result in the presentation of the profile $U(z) = U(\eta)$ in the form:

$$U(z) = [(1+M^2)\cos^2(p\eta - \mathrm{arctg}M)]^{-1}; \quad 0 \le p\eta \le 2\mathrm{arctg}M. \qquad (3.63)$$

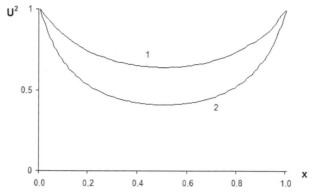

Fig. 3.5. Exactly solvable models of symmetric gradient photonic barriers $U^2(x)$, described by the parametric representations (3.60), (3.63), are shown by profiles 1 and 2, corresponding respectively to the values $M = 0.5$ and $M = 0.75$ in representation (3.60), $x = z/d$.

Expressions (3.60) and (3.63) determine the parametric representation of profile $U(z)$ for symmetric gradient barrier with thickness d and minimum value $U_{\min}$ (3.58) shown in Fig. 3.5.

To emphasize the differences between profiles (3.63) and (2.16) even in the case their thicknesses d and minimum values $U_{\min}$ are equal, one can compare the slope factors $\xi_{1,2}$ (3.1), calculated for the profiles (2.16) and (3.63) respectively. Setting in this case the values y (2.17) and M (3.58) to be equal and, using (3.2), we can find the ratio

$$\frac{\xi_1}{\xi_2} = 2[1 + (M + M^{-1})\operatorname{arctg} M]^{-1} < 1. \qquad (3.64)$$

Inequality (3.64) shows that the profile (3.63) is more steepened near the ends, and, thus, flatter in the central part, than profile (2.16).

To calculate the field components, determined by this generating function θ, one has to represent the field θ inside the gradient layer as a sum of forward and backward waves in η-space:

$$\theta = \sqrt{U}[\exp(iq\eta) + Q\exp(-iq\eta)]\exp(-i\omega t). \qquad (3.65)$$

Here the wave number q is defined by analogy with (2.14) as $q = \sqrt{\omega^2 n_0^2 c^{-2} - p^2}$; however, the value of the scale factor p is given, unlike (2.18), by formula (3.62).

Substitution of the solution (3.65) into definitions (3.37) yields explicit expressions for the E_x and H_y components of the EM fields inside the gradient barrier. The usual continuity conditions for these components produce the expression for the complex reflection coefficient R. Thus, despite the parametric representation of the profile $U(z)$, this representation proves to be an exactly solvable model of a gradient photonic barrier.

It is remarkable that the approach used here for implicit and parametric presentations of exactly solvable models of photonic barriers $U(z)$ can also be used for finding new explicit models of these barriers. Thus, returning to Eq. (3.43) and setting $U(\tau) = (\cos \tau)^{-1}$, we find from (3.43), that $z = \text{tg}\tau$. Substitution of this result into relation (3.41) yields an explicit exactly solvable model of a of photonic barrier, that coincides with the well known Caushy distribution:

$$U^2(z) = \left[1 + \left(\frac{z}{L}\right)^2\right]^{-1}. \tag{3.66}$$

Equation (3.42), describing wave propagation through the barrier (3.66), reads

$$\frac{d^2\theta}{d\tau^2} + \frac{\omega^2 n_0^2}{c^2} \frac{\theta}{\cos^2 \tau} = 0. \tag{3.67}$$

Solutions of this type of equation, useful for analysis of oblique propagation of waves through gradient media, will be examined below in Ch. 6.

Finally, let us note that solution (3.56)–(3.65) relates to the case $p^2 > 0$ (concave profile $U(z)$, characterized by $U_{\min}$ (3.58)). In the opposite case $p^2 < 0$ the function $W(\eta)$ (3.56) can be written as.

$$W(\eta) = \cosh(p\eta) + M \sinh(p\eta). \tag{3.68}$$

The representation (3.68) corresponds to a convex profile $U(z)$ with a maximum $U_{\max} = (1 - M^2)^{-1}$. Here the additional condition $M^2 < 1$ arises, and all the subsequent analysis has to be carried out according to the scheme (3.59)–(3.63). It is worth emphasizing, that profiles (3.56) and (3.68) correspond to the conditions $p^2 > 0$ and $p^2 < 0$, respectively. However, the special case $p = 0$ deserves to be mentioned too. In this case it follows from Eq. (3.54), that $W(\eta) = 1 + s_1\eta L^{-1}$; here $s_1 = \pm 1$, L is some unknown spatial scale. Substitution of this function $W(\eta)$ into Eq. (3.59) and use of (3.52) yields an explicit representation of the photonic barrier. Thus in the case $s_1 = 1$ we have

$$U^2(z) = \left(1 + \frac{3z}{L}\right)^{-\frac{4}{3}}. \quad (3.69)$$

The unknown parameter L can be found from the value of function (3.69) at the boundary of the barrier $z = d$. The solution of Eq. (3.55) under the condition $p = 0$ reads

$$f = \exp\left(\pm\frac{i\omega n_0 \eta}{c}\right); \quad \frac{\eta}{L} = \left(1 + \frac{3z}{L}\right)^{\frac{1}{3}} - 1. \quad (3.70)$$

The solution (3.70) represents a wave traveling in the heterogeneous dielectric, that does not possess heterogeneity-induced dispersion. Side by side with profile (2.63) model (3.69) exemplifies a dispersionless gradient photonic barrier.

Comments and Conclusions to Chapter 3

1. The thicknesses of gradient photonic barriers, intended for operating in some spectral range, are several times thinner than the wavelengths from the corresponding spectral range; this property is important for the design of miniaturized subwavelength dispersive elements.
2. The solution of wave equation (3.40) with the distribution of dielectric permittivity, represented by the implicit function (3.46), can be generalized by the replacement of model (3.44) by the more flexible one, obtained due to mapping the model $U(z)$ (2.16) to τ-space: $U(\tau) = (1 + \tau/L_1 - \tau^2/L_2^2)^{-1}$. By using the same algorithm, one can find that integration of Eq. (3.45) leads in this case to the convex profile, given by an implicit function, containing both ascending and descending branches, shown in Fig. 3.4. by curve 2. The ascending branch

$$x = x_m - x_1(\xi); \quad \xi = \frac{U}{U_m}; \quad \xi_0 \leq \xi \leq 1; \quad \xi_0 = U_m^{-1}, \quad (3.71)$$

is located in the interval $0 \leq x \leq x_m$; the value x_m, corresponding to the maximum of profile $U(z)$, is $x_m = x_1(\xi_0)$,

$$x_1(\xi) = \frac{U_m^{-\frac{3}{2}}}{2}\left[\frac{\sqrt{1-\xi}}{\xi} - \ln\left(\frac{1-\sqrt{1-\xi}}{\sqrt{\xi}}\right)\right]. \qquad (3.72)$$

The descending branch, located in the segment $x \geq x_m$ is

$$x = x_m + x_1(\xi); \quad 0 < \xi \leq 1. \qquad (3.73)$$

This profile, given by Eqs. (3.71)–(3.73), will be used in Sec. 5.2. for the analysis of spectra of graded left-handed metamaterials.

3. Note, that the exactly solvable models of gradient barriers, examined above, are based on continuous distributions of the dielectric permittivity $\varepsilon(z)$. However, these distributions, owing to technological conditions, are actually formed by plane layers of different optical and geometrical thicknesses. Thus, it is necessary to check the accuracy of the reflectance/transmittance spectra calculated for continuous profiles $\varepsilon(z)$, by comparing them with spectra, found due to the approximation of $\varepsilon(z)$ by multilayer structures. The goal of such a comparison is the optimization of parameters of a technologically controlled multilayer structure, approaching continuous profile $\varepsilon(z)$. An example of such optimization is shown below for the reflectance of a nanofilm with the linear profile $U^2 = 1 + z/L$, which is known to admit an exact analytical solution, expressed via Airy functions [3.8]. The reflection coefficients $|R|^2$ in the visible spectral range for this film were calculated analytically by means of these functions. Other results for the coefficients $|R|^2$ in the same spectral range were obtained by approximating this profile by a multilayer structure, containing n layers; the difference of the results as a function of the number of layers is depicted in Fig. 3.6.

Inspection of Fig. 3.6. shows that the discrepancy between the mean square deviation of $|R|^2$, found by means of Airy functions, from $|R|^2$, calculated in the framework of a multilayer model, does not exceed 1% for discretizing into 5 layers. Discretizing into 6 layers in this case seems to be optimized, since the subsequent increase of the number

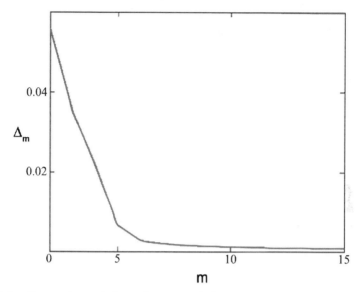

Fig. 3.6. Comparison of the reflectance of the gradient barrier, calculated by means of the exactly solvable model of the dielectric permittivity $U^2 = 1 + z/L$, and by an n-layer step-like distribution approximating this profile. The discrepancy Δ_n between the mean square deviation of reflection coefficients $|R|^2$ found in the framework of the exactly solvable model $U^2 = 1+z/L$, from $|R|^2$ calculated by means of a multilayer model, is plotted as a function of the number of layers n; parameters of the barrier are $n_0 = 1.5$, $n = 1.47$, $d = 500$ nm, $L = 10^4$ cm^{-1}. The optimized discretization, providing a discrepancy smaller than 1%, contains 6 layers.

of layers does not result in a significant decrease of the discrepancy. However, in the case of a more complicated profile $\varepsilon(z)$ the number of layers in the equivalent multilayer structure can be increased.

Bibliography

[3.1] M. Abramowitz and I. Stegun, *Handbook of Mathematical Functions* (Dover Publications, NY, 1968).
[3.2] P. Han and H. Wang, *JOSA B* **22**, 1571–1575 (2005).
[3.3] M. Deopura, C. K. Ullal, B. Temelkuran and Y. Fink, *Opt. Lett.* **26**, 1197–1199 (2001).
[3.4] L. Martin-Moreno, F. J. Garsia-Vidal, H. J. Lezek, K. M. Pellerin, T. Thio, J. B. Pendry and T. B. Ebbesen, *Phys. Rev. Lett.* **86**, 1114–1117 (2001).
[3.5] A. Shvartsburg, V. Kuzmiak and G. Petite, *Physics Reports*, **452**, 33–88 (2007).

[3.6] A. Shvartsburg, *Physics — Uspekhi*, **170**, 1297–1324 (2000).
[3.7] A. Shvartsburg, V. Kuzmiak and G. Petite, *Eur. Phys. J. B*, **72**, 77–88 (2009).
[3.8] P. Yeh, *Optical Waves in Layered Media, Wiley Series in Pure and Applied Optics*, 1997.

CHAPTER 4

RESONANT TUNNELING OF LIGHT THROUGH GRADIENT DIELECTRIC NANOBARRIERS

Tunneling is one of the fundamental processes in the dynamics of waves of different physical nature. The first steps in studies of these phenomena were taken in optics more than a century ago in the theoretical analysis of total internal reflection (TIR) of light, performed by A. Eikhenwald as long ago as in 1908 [4.1]. Using Maxwell equations, Eikhenwald showed, that the light wave, while incident on the boundary between two transparent dielectrics under an angle exceeding the critical angle for total internal reflection, does not vanish on this boundary, but penetrates partially into the subsurface layer of the reflecting medium. This penetration is accompanied by an exponential attenuation of the wave, the characteristic spatial scale of attenuation being about one wavelength. Unlike this effect of total internal reflection of a wave from the dispersiveless transparent half-space, the partial reflection from a non-transparent barrier of finite thickness, named "frustrated total internal reflection" (FTIR), began to attract attention in wave dynamics of dispersive media. However, the burst of interest in these intriguing phenomena of wave penetration into the forbidden area (tunneling effect) arose after Gamow's work [4.2], devoted to nuclear alpha-decay, connected with the penetration of an alpha-particle with energy E through a potential barrier a with maximum U_0, surrounding the nucleus, under the paradoxical condition $E < U_0$. In the framework of Gamow's approach alpha-decay of atomic nucleus was explained namely by tunneling of de Broglie waves, describing the alpha-particle, through this potential barrier. The exponentially small probability of this effect was calculated in this work by means of the formal analogy between equations governing electromagnetic and de Broglie waves.

Later on this approach catalyzed the applications of the tunneling concept to numerous EM wave phenomena in different spectral ranges — from optics [4.3] up to electromagnetics of a heterogeneous plasma [4.4]. These applications were based on the formal similarity between the stationary Schrodinger equation in quantum mechanics and the Helmholtz equation in the classical wave theory; tunneling of quantum particles through the forbidden zone was confronted with the propagation of an EM wave through a dispersive medium [4.5], the wave frequency being smaller than the cut-off frequency of the medium. Based on this equivalence, the electronic tunnel effect and the frustrated optical transmission phenomenon were shown to be related [4.6]. This opened up the opportunity to perform experiments with light beams, easier to perform and interpret than those with electron waves, since optical tunneling requires a micrometer or even hundreds of nanometers sized barrier. In these problems the tunneling effects were discussed from the viewpoint of wave propagation in heterogeneous media with so-called photonic barriers formed by metal nanofilms or heterogeneties of the dielectric permittivity ε in dielectric layers. However, the standard models of wave barriers, used in these topics (box-like barrier and linear potential), resulted in an exponentially small transmission; processes of highly effective energy transfer due to tunneling through more complicated barriers remained beyond of the scope of these models. The advent of nanotechnologies opened the avenue to creation of gradient dielectric films with feasible controlled spatial distributions of ε, providing, in particular, a series of unusual tunneling phenomena for EM waves [4.7]. The researches in this field are focused mainly on the effects of frustrated total internal reflection (FTIR) and penetration of evanescent waves through dispersive photonic barriers of finite thickness, demonstrating the growing potential for applications of gradient metamaterials in electromagnetics.

To compare and contrast the tunneling of waves through homogeneous and gradient photonic barriers it is worthwhile to recall some results of the theory of the interaction of an EM wave with a homogeneous layer of a metallic plasma with electron density N; this layer can be characterized by a plasma frequency ω_p and dielectric permittivity ε:

$$\omega_p^2 = \frac{4\pi e^2 N}{m}; \quad \varepsilon(\omega) = 1 - \frac{\omega_p^2}{\omega^2}, \qquad (4.1)$$

where m is the electron mass. Considering the normal incidence of a linearly polarized wave on a lossless plasma layer, one can find the generating function Ψ for the EM field inside the layer by means of Eq. (2.9). This layer is known to act as a high-pass filter: a high frequency wave ($\omega > \omega_p$) propagates through the plasma, while a low frequency wave ($\omega < \omega_p$) displays significant reflection and attenuation inside the plasma; this regime, considered to be the tunneling of the wave, is accompanied by a small transmission. To examine the last case, let us solve Eq. (2.9), by putting there $U^2 = \varepsilon$ (4.1). The generating function Ψ in this case ($\mathrm{Re}\,\varepsilon < 0$), written as

$$\Psi = A[\exp(-pz) + Q\exp(pz)]\exp(-i\omega t),$$
$$p = \frac{\omega}{c}\sqrt{u^2 - 1}; \quad u = \frac{\omega_p}{\omega} \geq 1, \qquad (4.2)$$

which represents the result of the interference of two monochromatic fields with imaginary wave numbers and A is the normalization constant. Substitution of (4.2) into (2.8) yields the field components E_x and H_y inside the plasma layer:

$$E_x = \frac{i\omega A}{c}[\exp(-pz) + Q\exp(pz)],$$
$$H_y = -pA[\exp(-pz) - Q\exp(pz)]. \qquad (4.3)$$

Denoting the boundaries of the layer as the planes $z = 0$, bordering with air, and $z = d$, bordering with a homogeneous dielectric substrate with refractive index n, and using the standard boundary conditions of continuity of E_x and H_y on these planes, one can calculate the parameter Q in (4.3); after this the complex reflection/transmission coefficients R and T of the plasma layer in the tunneling regime can be found by means of the standard procedure, used repeatedly in Chs. 2 and 3:

$$R = \frac{1 - in_e\Lambda}{1 + in_e\Lambda}, \quad \Lambda = \frac{1-Q}{1+Q},$$

$$Q = -\exp(-2pd)\left(\frac{n - in_e}{n + in_e}\right), \quad n_e = \sqrt{u^2 - 1}. \qquad (4.4)$$

Manipulations with the quantities Q and Λ in (4.4) results in the representation of R in the form (2.31):

$$R = \frac{\sigma_1 + i\sigma_2}{\chi_1 + i\chi_2}, \quad \sigma_1 = t(n + n_e^2), \quad \sigma_2 = -n_e(n-1),$$
$$\chi_1 = t(n - n_e^2), \quad \chi_2 = n_e(n+1), \quad t = \text{th}(pd). \qquad (4.5)$$

The complex transmission coefficient T_E may be written by analogy with (2.34):

$$T_E = \frac{2in_e\sqrt{1-t^2}}{\chi_1 + i\chi_2} = |T_E|\exp(i\phi_t), \qquad (4.6)$$

where ϕ_t is the phase shift of the transmitted wave. Finally, the transmission coefficient with respect to intensity $|T|^2 = T_E T_H^* = n|T_E|^2$ is

$$|T|^2 = 1 - |R|^2 = \frac{4nn_e^2(1 - t^2)}{|\chi|^2}. \qquad (4.7)$$

The phase shift of the transmitted wave is

$$\phi_t = \text{arctg}\left[\frac{(n - n_e^2)\text{th}(pd)}{(n+1)n_e}\right]. \qquad (4.8)$$

Inspection of formulae (4.6) and (4.7) permits outlining the main features of EM wave tunneling through a homogeneous layer of electronic plasma:

1. The transmittance decreases exponentially, when the barrier thickness d increases: when $pd \gg 1$, one has from (4.7): $|T|^2 \propto \exp(-2pd)$; hence $|R|^2 \to 1$, and a weakly attenuated (reflectionless) regime cannot arise.

2. The velocity of energy transfer through the barrier may be found from the z-component of the Poynting vector P_z and the energy density W for a dispersive ($\varepsilon = \varepsilon(\omega)$) non-magnetic medium [4.8]:

$$v_g = \frac{P_z}{W}, \quad P_z = \frac{c}{4\pi}\mathrm{Re}(E_x H_y^*),$$

$$W = \frac{1}{8\pi}\left\{\frac{\partial}{\partial\omega}[\omega\varepsilon(\omega)]|E|^2 + |H|^2\right\}. \qquad (4.9)$$

Calculating P_z and W by means of the expressions for Q (4.4) and $\varepsilon(\omega)$ (4.1), we find

$$Q + Q^* = -\frac{2(n^2 - n_e^2)\exp(-2pd)}{n^2 + n_e^2},$$

$$Q - Q^* = \frac{4inn_e \exp(-2pd)}{n^2 + n_e^2}, \quad |Q|^2 = \exp(-4pd).$$

$$P_z = \frac{cMnn_e^2}{n^2 + n_e^2}; \quad W = \frac{M}{2}\left(u^2 - \frac{n^2 - n_e^2}{n^2 + n_e^2}\right); \qquad (4.10)$$

$$M = \frac{1}{\pi}\left|\frac{\omega A \exp(-pd)}{c}\right|^2.$$

thus, both P_z and W are decaying exponentially in the depth of the plasma layer. Substitution of P_z and W (4.10) into the expression for v_g (4.9), and use of the definition $n_e^2 = u^2 - 1$ (4.4) produces a simple expression for the velocity of energy transfer in this tunneling regime:

$$\frac{v_g}{c} = \frac{2n}{1 + n^2 + u^2}. \qquad (4.11)$$

The result obtained shows, that the velocity v_g is always subluminal ($v_g < c$) and possesses a constant value at any point inside the homogeneous plasma layer.

3. Analysis of the expressions for R (4.5) and T_E (4.6) permits establishing a link between the phases of the reflected (ϕ_r) and transmitted (ϕ_t) waves. It is worthwhile to present the complex

reflection coefficient R in the form

$$R = |R|\exp(i\phi_r), \quad |R| = \sqrt{\frac{|\sigma|}{|\chi|}}, \quad \sigma = \sqrt{\sigma_1^2 + \sigma_2^2}. \qquad (4.12)$$

The quantities $\sigma_{1,2}$ and $|\chi|$ are defined in (4.5) and (4.6), respectively, while the phase ϕ_r is given by the relations:

$$\cos\phi_r = \frac{\sigma_1\chi_1 + \sigma_2\chi_2}{\sigma|\chi|}, \quad \sin\phi_r = \frac{\chi_1\sigma_2 - \chi_2\sigma_1}{\sigma|\chi|}. \qquad (4.13)$$

Introducing an auxiliary phase ϕ_σ via the equalities

$$\sin\phi_\sigma = \frac{\sigma_1}{\sigma}, \quad \cos\phi_\sigma = \frac{\sigma_2}{\sigma}, \qquad (4.14)$$

and substituting it together with the definition of the phase of the transmitted wave ϕ_t (4.8) into (4.13), we find the relation between the phases ϕ_r and ϕ_t: $\phi_r = \phi_t - \phi_\sigma$. In the simplest case, when the plasma layer is considered without a substrate ($n = 1$, $\sigma_2 = 0$), it follows from (4.14), that $\phi_\sigma = \pi/2$ and, thus [4.9],

$$\phi_t - \phi_r = \frac{\pi}{2}. \qquad (4.15)$$

Note, that under the barrier no phase shift is accumulated, and the entire phase shift comes from the boundaries [4.10].

A widely used model of a gradient layer, providing a continuous transition between regions with $\varepsilon > 0$ and $\varepsilon < 0$ is given by a linear profile of the normalized dielectric permittivity $U^2(z) = 1 - z/L$. The exact analytical solutions of wave equation (2.9) for this profile $U^2(z)$ are given by Airy functions [4.11], presenting the EM field both in traveling ($\varepsilon > 0$) and tunneling ($\varepsilon < 0$) regions. These solutions are discussed in details in many text books, devoted to geophysics [4.12] and plasma electromagnetics [4.13] and, therefore, this standard model of a linear distribution of dielectric susceptibility remains beyond of the scope of this book.

In contrast, this chapter is focused on the salient features of EM waves tunneling through gradient transparent dielectric nanofilms without free carriers. The effect of non-local dispersion, described above in Chs. 2–3, is shown here to provide a new mechanism of tunneling in dielectrics with $\mathrm{Re}\,\varepsilon > 0$; this condition reveals the principal distinction between the gradient and abovementioned homogeneous layers. Such an unusual situation, examined in Sec. 4.1, is shown to arise for photonic barriers (2.16) with a concave profile $U^2(z)$ (Fig. 2.1). Transmission spectra $T(u)$ for such photonic barriers illustrate the possibility of weakly attenuated, almost reflectionless tunneling of light ($|T|^2 \to 1$, $|R|^2 \to 0$) in some spectral ranges below the cut-off frequency $\omega = \Omega_1$. These transmission spectra, as well as the velocity of EM energy transmission through gradient barriers by evanescent waves, are examined in Sec. 4.2. Note, that these phenomena of weakly attenuated tunneling of light are caused by the interference of waves, reflected from every point inside the gradient layer. In contrast, Sec. 4.3. is devoted to the propagation of waves through a homogeneous transparent plane layer, located in the curvilinear narrowing of a waveguide; the interference of waves, reflected from every point of these curvilinear boundaries, is shown to provide the weakly attenuated tunneling through the subwavelength narrowing.

4.1. Transparency Windows for Evanescent Modes: Amplitude — Phase Spectra of Transmitted Waves

It was emphasized in Sec. 2.1, that the gradient dielectric photonic barriers with a concave profile of dielectric permittivity $U^2(z)$ (Fig. 2.1) possess the normal waveguide-like dispersion and cut-off frequency Ω_1 (2.19), determined by non-local dispersion of the dielectric nanobarrier. This feature can change drastically the reflectance/transmittance spectra for EM waves in the low frequency spectral range ($\omega < \Omega_1$). Such spectra are examined below for waves incident normally from the air ($z \leq 0$) on the plane interface of the photonic barrier ($z = 0$). The EM field inside this barrier is determined, instead of (2.21), by the generating function Ψ_t and imaginary wave number $q = ip$ [4.14]:

$$\Psi_t = \frac{A[\exp(-p\eta) + Q\exp(p\eta)]\exp(-i\omega t)}{\sqrt{U(z)}},$$

$$p = \frac{\omega n_e}{c}, \quad n_e = n_0\sqrt{u^2 - 1}, \quad u = \frac{\Omega_1}{\omega} > 1.$$

(4.16)

Considering a barrier with thickness d, located on a homogeneous substrate with refractive index n, one can calculate the parameter Q, describing the contribution of the backward wave to the field inside the barrier:

$$Q = -\exp(-2p\eta_0)\left(\frac{n - \frac{i\gamma}{2} - in_e}{n - \frac{i\gamma}{2} + in_e}\right), \quad \gamma = \frac{2un_0 y}{\sqrt{1+y^2}}. \quad (4.17)$$

Proceeding by analogy with the derivation of formulas for the reflection and transmission coefficients R and T_E in the travelling regime (Ch. 2.2), we obtain these coefficients for a gradient barrier in the tunneling regime, written in forms (4.5) and (4.6), respectively:

$$R = \frac{\sigma_1 + i\sigma_2}{\chi_1 + i\chi_2},$$

$$\sigma_1 = t\left(n + \frac{\gamma^2}{4} + n_e^2\right) - \gamma n_e, \quad \sigma_2 = -(n-1)\xi, \quad (4.18)$$

$$\chi_1 = t\left(n - \frac{\gamma^2}{4} - n_e^2\right) + \gamma n_e, \quad \chi_2 = (n+1)\xi, \quad \xi = n_e - \frac{\gamma t}{2},$$

$$n_e^2 = n_0^2(u^2 - 1), \quad t = \text{th}(l\sqrt{1-u^{-2}}),$$

$$l = \ln\left(\frac{\sqrt{1+y^2}+y}{\sqrt{1+y^2}-y}\right), \quad \gamma = \frac{2un_0 y}{\sqrt{1+y^2}}. \quad (4.19)$$

Expression (4.18), determining the reflection coefficient for an asymmetric optical structure (air-gradient layer-substrate), can be easily generalized for a symmetric structure, containing the gradient layer between two similar homogeneous substrates with equal refractive indices n. The reflection coefficient for this symmetrical structure can be presented in the standard form (4.18), where

$$\sigma_1 = t\left(n^2 + \frac{\gamma^2}{4} + n_e^2\right) - \gamma n_e, \quad \sigma_2 = 0,$$

$$\chi_1 = t\left(n^2 - \frac{\gamma^2}{4} - n_e^2\right) + \gamma n_e, \quad \chi_2 = 2n\xi. \quad (4.20)$$

In view of subsequent applications it is worthwhile to rewrite the expression for the complex transmission coefficient T_E by separating its real and imaginary parts:

$$T_E = |T_E|\exp(i\phi_t), \quad |T_E| = \frac{2n_e\sqrt{1-t^2}}{|\Delta|},$$

$$\mathrm{tg}\phi_t = \frac{\chi_1}{\chi_2}, \quad |\Delta|^2 = |\chi_1|^2 + |\chi_2|^2. \tag{4.21}$$

Now one can find the complex transmission coefficient of the gradient barrier with respect to intensity $|T|^2 = T_E T_H^* = n|T_E|^2$. Examples of spectra $|T|^2$ and the phase of the transmitted wave ϕ_t, characterizing the single photonic barrier, are depicted in Fig. 4.1. These spectra show a high transmittance ($|T|^2 \geq 0.85$), decreasing monotonically at the longer wavelengths. Note, that, putting in expressions for $\sigma_{1,2}$ and $\chi_{1,2}$ (4.18) the value $y = 0$ we obtain again formulae (4.5) and (4.6) for the homogeneous layer.

The multilayer gradient barrier, containing m ($m > 1$) similar adjusted barriers, can be examined in the same way. Attributing the number $m = 1$ to the first layer at the far side of this stack, we write the reflection coefficient for this stack R_m by means of continuity conditions on its near side $z = 0$

$$R_m = \frac{1 + \frac{i\gamma}{2} - in_e\Lambda_m}{1 - \frac{i\gamma}{2} + in_e\Lambda_m}, \quad \Lambda_m = \frac{1-Q_m}{1+Q_m}. \tag{4.22}$$

The parameters Λ_m in (4.22) are linked by the chain of recursive relations, obtained from the continuity conditions on the interfaces between the $(m+1)$-th and (m)-th barriers ($m > 1$):

$$\Lambda_m = \frac{n_e(\Lambda_{m-1}+t)-\gamma}{n_e(1+t\Lambda_{m-1})-\gamma t}. \tag{4.23}$$

The first term in this chain Λ_1, related to the far layer in the stack, is defined via the quantity Q (4.17):

$$\Lambda_1 = \frac{1-Q}{1+Q} = \frac{n - \frac{i\gamma}{2} + in_e t}{t\left(n - \frac{i\gamma}{2}\right) + in_e}. \tag{4.24}$$

Transmittance spectra for a periodic structure, containing several similar adjacent gradient barriers with concave profiles $n(z)$ (2.16),

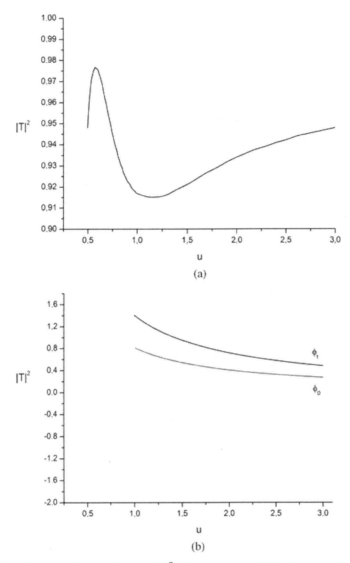

Fig. 4.1. Transmission coefficient $|T|^2$ (a) and phase shift of transmitted wave ϕ_t (b) are plotted vs $u = \Omega/\omega$ for a gradient barrier with the profile $n(z)$, shown in Fig. 2.1; $n_0 = 2.3$; $n_m = n = 1.47$. Curve ϕ_0 shows the phase shift accumulated by a wave with the same frequency ω, traveling in a free space along the distance d, equal to the barrier's thickness.

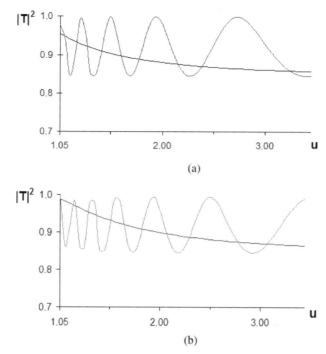

Fig. 4.2. Resonant transmission spectra $|T|^2$ for waves tunneling through a periodic structure, containing m gradient nanobarriers (2.16), supported by a homogeneous thick substrate with refractive index $n = 2.3$; the nanobarrier thickness $d = 100$ nm. (a): $n_0 = 1.8928$, $y = 0.577$, cut-off frequency $\Omega = 2.1210^{15}$ rad/s; (b): $n_0 = 2.2187$; $y = 0.75$, cut-off frequency $\Omega = 2.45\ 10^{15}$ rad/s; spectra 1 and 2 relate to nanostructures, containing $m = 1$ and $m = 20$ nanobarriers respectively.

are shown in Fig. 4.2. The following features of these spectra have to be emphasized:

1. The photonic barriers with cut-off frequencies Ω_1, restricting the spectral range of the tunneling regime, can be formed in transparent gradient dielectric structures ($\mathrm{Re}\,\varepsilon > 0$). The wave energy is transmitted through these structures by means of evanescent modes.
2. Interference of evanescent modes in the stack of gradient layers results in the complication of their transmittance spectra subject to the number of layers: thus, the monotonic decrease of

the transmittance of a single layer, shown by curves 1 in the Figs. 4.2(a) and 4.2(b), is replaced by periodic transmittance spectra, inherent to the stack containing 10 layers. Thus, instead of frustrated total internal reflection in the homogeneous plasma layer, arising for low frequencies ($\omega < \omega_p$), this structure possesses complete transmittance.

3. The reflectionless tunneling ($|R_m|^2 \to 0, |T_m|^2 \to 1$) is shown to arise for a series of frequencies in a periodic gradient structure containing several nanofilms (Fig. 4.2). Comparison of Figs. 4.2(a) and 4.2(b) shows that the increase of refractive index n_0 as well as the increase of the depth of this index modulation in the gradient structure, characterized by the parameter $y = \sqrt{n_0/n_{\min} - 1}$, results in an increase in the number of peaks with $|T|_m^2 \to 1$, accompanied by a narrowing of these peaks.

4.2. Energy Transfer in Gradient Media by Evanescent Waves

Inspection of amplitude-phase spectra of waves traversing the gradient barrier in the tunneling regime (4.16), shows the link between the amplitude and phase of the transmitted wave. To optimize the properties of the transmitted radiation it is worthwhile to consider its amplitude and phase spectra separately. Let us outline first some salient features of amplitude spectra:

1. Until now we were discussing the transmittance spectra for travelling ($u < 1$) and tunneling ($u > 1$) monochromatic waves separately. The expressions, describing these spectra were shown to possess the transition from travelling to tunneling regime due to the simple replacement $\sqrt{1 - u^2} \to i\sqrt{u^2 - 1}$ [4.14] in these expressions. However, to examine the energy transfer by polychromatic waves and pulses, containing both subcritical ($u < 1$) and supercritical ($u > 1$) frequencies, it makes sense to consider their transmittance spectra in the vicinity of the transition point ($u = 1$). Recalling the expressions for the complex transmission coefficients T_E for traveling (2.34) and tunneling (4.21) waves, one can see, that the transmission coefficient with respect to energy

$|T|^2$ is continuous at the transition point; its value is:

$$|T(u=1)|^2 = \frac{4nn_0^2}{\left[l\left(n - \frac{\gamma^2}{4}\right) + \gamma n_0\right]^2 + (n+1)^2\left(n_0 - \frac{\gamma l}{2}\right)^2}. \quad (4.25)$$

2. The components of a tunneling wave inside the gradient layer possess a complicated spatial structure and coordinate-dependent mutual phase shift. To examine the spatial structure of low frequency ($u > 1$) E and H components of tunneling fields, represented inside the barrier ($0 \leq z \leq d$) by evanescent modes, one has to substitute the generating function Ψ_t (4.16) into equations (2.8):

$$E = \frac{E_0 M}{\sqrt{U(z)}}[\exp(-p\eta) + Q\exp(p\eta)], \quad M = \frac{1+R}{1+Q}, \quad x = \frac{z}{d},$$

$$H = -iE_0 M \sqrt{U(z)} \left\{ \left[\frac{\gamma}{2}(1-2x) - n_e\right]\exp(-p\eta) \right.$$

$$\left. + Q\left[\frac{\gamma}{2}(1-2x) + n_e\right]\exp(p\eta) \right\}. \quad (4.26)$$

For simplicity the temporal factor $\exp(-i\omega t)$ is omitted in Eq. (4.26). Factors Q and R, determining the amplitude of evanescent mode, were calculated in (4.16) and (4.17) from the usual continuity conditions on the boundaries $z = 0$ and $z = d$. Equations (4.26) illustrate the heterogeneous distribution of the field energy density inside the gradient layer, which is needed for the calculation of the energy transfer velocity.

3. The velocity of energy transfer by evanescent modes v_g can be found by substitution of (4.26) into formulae (4.9). Calculation of the energy flux P_z and energy density W yields the expressions:

$$P_z = \frac{c}{\pi} \frac{|E_0|^2 |M|^2 nn_e^2 \exp(-2p\eta_0)}{|\Delta|^2},$$

$$W = \frac{|E_0|^2 |M|^2 n_e^2 U \exp(-2p\eta_0)}{2\pi |\Delta|^2}[n^2 + n_0^2 + \gamma^2(1-x)^2], \quad (4.27)$$

$$0 \leq x \leq 1, \quad U = U(z) = [1 + 4y^2 x(1-x)]^{-1}.$$

The quantity $|\Delta|^2$ was determined in (4.21). The definition of the energy flux P_z (4.9) shows that the forward as well the as backward wave alone produces no energy flux; only their superposition (4.15) represents the tunneling electromagnetic flux. Substitution of (4.27) into Eq. (4.9) brings the value of the normalized velocity V

$$V = \frac{v_g}{c} = \frac{2n[1 + 4y^2 x(1-x)]}{n^2 + n_0^2 + \gamma^2 (1-x)^2}. \qquad (4.28)$$

In the limiting case, where the heterogeneity is vanishing ($y \to 0, \gamma \to 0$), Eq. (4.28) is reduced to a well-known expression for normalized the group velocity in a homogeneous transparent layer, located on a substrate:

$$V = \frac{2n}{n^2 + n_0^2}. \qquad (4.29)$$

The spatial distribution of V across the gradient layer (Fig. 4.3) shows that the velocity of energy transfer v_g by the tunneling mode is frequency- and coordinate-dependent; however, this velocity always remains subluminal ($V < 1$). Moreover, in the case, where the homogeneous layer and the substrate are produced from the same material ($n = n_0$), Eq. (4.29) reduces to the well known expression for the

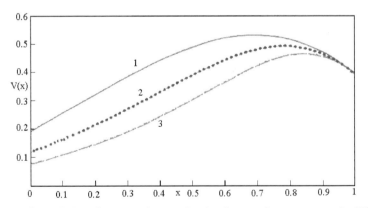

Fig. 4.3. Spatial distributions of normalized velocity of energy transfer $V(x) = v_g/c$ by waves tunneling through the gradient barrier ($y^2 = 0.5646$; $n_0 = 2.3$) located on a homogeneous substrate with refractive index $n = 1.47$; $x = z/d$ — the normalized coordinate across the barrier. Curves 1, 2, 3 relate to the normalized frequencies $u = 1; 1.5; 2$.

normalized group velocity in a homogeneous non-dispersive medium: $V = n^{-1}$.

Now let us consider some tunneling-related phase effects in the transmitted waves:

1. It is seen that tunneling of a wave through a homogeneous layer is always accompanied by a non-vanishing reflection; this means, that the condition $R = 0$, where the reflection coefficient R is defined in (4.5), never can be fulfilled. In contrast, this condition can be fulfilled for some gradient structures, shown, e.g. in Fig. 4.2, for the frequencies, related to the cases $|T| = 1$. In the case, corresponding to tunneling through the simple symmetric structure, containing one gradient layer, located between two homogeneous substrates (4.20), the condition of reflectionless tunneling $R = 0$ is reduced to the equation

$$\gamma n_e = t \left(n^2 + \frac{\gamma^2}{4} + n_e^2 \right)^{-1}. \qquad (4.30)$$

Substitution of Eq. (4.30) into the definition of the phase shift ϕ_t (4.21) permits calculating ϕ_t in the case $R = 0$:

$$\phi_t = \text{arctg}\left(\frac{\gamma t - 2n_e}{nt}\right). \qquad (4.31)$$

Thus, the structure discussed can be viewed as a phase shifter, leaving the wave amplitude unchanged.

2. In the tunneling spectral range phase shift ϕ_t can exceed the phase shift ϕ_0, accumulated by the wave with the same frequency w, traversing the same distance d in free space; according to the definition (2.19) the phase shift ϕ_0 can be found as

$$\phi_0 = \frac{\omega d}{c} = \frac{2y\sqrt{1+y^2}}{n_0 u}. \qquad (4.32)$$

this superluminal effect ($\phi_t > \phi_0$) is shown in Fig. 4.1(b).

3. The phase shift ϕ_t is linked with an another important characteristic of tunneling phenomena — the so-called phase time τ_s [4.15]

$$\tau_s = \frac{\partial \phi_t}{\partial \omega}. \qquad (4.33)$$

It is worthwhile to compare this time τ_s with another time scale $t_0 = d/c$, determining the travel time of radiation with the free space light velocity through the distance d, equal to the width of gradient layer [4.16]. Bringing together the quantities τ_s and t_0 we find from Eq. (4.33)

$$\frac{\tau_s}{t_0} = -\frac{u}{\phi_0}\frac{\partial \phi_t}{\partial u}. \tag{4.34}$$

Here the phase ϕ_0 is defined in (4.32). Using the phase spectra ϕ_t and ϕ_0, depicted in Fig. 4.1, one can see that in the spectral range $1.5 < u < 3$ we have $\tau_s > t_0$, namely, $\tau_s \approx (1.2 \div 2.5)t_0$. Note, that this time does not relate to any "phase velocity of the tunneling wave", since no phase shift of an evanescent wave is accumulated inside the gradient barrier [4.16], and the phase shift ϕ_t of the transmitted wave is governed by the boundary conditions on the interfaces of the gradient layer. However, the subluminal speed of energy transfer in the tunneling region (4.28) may be considered as the tunneling speed of light [4.17].

4.3. Weakly Attenuated Tunneling of Radiation Through a Subwavelength Slit, Confined by Curvilinear Surfaces

Another family of heterogeneous wave barriers, producing the resonant tunneling of light, may be formed in directional systems, containing a slit, confined by curvilinear interfaces. This peculiar type of heterogeneity-induced tunneling proves to be controlled by the geometrical parameters of the confining interfaces. To examine this effect, let us consider the propagation of the TE_{10} mode with EM field components E_y, H_x, H_z along the single-mode waveguide with axis z and thickness $d(-d/2 \leq x \leq d/2)$, filled by a dielectric with dielectric permittivity ε. The waveguide contains a symmetric smoothly shaped narrowing in the area $(-b/2 \leq z \leq b/2)$, and the planes $x = d/2$ and $x = -d/2$ are assumed to be perfectly conducted (Fig. 4.4). Our goal is to examine the wave propagation through the slit, formed by these convex curves, determining its complex reflection/transmission coefficients [4.18].

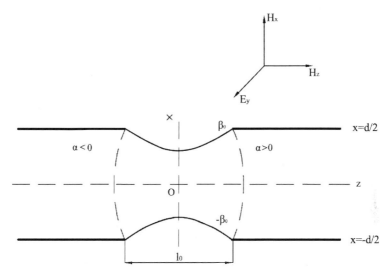

Fig. 4.4. Geometry of the narrowed waveguide. The following geometrical parameters are indicated: the narrowing length b, the distance between the waveguide walls d; the distance between the tops of the coordinate lines β_0 and $-\beta_0$ determines the minimal width s.

To calculate these coefficients let us consider the narrowings, formed by hyperbolic or elliptical surfaces. To do this its convenient to introduce in the (x, z) plane the curvilinear coordinate system (α, β), formed by mutually orthogonal ellipses α and hyperbolas β [4.19]:

$$x = a \operatorname{ch} \alpha \sin \beta, \quad z = a \operatorname{sh} \alpha \cos \beta. \tag{4.35}$$

The values $\alpha < 0$ ($\alpha > 0$) relate to the half-plane $z < 0$ ($z > 0$), the line $\alpha = 0$ corresponds to the x-axis. Analogously, the values $\beta > 0$ ($\beta < 0$) relate to the upper (lower) half-planes, the line $\beta = 0$ coincides with the z-axis. Let us consider, e.g. the slit, formed by symmetrical hyperbolas, located between the points $x = d/2, z = \pm b/2$ and $x = -d/2, z = \pm b/2$. In the coordinate system (α, β) these hyperbolas coincide with the coordinate lines β_0 and $-\beta_0$. Designating the minimal width of the slit (the distance between the tops of hyperbolas) as s, one can define the parameters β_0 and a:

$$\beta_0 = \pm \operatorname{arctg}\left(\frac{\sqrt{d^2 - s^2}}{b}\right), \quad a = \frac{s}{2|\sin \beta_0|}. \tag{4.36}$$

The components of a TE_{10} mode travelling in this waveguide are known to be expressed via the generating function Ψ:

$$E_y = -\frac{1}{c}\frac{\partial \Psi}{\partial t}, \quad H_x = \frac{\partial \Psi}{\partial z}, \quad H_z = -\frac{\partial \Psi}{\partial x}. \tag{4.37}$$

The wave equation governing the generating function Ψ inside the slit, is derived from the Maxwell equations. Using the (α, β) coordinates and separating the variables $\Psi = F(\alpha)f(\beta)$, one can find the equations, determining the unknown functions $F(\alpha)$ and $f(\beta)$:

$$\frac{d^2 F}{d\alpha^2} + p^2(\text{ch}^2\alpha - A)F = 0, \quad p^2 = \frac{\omega^2 a^2 \varepsilon}{c^2}, \tag{4.38}$$

$$\frac{d^2 f}{d\beta^2} + p^2(A - \sin^2 \beta)f = 0. \tag{4.39}$$

Here A is some dimensionless constant, which will be determined below from the boundary conditions.

Rigorously speaking, the regular solutions of Eqs. (4.38) and (4.39) can be written in terms of Mathieu functions, which are known to be expressed via power series. However, for our goals it makes sense to present the solutions of these equations directly by power series, defining simultaneously the eigenvalues A. The linearly independent solutions of Eq. (4.38) may be written by means of even F_1 and odd F_2 functions

$$F_1 = \sum_{n=0}^{\infty} F_{1n}\alpha^{2n}, \quad F_2 = \sum_{n=0}^{\infty} F_{2n}\alpha^{2n+1}. \tag{4.40}$$

Substitution of the function $\text{ch}^2\alpha$, expanded in a Taylor series, into Eq. (4.38) yields the coefficients F_{1n} and F_{2n} in the solutions (4.40). The values of the first coefficients are:

$$F_{10} = 1, \quad F_{11} = -\frac{p^2(1-A)}{2},$$

$$F_{12} = \frac{p^2}{12}\left[\frac{p^2(1-A)^2}{2} - 1\right], \tag{4.41}$$

$$F_{20} = 1, \quad F_{21} = -\frac{p^2(1-A)}{6},$$
$$F_{22} = \frac{p^2}{20}\left[\frac{p^2(1-A)^2}{6} - 1\right]. \tag{4.42}$$

Since, according to (4.37), the electric field component E_y is given by $E_y = E_0 F(\alpha) f(\beta)$, the function $f(\beta)$ satisfies the boundary condition:

$$f(\beta_0) = f(-\beta_0) = 0. \tag{4.43}$$

Due to condition (4.43) one has to use below only an even solution of Eq. (4.39):

$$f = \sum_{n=0}^{\infty} f_n \beta^{2n}, \quad f_0 = 1, \quad f_1 = -\frac{p^2 A}{2}, \quad f_2 = \frac{p^2}{12}\left(1 + \frac{p^2 A^2}{2}\right). \tag{4.44}$$

The parameter A, indicating the separation constant, till now remains unknown. Substitution of solution f (4.44) to condition (4.43) yields an infinite set of values of A; each of these values relates to some mode of the field in the range $(-b/2 \leq z \leq b/2)$. Thus, using the series

$$\sin^2\beta = \beta^2 - \frac{\beta^4}{3} + \frac{2\beta^6}{45} \cdots \tag{4.45}$$

and restricting ourselves to the first two terms in (4.45), we obtain from (4.43) for a single mode approximation for the field inside the slit:

$$A_0 = \frac{2}{(p\beta_0)^2}. \tag{4.46}$$

To evaluate the accuracy of this approximation one has to take into account that the rectangular waveguide with walls d and $l(d > l)$ supports the propagation of waves in the TE_{10} mode in the following spectral range of wavelengths λ

$$d\sqrt{\varepsilon} < \lambda < 2d\sqrt{\varepsilon}. \tag{4.47}$$

Choosing the waveguide's side d, obeying condition (4.47) for the spectral range under discussion, we can find from (4.36) the value β_0, related to the width of the slit s and the size of narrowing b (Fig. 4.4). The field inside the slit is characterized by values of the variable β, located in the interval $\beta_0 > \beta > -\beta_0$.

Determining the value β_0, we can neglect the β^4 and subsequent terms in (4.45), if $\beta_0^2 \ll 3$; putting, e.g. $\beta_0^2 = 0.3$ ($|\beta_0| = 0.55$), we can represent the field in the narrowing by one mode (4.46).

To examine the reflectance and transmittance of the narrowed section of waveguide, one has to use the continuity conditions on the planes $z = -b/2$ and $z = b/2$, bounding the narrowed section. The components of the TE_{10} mode, incident on the slit from the area $z = -b/2$, are described by the generating function Ψ_1

$$\Psi_1 = B_1 \cos(k_\perp x) \exp[i(\gamma z - \omega t)], \quad k_\perp = \frac{\pi}{d}, \quad \gamma = \sqrt{\frac{\omega^2}{c^2}\varepsilon - k_\perp^2}, \tag{4.48}$$

where B_1 is the normalization constant. Substitution of function Ψ_1 into definitions (4.37) yields the expressions for the mode components. Omitting for simplicity the phase factor, we obtain:

$$E_y = \frac{i\omega B_1}{c} \cos(k_\perp x), \quad H_x = i\gamma B_1 \cos(k_\perp x),$$
$$H_z = -k_\perp B_1 \sin(k_\perp x). \tag{4.49}$$

For simplicity we will examine below the salient features of wave propagation through the slit in the framework of a one mode approximation. In this case the generating function for the field in the narrowing can be written as

$$\Psi_2 = B_2[F_1(\alpha) + QF_2(\alpha)]f(\beta). \tag{4.50}$$

The functions F_1, F_2 and f are determined in (4.40)–(4.44), the constant Q will be defined from the continuity conditions on the plane $z = b/2$. To use these conditions, one has to express the values of the variables α and β as functions of the variable x on the planes

$z = \pm b/2$. The relations needed can be derived directly from (4.35):

$$\alpha(x) = \pm \operatorname{arcch}\left(\sqrt{\frac{q+g}{2}}\right),$$

$$\beta(x) = \pm \arcsin\left(\sqrt{\frac{q-g}{2}}\right), \quad x = \frac{ud}{2}, \tag{4.51}$$

$$q = 1 + \left(\frac{b}{2a}\right)^2 + \left(\frac{ud}{2a}\right)^2, \quad g = \sqrt{q^2 - \left(\frac{ud}{a}\right)^2}. \tag{4.52}$$

Substitution of (4.51) and (4.52) into (4.49) yields the function $\Psi_2(x)$ on the planes $z = \pm b/2$, the dimensionless variable u varies here in the range $-1 \leq u \leq 1$.

To find the complex reflection coefficient R of the TE_{10} mode on the slit one has to represent the field components on the planes $z = \pm b/2$ by means of Fourier transforms of the eigenfunctions of a rectangular waveguide. In a single-mode waveguide its spectrum is characterized by only one eigenfunction $\cos k_\perp x$. The continuity condition for electric field E_y on the plane $z = b/2$ can be written due to the Fourier transform of the function $\Psi_2(x)$ (4.50) as (according to (4.48) and (4.51) $k_\perp x = \frac{\pi}{2} u$):

$$B_1(1+R) = \frac{B_2 d}{2}(\chi_1 + Q\chi_2). \tag{4.53}$$

$$\chi_1 = \int_{-1}^{1} F_1(\alpha) f(\beta) \cos\left(\frac{\pi}{2} u\right) du,$$

$$\chi_2 = \int_{-1}^{1} F_2(\alpha) f(\beta) \cos\left(\frac{\pi}{2} u\right) du. \tag{4.54}$$

The continuity condition for the longitudinal magnetic component H_z (4.49) coincides with (4.53); however, such a condition for the H_x component, defined, according to (4.49), via the derivative $\partial/\partial z$, needs special consideration. Making use of (4.35), we can express this derivative via the derivatives $\partial/\partial \alpha$ and $\partial/\partial \beta$:

$$\frac{\partial}{\partial z} = \frac{1}{ag}\left(\operatorname{ch}\alpha \cos\beta \frac{\partial}{\partial \alpha} - \operatorname{sh}\alpha \sin\beta \frac{\partial}{\partial \beta}\right). \tag{4.55}$$

The quantity g in (4.55) is determined in (4.52). Substitution of the generating function Ψ_2 (4.50) into (4.55) yields, after a Fourier transformation, the continuity condition for the H_x component on the plane $z = b/2$:

$$i\gamma B_1(1-R) = \frac{B_2 d}{2a}(\sigma_1 + Q\sigma_2). \qquad (4.56)$$

$$\sigma_1 = \int_{-1}^{1} \frac{du}{g}\left[\operatorname{ch}\alpha\cos\beta\frac{\partial F_1}{\partial\alpha}f(\beta) - \operatorname{sh}\alpha\sin\beta F_1(\alpha)\frac{\partial f}{\partial\beta}\right]. \qquad (4.57)$$

$$\sigma_2 = \int_{-1}^{1} \frac{du}{g}\left[\operatorname{ch}\alpha\cos\beta\frac{\partial F_2}{\partial\alpha}f(\beta) - \operatorname{sh}\alpha\sin\beta F_2(\alpha)\frac{\partial f}{\partial\beta}\right]. \qquad (4.58)$$

To carry out the integration in (4.54), (4.57) and (4.58), the variables α and β have to be expressed via the variable u by means of (4.51) and (4.52); the quantities $\chi_{1,2}$ and $\sigma_{1,2}$ are dimensionless. Finally, manipulating with Eqs. (4.53) and (4.56) one finds the reflection coefficient R

$$R = \frac{i\gamma a\chi_1 - \sigma_1 + Q(i\gamma a\chi_2 - \sigma_2)}{i\gamma a\chi_1 + \sigma_1 + Q(i\gamma a_2\chi_2 + \sigma_2)}. \qquad (4.59)$$

The factor Q in (4.59) can be calculated from the continuity conditions on the plane $z = -b/2$. This calculation is based on the following symmetry properties of the functions F_1, F_2 and f, following from their definitions (4.40)–(4.44):

$$F_1|_{z=-b/2} = F_1|_{z=b/2}, \quad \left.\frac{\partial F_1}{\partial\alpha}\right|_{z=-b/2} = -\left.\frac{\partial F_1}{\partial\alpha}\right|_{z=b/2},$$

$$F_2|_{z=-b/2} = -F_2|_{z=b/2}, \quad \left.\frac{\partial F_2}{\partial\alpha}\right|_{z=-b/2} = \left.\frac{\partial F_2}{\partial\alpha}\right|_{z=b/2}, \qquad (4.60)$$

$$\chi_1|_{z=-b/2} = \chi_1|_{z=b/2}, \quad \chi_2|_{z=-b/2} = -\chi_2|_{z=b/2},$$

$$\sigma_1|_{z=-b/2} = -\sigma_1|_{z=b/2}, \quad \sigma_2|_{z=-b/2} = \sigma_2|_{z=b/2}.$$

By denoting the complex amplitude of the transmitted mode as B_3, and using these properties, one can write the continuity conditions for field components E_y and H_x by analogy with (4.53) and (4.56), respectively:

$$\frac{B_2 d}{2}(\chi_1 - Q\chi_2) = B_3, \quad \frac{B_2 d}{2a}(-\sigma_1 + Q\sigma_2) = i\gamma B_3. \quad (4.61)$$

Manipulations with Eq. (4.61) yield the factor Q:

$$Q = \frac{i\gamma a \chi_1 + \sigma_1}{i\gamma a \chi_2 + \sigma_2}. \quad (4.62)$$

Finally, substitution of (4.62) into (4.59) yields the complex reflection coefficient R

$$R = \frac{(\gamma a)^2 \chi_1 \chi_2 + \sigma_1 \sigma_2}{\Delta}, \quad (4.63)$$

$$\Delta = (\gamma a)^2 \chi_1 \chi_2 - \sigma_1 \sigma_2 - i\gamma a(\chi_1 \sigma_2 + \chi_2 \sigma_1). \quad (4.64)$$

The complex transmission coefficient, defined as the ratio of the transmitted and incident amplitudes $T = B_3/B_1$, can be obtained by the substitution of (4.62)–(4.64) into Eq. (4.61)

$$T = \frac{-i\gamma a(\chi_1 \sigma_2 - \chi_2 \sigma_1)}{\Delta}. \quad (4.65)$$

The denominator Δ in (4.65) is defined in (4.64). It is worthwhile to present the coefficients R and T, visualizing their amplitudes and phases, in the forms $R = |R|\exp(i\phi_r)$ and $T = |T|\exp(i\phi_t)$:

$$|R| = \frac{|(\gamma a)^2 \chi_1 \chi_2 + \sigma_1 \sigma_2|}{|\Delta|}, \quad |T| = \sqrt{1 - |R|^2}. \quad (4.67)$$

$$\phi_r = \operatorname{arctg}\left[\frac{\gamma a(\chi_1 \sigma_2 + \chi_2 \sigma_1)}{(\gamma a)^2 \chi_1 \chi_2 - \sigma_1 \sigma_2}\right],$$

$$\phi_t = \operatorname{arctg}\left[\frac{\sigma_1 \sigma_2 - (\gamma a)^2 \chi_1 \chi_2}{\gamma a(\chi_1 \sigma_2 + \chi_2 \sigma_1)}\right]. \quad (4.68)$$

Comparison of the phases of reflected and transmitted waves in he waveguide shows the relation $\phi_t - \phi_r = \pi/2$, coincident with the same relation (4.15), established for waves tunneling through the gradient layer.

This analysis illustrates the influence of each of the geometric parameters describing the waveguide slit (distance d, slit length b

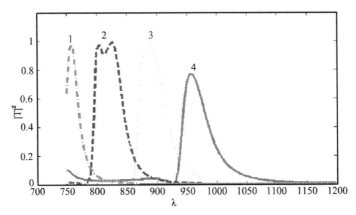

Fig. 4.5. Dependence of the transmittance on the slit width s in the range $225\,\text{nm} < s < 300\,\text{nm}$, while $b = 1400\,\text{nm}$, $d = 500\,\text{nm}$ are kept fixed and the wavelength range is determined according to Eq. (4.47); curves 1, 2, 3 and 4 relate to the values of width $s = 225$, 250, 275 and 300 nm, respectively.

and its width s) on the tunneling of waves through the slit. Thus, the transmittance spectra $|T|^2$ as functions of the variations of slit width, while the other parameters are kept fixed, are shown in Fig. 4.5. A narrow peak in transmittance with $|T|^2 \to 1$ belonging to the fundamental TE_{01} mode starts to appear at the lower end of the wavelength range $700\,\text{nm} < \lambda < 1200\,\text{nm}$, when $s = 225\,\text{nm}$ (curve 1). When the width s is increased a splitting of the peak occurs (curve 2). The subsequent increase of the slit width results in the decrease of the transmittance (curve 4).

The effect of the resonant tunneling of the fundamental mode of radiation through a subwavelength slit, formed by a smoothly shaped narrowing in the waveguide, is caused by the interference of waves, reflected from the different parts of curvilinear slit with the different phases. Unlike the exponential weakening of the fundamental mode, tunneling through the rectangular "undersized" segment in the waveguide [4.20], an effective energy transfer by evanescent waves proves to become possible due to the geometry of the curvilinear slit. This interference of evanescent and antievanescent waves is shown to provide the possibility of reflectionless transmittance for waves with wavelengths 2.5–3 times longer than the width of the slit.

Comments and Conclusions to Chapter 4

Side by side with the academic interest in the effects discussed, some trends, promising for the design of tunneling-assisted phenomena in photonic crystals, can be outlined.

1. Tunneling of light through a gradient nanolayer, fabricated from a dielectric with $\varepsilon(z) > 0$ proves to be possible for some concave spatial distributions of $\varepsilon(z)$.
2. Tunneling phenomena in gradient nanophotonics open the potential of subwavelength nanofilms for the design of broadband antireflection coatings.
3. The speed of energy transfer through the gradient medium is shown to be subluminal.
4. Unlike the tunneling of light through nanostructured metallic films, caused by narrow-banded plasmon-polariton resonances [4.21], the weakly attenuated tunneling propagation in gradient dielectrics, caused by artificial non-local dispersion, can be realized in a wide spectral range, determined by technologically controlled parameters of the metamaterial. The effect of tunneling of EM waves, based on non-local dispersion in gradient dielectrics, can result in a broadening of the list of materials, promising for gradient nanophotonics.
5. To increase the accuracy of the calculation of the tunneling-assisted transparency of the slit, one can take into account the term with β^4 in (4.45); in this case larger values of β_0^2 can be considered ($\beta_0^2 \ll 15/2$). This approximation results in two eigenvalues A for the two modes regime:

$$A_{1,2} = \frac{6}{(p\beta_0)^2}\left(1 \pm \sqrt{\frac{1}{3} - \frac{p^2\beta_0^4}{18}}\right). \tag{4.69}$$

Substitution of $|\beta_0|$ into (4.36) leads to a more precise value of the hyperbolic parameter a. Following the procedure, developed in Sec. 4.3, one can specify the values of the transmission coefficients; the same algorithm is applicable to the subsequent approximations as well [4.18].

Bibliography

[4.1] A. A. Eikhenwald, *J. Russ. Phys. — Chem. Soc.* **41**, 131–157 (1909).
[4.2] G. A. Gamow, *Z. Phys.* **51**, 204–212 (1928).
[4.3] F. de Fornel, Evanescent waves: From newtonian optics to atomic optics, in *Opt. Sciences*, Springer Series, Vol. 73 (Springer, Berlin, 2001).
[4.4] A. B. Mikhailovskii, *Electromagnetic Instabilities in an Inhomogeneous Plasma* (A. Hilger, 1992).
[4.5] V. S. Olkhovsky, E. Recami and J. Jakiel, *Phys. Rep.* **398**, 133–178 (2004).
[4.6] V. Laude and P. Tournois, *JOSA B* **16**(1), 194–198 (1999).
[4.7] A. B. Shvartsburg and G. Petite, *Opt. Lett.* **31**, 1127–1130 (2006).
[4.8] M. Born and E. Wolf, *Principles of Optics*, 7th edn. (Cambridge University Press, 1999).
[4.9] R. M. A. Azzam, *JOSA A* **23**(4), 960–965 (2006).
[4.10] Y. Aharonov, N. Erez and B. Reznik, *Phys. Rev. A* **65**(1–12), 052124 (2002).
[4.11] M. Abramowitz and I. Stegun, *Handbook of Math. Functions* (Dover Publications, NY, 1968).
[4.12] L. M. Brekhovskikh and O. A. Godin, *Acoustics of Layered Media*, I (Berlin, Springer-Verlag, 1990).
[4.13] V. L. Ginzburg, *Propagation of Electromagnetic Waves in a Plasma* (Pergamon, 1967).
[4.14] A. B. Shvartsburg and G. Petite, *European Phys. J. B* **36**, 111–118 (2005).
[4.15] E. P. Wigner, *Phys. Rev.* **98**, 145–147 (1955).
[4.16] Ph. Balcou and L. Dutriaux, *Phys. Rev. Lett.* **78**(5), 851–854 (1997).
[4.17] C.-F. Li and Q. Wang, *JOSA B* **18**(8), 1174–1179 (2001).
[4.18] A. B. Shvartsburg and V. Kuzmiak, *JOSA B*, **27**(12), 2766–2773 (2010).
[4.19] M. Friedman and R.F. Fernsher, *Appl. Phys. Lett.* **74**, 3468–3470 (1999).
[4.20] A. Ranfagni, P. Fabeni, G. P. Pazzi and D. Mugnai, *Phys. Rev. E* **48**, 1453–1460 (1993).
[4.21] A. V. Zayatz, I. I. Smolyaninov and A. A. Maradudin, *Phys. Rep.* **408**, 131–314 (2005).

CHAPTER 5

INTERACTION OF ELECTROMAGNETIC WAVES WITH CONTINUOUSLY STRUCTURED DIELECTRICS

The analyses in Chs. 2–4 were focused on calculations of reflectance/transmittance spectra of thin gradient dielectric films. To emphasize the peculiar effects of technologically controlled heterogeneity-induced dispersion on these spectra, a series of exactly solvable models of $\varepsilon(z)$ was examined. These models demonstrate the influence of the geometrical parameters of the profiles $\varepsilon(z)$, e.g. their steepness, curvature, symmetry, on the amplitude-phase structure of propagating and evanescent waves, traversing the film. However, the dependence of the spectra discussed on such physical features of gradient media as absorption, natural dispersion, resonances, bordering the passbands, remained beyond of scope of these models. To extend the physical insight on the diversity of nanogradient optical phenomena in photonic barriers, the interplay of heterogeneity-induced dispersion with the aforesaid features is discussed in this Chapter.

Absorption spectra of radiation in sputtered gradient dielectric films are known to possess some narrow-banded peaks in the IR range [5.1]. Thus, the X-rays measurements of gradient films had revealed the cluster structure of a series of typical materials (TiO_2, Ta_2O_5, SiO_2), used in these films. These clusters are characterized by resonant absorption peaks in the IR range, e.g., for $(SiO_2)_3$ the resonant frequencies are $6.531 \ 10^{13}$ rad/s and $8.291 \ 10^{13}$ rad/s [5.2]. Outside this resonant IR range, stipulated by eigenfrequencies of intermolecular vibrations, the absorption in dielectrics usually can be assumed to be frequency-independent [5.3]. To generalize the expressions, obtained above for spectra of reflectance R and transmittance T for lossless films, let us note, that in the course of the derivation of these expressions the refractive indices of gradient films and substrates were not assumed to be purely real. Thus, while considering the spectrum of a lossy film with a complex value of its dielectric permittivity $\varepsilon = \varepsilon_1 + i\varepsilon_2$, one can use the obtained formulae for R and T, replacing the refractive index $n_0 \to n_0(1 + i\kappa)$. The values n_0 and κ are linked with the real and imaginary parts of dielectric permittivity ε:

$$n_0 = \frac{\varepsilon_2}{\sqrt{2(\sqrt{\varepsilon_1^2 + \varepsilon_2^2} - \varepsilon_1)}}, \quad \kappa = \frac{\sqrt{\varepsilon_1^2 + \varepsilon_2^2} - \varepsilon_1}{\varepsilon_2}. \quad (5.1)$$

This approach is used in the analysis of the reflectance and transmittance of a lossy gradient barrier in Sec. 5.1. Here no assumptions about the smallness of radiation losses in the barrier are made.

The natural dispersion of materials, as well as the absorption, was ignored in the above analysis. Many transparent non-polar dielectric materials do not possess cut-off frequencies in the visible and near IR spectral ranges, meanwhile the dependence of the refractive index n on the wavelength λ is smooth, and its variations don't exceed several percents. For example, the values of n for fused silica are known to decrease monotonically in the visible and near infrared ranges from $n(\lambda = 400\,\text{nm}) = 1.47012$ to $n(\lambda = 2010\,\text{nm}) = 1.43794$, so that the relative variation of n in this spectral range does not exceed 2–2.5% [5.4]. Thus, the natural dispersion of optical materials in these ranges is usually weak and negative. In contrast, the artificial non-local dispersion in gradient media was shown to provide drastic changes in the function n(λ) as well as controlled formation of both positive or negative dispersion in the given host material. This way to strengthen and control dispersion seems to be especially useful for the visible spectral range, where the dispersion of a natural material proves often to be weak.

However, the interplay between natural (local) and artificial (non-local) dispersive effects may become important in some media for definite spectral ranges in the vicinity of resonances, caused, e.g. by plasmon or polariton excitations. Propagation of light through such barriers, influenced simultaneously by both natural and artificial dispersion, is considered in Sec. 5.2 by means of exactly solvable models of $\varepsilon(z)$, describing the saturation of dielectric functions in the depth of gradient dielectrics. Simultaneous action of natural and artificial dispersion in these media is illustrated by the use of generalizations of dielectric functions, widely used for analysis of both plasmon (5.2) and polariton (5.3) resonances [5.5, 5.6]:

$$\varepsilon = \varepsilon_\infty - \frac{\omega_p^2}{\omega^2}, \quad \omega_p^2 = \frac{4\pi e^2 N_e}{m_e}. \tag{5.2}$$

$$\varepsilon = \varepsilon_\infty \left(\frac{\omega^2 - \omega_L^2}{\omega^2 - \omega_T^2} \right). \tag{5.3}$$

Here ε_∞ is the frequency-independent background dielectric constant, the plasma frequency ω_p in (5.2) depends upon the electron density N_e and electron charge e and effective mass m_e; ω_T and ω_L in (5.3) are the frequencies of transverse and longitudinal vibration modes, forming the edges of the stop band for radiation in the spectral range between ω_L and ω_T.

Another effect of heterogeneity-induced dispersion in artificial resonant media is connected with a frequency-dependent magnetic permeability $\mu(\omega)$, produced by an array of non-magnetic conducting elements, which exhibit a strong resonant response on the magnetic component of electromagnetic field. These elements, so-called split-ring resonators (SRR), formed by a pair of plane opened contours, embedded in a plastic plate, were shown to provide the effective permeability [5.7, 5.8]

$$\mu(\omega) = 1 - \frac{Y\omega^2}{\omega^2 - \omega_0^2}. \tag{5.4}$$

Here ω_0 is the resonance frequency, determined by the split-ring resonator geometry, Y is a the geometrical factor; for convenience damping effects are neglected in the model (5.4). Note, that a medium with a positive constant value of ε supports the propagation of a wave with frequency ω in the spectral ranges $\omega^2 > \omega_0^2(1-Y)^{-1}$ and $\omega^2 < \omega_0^2$ in which $\mu(\omega)$ is positive. The interval $\omega_0^2 < \omega^2 < \omega_0^2(1-Y)^{-1}$, where $\mu(\omega) < 0$, relates to the stop band; owing to technical limitations the values of ω_0 are reported recently to be about 1 THz ($\lambda \approx 3$ nm) [5.9]. An array of SRR, characterized by spatial dimensions about 300 nm, thickness about 90 nm, and the distance between SSR about 600 nm [5.10], embedded in a plastic plate, may be viewed as an effective continuous medium for the mid-IR radiation; heterogeneity of this structure can be caused by varying the spacing between the adjacent elements. The influence of heterogeneties of $\varepsilon(z)$ and $\mu(z)$ on spectral properties and skin layers of these heterogeneous structures, modeled by distributions

$$\varepsilon(z) = \varepsilon(\omega)U^2(z), \quad \mu(z) = \mu(\omega)\Phi^2(z), \qquad (5.5)$$

is examined in Sec. 5.2.

The potential of gradient nanophotonic barriers for control of radiation flows stimulated interest in periodic arrays of such elements — so-called gradient superlattices. Historically, the first superlattices were composed from a sequence of ultrathin alternating semiconductor layers, consisting either of a sequence of two different semiconductor materials, or of a sequence of n- and p-doped layers. The reflectance and transmittance spectra of these structures, unattainable in natural materials, were designed by an appropriate choice of superlattice period and semiconductor species. Later on progress in fabrication of gradient nanofilms has made feasible the realization of dielectric superlattices, based on controlled heterogeneity-induced dispersion and formed by periodic distributions of dielectric permittivity; some examples of such structures, possessing periodic discontinuities of the gradient $\varepsilon(z)$ in nonmagnetic media, were considered in Sec. 2.3 (Fig. 2.4). In contrast, Sec. 5.3 is focused on more flexible models of gradient superlattices, providing continuous periodic profiles of the refractive index $n(z)$ with smooth transitions between its alternating maxima and minima. Interaction of radiation with these structures can be viewed as the optical analogies of electron scattering in the Kronig–Penny model in the solid state physics [5.11]. Attention is given to periodic structures with a distributed negative magnetic response $\mu(z)$, when combined with plasmonic wires, which exhibit a negative dielectric permittivity [5.12]; these structures should produce a negative refractive index material, a so-called left-handed material, distinguished by intriguing parameters: $\varepsilon(\omega) < 0$, $\mu(\omega) < 0$, $n < 0$. A negative $\varepsilon(\omega)$, occurring in some spectral range, being combined with a negative $\mu(\omega)$, allows the formation of a pass band with negative n [5.13]. Strongly dispersive gradient superlattices, fabricated both from these structures and right-handed materials ($\varepsilon > 0, \mu > 0, n > 0$), are examined by means of new exactly solvable models in Sec. 5.3.

5.1. Reflectance/Transmittance Spectra of Lossy Gradient Nanostructures

Analysis of reflectance/transmittance spectra for lossless photonic nanobarriers, carried out in Chs. 2–4, was simplified due to the relation between intensity reflection and transmission coefficients $|R|^2 + |T|^2 = 1$. However, this relation becomes invalid for lossy media, and one has to derive the formulae, describing each of the complex coefficients R and T for gradient absorbing barriers independently [5.14]. To illustrate the interplay between absorption and heterogeneity-induced dispersion effects it makes sense to consider first the case of normal dispersion, examining the tunneling and propagation regimes in gradient barriers with concave profile $U(z)$ separately. For simplicity we will restrict ourselves below to an analysis of the simplest configuration, containing one barrier, located on a substrate; here both the real and imaginary parts of the complex refractive index $n(z)$ are assumed to be modulated by the same function $U(z)$: $n(z) = n_0 U(z)(1 + i\kappa)$ [5.15].

Tunneling through a lossy barrier can be described by means of the replacement of the refractive index n_0 in the expression for the effective refractive index n_e (4.18) by the model representation of n_0 (5.1):

$$n_e = n_0 N, \quad N = \sqrt{u^2 - (1 + i\kappa)^2} = a - ib. \tag{5.6}$$

The factor N in (5.6) becomes complex, its real and imaginary parts are:

$$a, b = \sqrt{\frac{1}{2}\left[\sqrt{(u^2 - 1 + \kappa^2)^2 + 4\kappa^2} \pm (u^2 - 1 + \kappa^2)\right]}. \tag{5.7}$$

The subsequent substitutions of (5.6) and (5.7) into the formula for the reflection coefficient R (4.17) permits one to exploit (4.17) for lossy barriers, using the generalized values of the factors $\sigma_{1,2}$ and $\chi_{1,2}$:

$$\sigma_1 = \left[n + \frac{\gamma^2}{4} + n_0^2(a-ib)^2\right] \text{sh}\left(\frac{lN}{u}\right) - \gamma n_0(a-ib)\text{ch}\left(\frac{lN}{u}\right),$$

$$\sigma_2 = -(n-1)\left[n_0(a-ib)\text{ch}\left(\frac{lN}{u}\right) - \frac{\gamma}{2}\text{sh}\left(\frac{lN}{u}\right)\right], \quad (5.8)$$

$$\chi_1 = \left[n - \frac{\gamma^2}{4} - n_0^2(a-ib)^2\right]\text{sh}\left(\frac{lN}{u}\right) + \gamma n_0(a-ib)\text{ch}\left(\frac{lN}{u}\right),$$

$$\chi_2 = (n+1)\left[n_0(a-ib)\text{ch}\left(\frac{lN}{u}\right) - \frac{\gamma}{2}\text{sh}\left(\frac{lN}{u}\right)\right], \quad (5.9)$$

$$\text{sh}\left(\frac{lN}{u}\right) = \text{sh}\left(\frac{la}{u}\right)\cos\left(\frac{lb}{u}\right) - i\text{ch}\left(\frac{la}{u}\right)\sin\left(\frac{lb}{u}\right),$$

$$\text{ch}\left(\frac{lN}{u}\right) = \text{ch}\left(\frac{la}{u}\right)\cos\left(\frac{lb}{u}\right) - i\text{sh}\left(\frac{la}{u}\right)\sin\left(\frac{lb}{u}\right). \quad (5.10)$$

Formula for complex reflection coefficient T_E (4.19) and intensity coefficient $|T|^2$ prove to be valid for

The expression for the complex reflection coefficient T_E (4.19) and the intensity coefficient $|T|^2$ prove to be valid for tunneling through lossy barriers as well, the expressions (5.6) for n_e and (5.9) for $\chi_{1,2}$ are used:

$$T_E = \frac{2in_0(a-ib)}{\chi_1 + i\chi_2}; \quad |T|^2 = n|T_E|^2; \quad (5.11)$$

To carry out a similar analysis for the propagation regime in the case of normal dispersion one has to represent the effective refractive index by $n_e = n_0 N$, where the factor N differs, according to (2.32), from the expression given by (5.6):

$$N = \sqrt{(1+i\kappa)^2 - u^2} = a + ib. \quad (5.12)$$

The values a and b in (5.12) differ from those given by (5.7) and are:

$$a, b = \sqrt{\frac{1}{2}\left[\sqrt{(1-u^2-\kappa^2)^2 + 4\kappa^2} \pm (1-u^2-\kappa^2)\right]}. \quad (5.13)$$

Substitution of (5.12) and (5.13) but now into the expression (2.31) for the complex reflection coefficient R in the propagation regime (2.31) yields formula (2.31), valid for the following values of the factors $\sigma_{1,2}$ and $\chi_{1,2}$:

$$\sigma_1 = \left[n + \frac{\gamma^2}{4} - n_0^2(a+ib)^2\right]\sin\left(\frac{lN}{u}\right) - n_0\gamma(a+ib)\cos\left(\frac{lN}{u}\right),$$

$$\sigma_2 = -(n-1)\left[n_0(a+ib)\cos\left(\frac{lN}{u}\right) - \frac{\gamma}{2}\sin\left(\frac{lN}{u}\right)\right]. \quad (5.14)$$

$$\chi_1 = \left[n - \frac{\gamma^2}{4} + n_0^2(a+ib)^2\right]\sin\left(\frac{lN}{u}\right) + n_0\gamma(a+ib)\cos\left(\frac{lN}{u}\right),$$

$$\chi_2 = (n+1)\left[n_0(a+ib)\cos\left(\frac{lN}{u}\right) - \frac{\gamma}{2}\sin\left(\frac{lN}{u}\right)\right]. \quad (5.15)$$

$$\sin\left(\frac{lN}{u}\right) = \sin\left(\frac{la}{u}\right)\operatorname{ch}\left(\frac{lb}{u}\right) + i\cos\left(\frac{la}{u}\right)\operatorname{sh}\left(\frac{lb}{u}\right),$$

$$\cos\left(\frac{lN}{u}\right) = \cos\left(\frac{la}{u}\right)\operatorname{ch}\left(\frac{lb}{u}\right) - i\sin\left(\frac{la}{u}\right)\operatorname{sh}\left(\frac{lb}{u}\right). \quad (5.16)$$

Formula (2.34) for the transmission coefficient can be generalized for lossy medium as

$$T_E = \frac{2in_0(a+ib)}{\chi_1 + i\chi_2}, \quad |T|^2 = n|T_E|^2. \quad (5.17)$$

Note, that the frequency $u = u_0 = \sqrt{1-\kappa^2}$ separates the spectral ranges, corresponding to the tunneling $(u > u_0)$ and propagating $(u < u_0)$ regimes; the values of a and b, determined from (5.7) and (5.13), become equal at $u = u_0$: $a = b = \sqrt{\kappa}$. Substitution of these values into (5.8) and (5.14) and using the formulae

$$\operatorname{sh}[\vartheta(1-i)] = -i\sin[\vartheta(1+i)]; \quad \operatorname{ch}[\vartheta(1-i)] = \cos[\vartheta(1+i)], \quad (5.18)$$

valid for real values of factor ϑ (in the case discussed $\vartheta = l\sqrt{\kappa}u_0^{-1}$), shows that the values of reflection and transmission coefficients $|R|^2$ and $|T|^2$ are continuous at this frequency u_0.

Spectra $|R|^2$ and $|T|^2$ for lossy barriers are depicted in Fig. 5.1(a) and Fig. 5.1(b), respectively; for comparison the spectra $|R|^2$ and $|T|^2$ for lossless barriers, other parameters of these barriers being the same, are also presented in these figures; according to the given value of κ the normalized frequency $u_0 = 1.054$. Inspection of these figures shows, that absorption provides an increase of the barrier's reflectance and a decrease of its transmittance. The phase spectrum $\phi_t(u)$ of a wave, transmitted through the lossy barrier (Fig. 5.1(c)), is characterized by a non-monotonic variation of ϕ_t. The characteristic frequency ω_{cr}, separating tunneling and propagation ranges in these spectra $|R|^2$ and $|T|^2$, is increased, due to the influence of losses, from the value $\omega_{cr} = \Omega_1$, corresponding to the lossless case, up to the value $\omega_{cr} = \Omega_1/\sqrt{1-\kappa^2}$.

Proceeding in a similar fashion one can examine the influence of absorption on the spectra of a photonic barrier with abnormal dispersion (2.20) as well. Moreover, the general formulae (2.31) and (2.34), describing the reflectance/transmittance spectra of a gradient film supported by a substrate, remain valid for a lossy substrate, characterized by a complex refractive index n.

The model discussed is characterized by heterogeneity-induced dispersion, the natural dispersion of the material is presumed to be insignificant. Note, that the spectral range, characterized by strong artificial dispersion, is determined by the technologically controlled profile of dielectric permittivity $\varepsilon(z)$ and, thus, can be chosen far from the absorption range of the barrier material. This possibility, as well as the miniaturized subwavelength thickness of the barrier, can provide a significant decrease of losses for waves passing through a gradient barrier.

5.2. Interplay of Natural and Artificial Dispersion in Gradient Coatings

To extend the physical insight on the diversity of nanogradient optical phenomena, the interplay of natural and artificial dispersion is illustrated below in the framework of simple examples of thin gradient coatings or transition layers. The dielectric functions of these layers are supposed to contain resonant frequencies, habitual to either

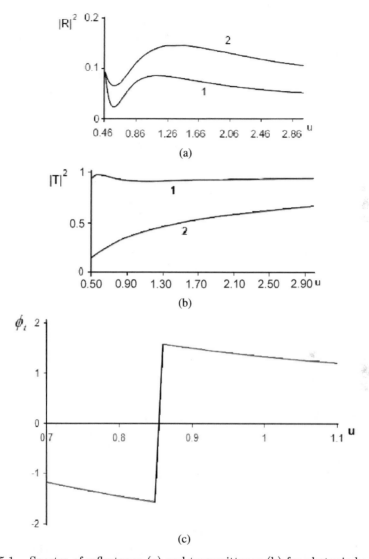

Fig. 5.1. Spectra of reflectance (a) and transmittance (b) for photonic barriers (2.16) with normal heterogeneity-induced dispersion ($n_0 = 2.3$, $y = 0.75$), located on a substrate with $n = 1.47$; u is the normalized frequency. Curves 1 and 2 correspond to the models of lossless ($\kappa = 0$) and lossy ($\kappa = 0.3162$) barriers. (c): continuous variation of the phase of transmitted wave ϕ due to transition from tunneling to propagation regime in the lossy barrier, characterized by curve 2 in (b).

plasmon or polariton mechanisms of dispersion in dielectrics. These frequencies are known to border the transmittance bands in the aforesaid dielectrics. The simultaneous influence of natural resonances and artificial dispersion on transmittance bands of a gradient transition layer are examined here by means of a dielectric function, that varies continuously across the subsurface layer, and saturates in the depth of medium. This approach is based on generalizations of two models of dielectric permittivity, widely used for analysis of a series of homogeneous lossless semiconductors and dielectrics. These models relate to stopbands of both semi-infinite and finite width.

1. Plasma, characterized by semi-infinite transparent and semi-infinite opaque spectral ranges, separated by the plasma frequency ω_p. This frequency, dependent on the electron density N_e, is determined by the well known formula (5.2). If the density N_e is coordinate dependent, $N_e(z) = N_0 U^2(z)$, one can write the dielectric function of this heterogeneous plasma for the frequency ω

$$\varepsilon(z) = \varepsilon_\infty - \frac{\omega_{p0}^2 U^2(z)}{\omega^2}, \quad \omega_{p0}^2 = \frac{4\pi e^2 N_0}{m_e}. \quad (5.19)$$

Here ε_∞ is the frequency-independent background dielectric constant; the dimensionless function $U^2(z)$, describing the distribution of electron density in the transition layer, can be represented, e.g. as [5.16]

$$U^2(z) = 1 + \frac{1}{g} - \frac{W^2(z)}{g}, \quad W(z=0) = 1, \quad W(z \to \infty) \to 0. \quad (5.20)$$

To outline the effect produced by the plasma heterogeneity (5.19), one can use the simple one-parameter models of the distribution $W(z)$:

$$W(z) = \left(1 + \frac{z}{L}\right)^{-1}, \quad W(z) = \exp\left(-\frac{z}{L}\right). \quad (5.21)$$

The spatial scale L and dimensionless factor g are the free parameters of model (5.20), where, the growth (decrease) of the electron density N_e in the depth of plasma relates to a positive (negative) sign of g. Thus, in the case $g > 0$ both distributions (5.21) indicate the increase of N_e from the value $N_e = N_0$ at the interface $z = 0$ up to the same value $N_e = N_0(1 + g^{-1})$ in the depth of the medium ($z \gg L$). However, the spatial structures of the EM fields in the heterogeneous subsurface layers, described by models (5.21), are different. This difference may become drastic, when the medium is not transparent for the wave discussed, and the radiation field proves to be localized in a subsurface skin layer.

Let the wave be incident normally from vacuum on the interface $z = 0$. The generating function Ψ, determining the wave field inside the medium ($z \geq 0$), is governed by the equation, obtained by substitution of (5.20) into the wave equation (2.9):

$$\frac{d^2\Psi}{dz^2} + \frac{\Psi}{c^2}\left\{\varepsilon_\infty \omega^2 - \omega_{p0}^2\left[1 + \frac{1}{g} - \frac{W^2(z)}{g}\right]\right\} = 0. \quad (5.22)$$

By using Eq. (5.22) we can examine the influence of each of the distributions $W(z)$ on the structure of the skin layer. Considering first the model $W(z) = (1+z/L)^{-1}$ and introducing the new function f and new variables η and ς,

$$f = \Psi\sqrt{W(z)}, \quad \eta = \int_0^z W(z_1)dz_1, \quad \varsigma = \exp\left(\frac{\eta}{L}\right), \quad (5.23)$$

we obtain the equation for the function f in a standard form of the Bessel equation:

$$\frac{d^2f}{d\varsigma^2} + \frac{1}{\varsigma}\frac{df}{d\varsigma} + \left(q^2 - \frac{s^2}{\varsigma^2}\right)f = 0. \quad (5.24)$$

$$q^2 = \frac{L^2[\omega^2\varepsilon_\infty - \omega_{p0}^2(1+g^{-1})]}{c^2}, \quad s^2 = \frac{1}{4} - \frac{L^2\omega_{p0}^2}{c^2 g}, \quad n_0^2 = \varepsilon_\infty.$$

$$(5.25)$$

Inspection of the expression for the parameter q^2 (5.25) reveals the cut-off frequency $\omega_c = \omega_{p0}\sqrt{1+g^{-1}}/\sqrt{\varepsilon_\infty}$, that separates the transparent ($\omega > \omega_c, q^2 > 0$) and non-transparent ($\omega < \omega_c, q^2 < 0$) spectral ranges. Introducing in the last case the quantity $q^2 = -q_1^2$, one can write the solution of Eq. (5.24) that decreases in the depth of the plasma by means of a modified Bessel function [5.17] $f = K_s(q_1\varsigma)$.

Substitution of this function f into (5.23) and using the relation $\varsigma = 1 + z/L$ yields the generating function Ψ in an explicit form,

$$\Psi = A\sqrt{1 + \frac{z}{L}} K_s \left[q_1 \left(1 + \frac{z}{L} \right) \right] \exp(-i\omega t). \qquad (5.26)$$

A is the normalization constant; according to (2.8) the electric field $E(z)$ can be written as $E = i\omega c^{-1}\Psi$. To visualize the spatial distribution of field $E(z)$ it makes sense to calculate the normalized field $e(z) = \Psi(z)\Psi^{-1}(0)$:

$$e(z) = \sqrt{1 + \frac{z}{L}} K_s \left[q_1 \left(1 + \frac{z}{L} \right) \right] K_s^{-1}(q_1). \qquad (5.27)$$

The spatial distribution of $e(z)$ (Fig. 5.2(a)) illustrates the monotonic weakening of the field inside the skin layer, characterized by an e-folding length of weakening close to the spatial scale of heterogeneity L.

In contrast, a non-monotonic field structure in the skin layer can be produced by the distribution $W(z) = \exp(-z/L)$ (5.21). In this case the generating function Ψ is governed by the Bessel equation (5.24) after the replacement $f \to \Psi$, however, the parameters q and s are now different from those given in (5.25):

$$q^2 = \frac{1}{g}\left(\frac{\omega_{p0}L}{c}\right)^2; \quad s^2 = \frac{L^2[\omega_{p0}^2(1 + g^{-1}) - \omega^2\varepsilon_\infty]}{c^2}; \qquad (5.28)$$

It is remarkable, that solutions of (5.24) can describe low frequency fields ($\omega < \omega_c$) tending to zero in the depth of the medium, where the electron density is increasing, the values of both parameters q^2 and s^2 (5.28) are positive. In this case the field amplitude in the subsurface layer grows up to some maximum, that exceeds its value on the interface. The damping of the amplitude, habitual for a skin layer, arises after this maximum. Thus the field structure in the skin layer proves to be non-monotonic. To illustrate these structures one can recall, that solutions of (5.24) for half-integer values of order $s = m+0.5$ ($m = 0; 1; 2\ldots$) are known to be expressed via elementary functions [5.18]; choosing the solutions, tending to zero in the depth

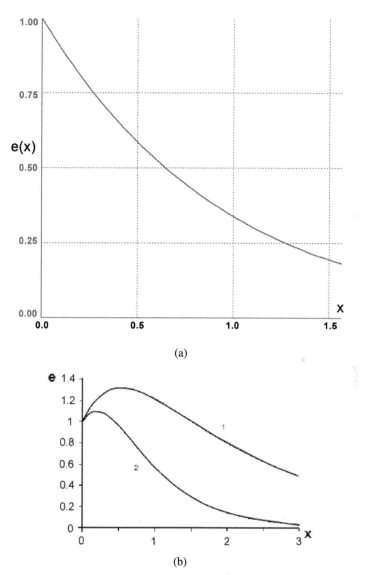

Fig. 5.2. Monotonic and non-monotonic distributions of normalized electric field $e(x)$ in the gradient transition layer (5.20), related to profiles $W(z) = (1+z/L)^{-1}$ (a) and $W(z) = \exp(-z/L)$ (b), are plotted vs. the dimensionless coordinate $z = x/L$. The values of the parameters s and q in Eq. (5.24), determining the distribution $e(x)$, are $s = 0.325$, $q = 0.75$ (a); curves 1 and 2 on (b) show the distributions of $e(x)$ (5.29) and (5.67), corresponding to the values $s = 0.5$, $q = 2$ and $s = 1.5$, $q = 3$, respectively.

of the medium ($\varsigma \to 0$) we have, e.g. the normalized distribution of the electric field in the skin layer in a simple case m = 0:

$$e(x) = \frac{\sin[q\exp(-x)]\exp\left(\frac{x}{2}\right)}{\sin q}, \quad x = \frac{z}{L}. \quad (5.29)$$

Examples of such non-monotonic field distributions in the skin layer, corresponding to the cases $m = 0$ and $m = 1$, are shown in Fig. 5.2(b).

2. Now let us turn to another spectral structure, related to a stopband of finite width. It is worthwhile to illustrate this model by means of a dielectric with a polariton gap, e.g. MgO or ZnSe [5.19], bounded by the resonant frequencies ω_T and ω_L (5.3). Recalling the relation between these frequencies $\omega_L^2 = \omega_T^2(1 + \beta/\varepsilon_\infty)$ [5.20], where β is the so-called "oscillator strength" that characterizes the polarization of molecules, one can use the distribution of $\omega_L^2(z)$, produced by a continuously varying content of polarized molecules in the subsurface layer of the medium:

$$\omega_L^2(z) = \omega_{L0}^2 U^2(z); \quad (5.30)$$

Using the distribution (5.30) one can represent the generalized profile $\varepsilon(z)$ (5.3) in the form

$$\varepsilon(z) = n_0^2 \left[\frac{\omega^2 - \omega_{L0}^2 U^2(z)}{\omega^2 - \omega_T^2}\right], \quad n_0^2 = \varepsilon_\infty. \quad (5.31)$$

Assuming the distribution $U^2(z)$ in (5.31) in the form (5.20), one can follow the approach developed above for the analysis of a gradient plasma layer in both models (5.21). Thus, in case of model $W(z) = \exp(-z/L)$ this approach brings the equation for the generating function into the form (5.24) with parameters

$$q^2 = \frac{1}{g}\left(\frac{\omega n_0 L}{c}\right)^2 \frac{\omega_{L0}^2}{\omega^2 - \omega_T^2},$$

$$s^2 = \left(\frac{\omega n_0 L}{c}\right)^2 \left[\frac{\omega_{L0}^2(1 + g^{-1}) - \omega^2}{\omega^2 - \omega_T^2}\right]. \quad (5.32)$$

This similarity opens the possibility of "double use" of each solution of Eq. (5.24) for treatment of EM wave processes in gradient dielectrics, possessing both plasmon and polariton resonances, if the values of the dimensionless parameters q and s, calculated for these dielectrics, coincide. Manipulations with the expressions for the parameters q^2 and s^2 (5.32) yields the relation

$$1 + g\left(1 - \frac{\omega^2}{\omega_{LO}^2}\right) = \frac{s^2}{q^2}. \tag{5.33}$$

Thus, referring to curve 1 in Fig. 5.2(b) corresponding to the values $q = 2, s = 0.5, g = 0.333$, we find from (5.33) the wave's frequency, $\omega = 1.95\omega_{LO}$, located in the broadened stop band $\omega_{LO}^2(1 + g^{-1}) > \omega^2 > \omega_T^2$. To illustrate the spectral properties of a spatially varying stop band one can consider the light wave, incident normally on the gradient subsurface layer in dielectric ZnSe, possessing a polariton bandgap between ω_{LO} and ω_T. Taking into account that this material in a natural state is characterized by the refractive index $n_0 = 2.365$ and a polariton gap, determined by frequencies from the near IR range $\omega_T = 0.47 \; 10^{15}$ rad/s, $\omega_{LO} = 0.88 \; 10^{15}$ rad/s [5.19], one can calculate $\omega = 1.715 \; 10^{15}$ rad/s ($\lambda = 1.08\,\mu$); here the condition $q^2 = 4$ yields the gradient scale $L = 160$ nm. On the other hand, to apply the same Fig. 5.2(b) for the gradient plasma layer in a semiconductor with free carriers, one has to use the expressions q^2 and s^2 from (5.28); in this case the replacement $\omega^2/\omega_{LO}^2 \to \omega^2\varepsilon_\infty/\omega_{p0}^2$ in Eq. (5.33) has to be done. Putting again $q = 2, s = 0.5, g = 0.333$, we find for $\varepsilon_\infty = 5$ the ratio $\omega = 0.875\omega_{p0}$; for an electron density $N_e = 6.5 \times 10^{17}$ cm^{-3} ($\omega_{p0} = 4.35 \times 10^{13}$ rad/s) the frequency ω ($\lambda = 50\,\mu$m) belongs to the far IR range, and the growth of the electron density is characterized by the spatial scale $L = 8\,\mu$m. Thus, the interaction of natural and artificial dispersion in a gradient dielectric can produce an enhancement of the electric field in the subsurface transition layer.

An interesting manifestation of this interaction can arise in the left-handed materials, whose magnetic permeability μ is modeled by the Lorentz-type oscillator (5.4). To take into account the values of μ, distinguished from unity, we introduce the generating function Ψ, generalizing the representation (2.8):

$$E_x = -\frac{1}{c}\frac{\partial \Psi}{\partial t}, \quad H_y = \frac{1}{\mu}\frac{\partial \Psi}{\partial z}. \tag{5.34}$$

Owing to representation (5.34) the set of Maxwell equations for a monochromatic wave is reduced to the equation

$$\frac{d^2\Psi}{dz^2} + \frac{\omega^2}{c^2}\varepsilon\mu\Psi = \frac{1}{\mu}\frac{d\mu}{dz}\frac{d\Psi}{dz}. \quad (5.35)$$

To emphasize the effects produced by a negative magnetic permeability $\mu(\omega)$, let us consider a so-called "single-negative metamaterial" [5.21], characterized by a positive value of ε, assuming, for simplicity, $\varepsilon = n_0^2 = $ const. In contrast, the spatial distribution of $\mu(z)$ is modeled, according to (5.5), as $\mu(z) = \mu(\omega)\Phi^2(z)$. It is worthwhile to use the analogy between Eqs. (5.35) and (3.38), which becomes obvious due to the replacement $\Phi(z) \to U(z)$ in (5.35). Continuing this analogy, introducing the variable τ (3.39), and following the algorithm (3.40)–(3.51), we present the solution of Eq. (5.35) for the profile (3.46), depicted in Fig. 3.4, in a form, similar to (3.48):

$$\Psi = A\sqrt{U(z)}[\exp(iq\varsigma) + Q\exp(-iq\varsigma)]. \quad (5.36)$$

$$q = \frac{\omega n_0 N}{c}, \quad N = \sqrt{\mu(\omega) - \frac{\Omega^2}{\omega^2}}, \quad \Omega = \frac{c}{2n_0 L}. \quad (5.37)$$

The generating function Ψ can be expressed via the propagating waves (5.36) when the wave number q (5.37) is real; the spectral ranges, providing this condition, are determined from the inequality $\mu(\omega) - \Omega^2\omega^{-2} > 0$. Substitution of the model $\mu(\omega)$ (5.4) into this inequality permits rewriting it in the form:

$$\frac{(\omega^2 - \omega_1^2)(\omega^2 - \omega_2^2)}{\omega^2(\omega^2 - \omega_0^2)} > 0,$$

$$\omega_{1,2}^2 = \frac{1}{2(1-Y)}\left[\omega_0^2 + \Omega^2 \pm \sqrt{(\omega_0^2 + \Omega^2)^2 - 4(1-Y)\omega_0^2\Omega^2}\right]. \quad (5.38)$$

Analysis of formula (5.38) reveals the relation between the frequencies ω_0, ω_1 and ω_2: $\omega_1 > \omega_0 > \omega_2$. Taking into account this relation it is convenient to examine the inequality (5.38) in four spectral ranges:

(1) $\omega > \omega_1$, (2) $\omega_1 > \omega > \omega_0$, (3) $\omega_0 > \omega > \omega_2$, (4) $\omega_2 > \omega$.

(5.39)

The condition $\mu(\omega) - \Omega^2 \omega^{-2} > 0$ is satisfied in spectral ranges (1) and (3); it means that the gradient structure with magnetic permeability (5.4), distributed according to (3.46), is transparent in these ranges. The opposite condition, $\mu(\omega) - \Omega^2 \omega^{-2} < 0$, satisfied in spectral ranges (2) and (4), indicates the non-transparent range. This analysis shows the drastic changes in reflectance/transmittance spectra of a dielectric with an artificial dispersive magnetic permeability, produced by its spatial heterogeneity.

Note, that negative $\mu(\omega)$ was shown to be possible, when a polariton resonance exists in the dielectric permeability such as in the antiferromagnetics MnF_2 and FeF_2 [5.21].

5.3. EM Radiation in Gradient Superlattices

Progress in crystal growth techniques has made feasible the realization of periodic dielectric structures, composed from ultrathin layers with alternating high and low refractive indices [5.22]. The periods of these artificial one-dimensional structures significantly exceed the natural periods of crystal lattices, constituting these structures, the so-called superlattices. The distribution of refractive index in these structures, modeled by broken straight lines, formed the basis for the design of systems for spectral filtration of radiation, e.g. transparency windows [5.23]. The flexibility of the parameters of these dielectric non-magnetic structures can be improved by the use of graded metamaterials, characterized by frequency dependent dielectric and magnetic parameters, providing both positive and negative values of $\varepsilon(\omega)$ and $\mu(\omega)$. These artificial materials attract growing attention now owing to their unusual electromagnetic properties.

To display these properties, a new exactly solvable model of a multilayer gradient photonic barrier, providing a smooth transition between adjacent layers is developed. This model can be viewed as a generalization of profile (5.20) containing, unlike (5.20), three free parameters L, M and g [5.11]:

$$n^2(z) = n_0^2 U^2(z), \quad U^2(z) = 1 - \frac{1}{g} + \frac{W^2(z)}{g},$$

$$W(z) = \left[\cos\left(\frac{z}{L}\right) + M\sin\left(\frac{z}{L}\right)\right]^{-1}. \qquad (5.40)$$

The values of L, M and g depend on the gap's width d, the minimum (maximum) refractive index $n_{\min,\max}$ and the slope ξ of the profile $n(z)$ near the gap's boundaries, the planes $z = 0$ and $z = d$ (3.1). Considering the symmetrical profile ($W(0) = W(d) = 1$), one can link these quantities with the physical and geometrical parameters of the gap:

$$\frac{d}{L} = 2\mathrm{arctg}M, \quad n_{\min,\max}^2 = n_0^2\left[1 - \frac{M^2}{g(1+M^2)}\right],$$

$$\frac{1}{n_0^2}\frac{dn^2}{dz}\bigg|_{z=0} = -\frac{2M}{gL} = \xi. \qquad (5.41)$$

Manipulations of Eq. (5.41) lead to an explicit expression for determining the quantity M via these parameters $n_0, n_{\min}, g,$ and ξ:

$$\frac{M}{(1+M^2)\mathrm{arctg}M} = -\frac{4}{\xi d}\left(1 - \frac{n_{\min}^2}{n_0^2}\right); \qquad (5.42)$$

Substitution of the quantity M, calculated from (5.42), into (5.41) yields the unknown values g and L.

Inspection of expressions (5.41) shows that positive (negative) values of the parameter g correspond to concave (convex) profiles of the refractive index $U^2(z)$. Moreover, the positive values of g are restricted by the condition $g > M^2(1+M^2)^{-1}$. The examples of the dimensionless profiles of refractive index $U^2(z)$ shown in Fig. 5.3, illustrate the profiles of equal width d and equal minimum/maximum values $U_{\min,\max}^2$, characterized, however, by different shapes.

By introducing the variable η and normalized variable ς,

$$\eta = \int_0^z W(z_1)dz_1, \qquad (5.43)$$

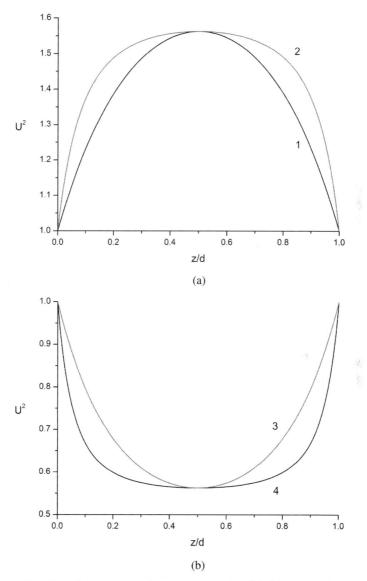

Fig. 5.3. Profiles of dielectric permittivity (5.40) $U^2(z/d)$, characterized by a variable half-width, their maximum (minimum) and width d being fixed; convex profiles 1 and 2 in (a) relate to the values $g = -0.5$, $M = 0.6255$ ($g = -1.5$, $M = 2.3238$), concave profiles 3 and 4 in (b) correspond to the values $g = 1.1$, $M = 0.9632$ ($g = 2.1$, $M = 3.3627$), respectively.

$$\frac{\eta}{L} = \frac{1}{\sqrt{1+M^2}} \ln\left[\frac{1 + m_+ \text{tg}\left(\frac{z}{2L}\right)}{1 - m_- \text{tg}\left(\frac{z}{2L}\right)}\right], \quad m_\pm = \sqrt{1+M^2} \pm M, \tag{5.44}$$

$$\varsigma = \frac{\eta\sqrt{1+M^2}}{L} - \varsigma_0, \quad \varsigma_0 = \ln(m_+), \quad \varsigma|_{z=0} = -\varsigma_0, \quad \varsigma|_{z=d} = \varsigma_0, \tag{5.45}$$

one can write an explicit expression for the function $W(\varsigma)$:

$$W(\varsigma) = \frac{\text{ch}\varsigma}{\sqrt{1+M^2}}, \quad W(\varsigma_0) = W(-\varsigma_0) = 1. \tag{5.46}$$

Let a wave with components E_x and H_y be incident normally on the interface ($z = 0$) of gradient barrier (5.40). To obtain an exact analytical solution of wave equation (2.9) for this barrier let us rewrite this equation by introducing the new function $f = \Psi\sqrt{W}$ and the new variable ς (5.45). After these transformations the equation governing the function f reads as

$$\frac{d^2 f}{d\varsigma^2} + f\left(q^2 - \frac{\Lambda}{\text{ch}^2\varsigma}\right) = 0. \tag{5.47}$$

$$q^2 = \frac{\omega^2 L^2 n_0^2}{c^2 g(1+M^2)} - \frac{1}{4}, \quad \Lambda = \frac{1}{4} - \frac{\omega^2 L^2 n_0^2}{c^2}\left(1 - \frac{1}{g}\right). \tag{5.48}$$

Thus, the equations for both concave and convex barriers are presented in similar forms [5.16]. This similarity simplifies the following analysis. To find the solutions of Eq. (5.47) it is worthwhile to introduce the new function F and new variable v by

$$\nu = \frac{1 - \text{th}\varsigma}{2}, \quad f = (\text{ch}\varsigma)^{-2p} F(v), \tag{5.49}$$

$$v(1-v)\frac{d^2 F}{dv^2} + [\gamma - (1+\alpha+\beta)v]\frac{dF}{dv} - \alpha\beta F = 0. \tag{5.50}$$

Equation (5.50), well known in quantum mechanics [5.24], is the standard form of the hypergeometric equation, where parameters α, β, γ

are defined via the quantities q^2 and Λ (5.48):

$$\gamma = 1 + 2p, \quad \alpha, \beta = \frac{1}{2} + 2p \pm \frac{\omega n_0 L}{c}\sqrt{1 - \frac{1}{g}}, \quad 2p = \pm\sqrt{-q^2}. \tag{5.51}$$

The hypergeometric equation (5.50) has two linearly-independent solutions. Since the parameters α, β, γ are linked by the relation $\text{Re}(\alpha + \beta + 1) = 2\gamma$, these solutions are given by hypergeometric functions F_1 and F_2 [5.18]:

$$F_1 = F(\alpha, \beta, \gamma, v), \quad F_2 = F(\alpha, \beta, \gamma, 1 - v). \tag{5.52}$$

The hypergeometric series F_1 and F_2 are known to converge absolutely inside the circle $|v| = 1$ under the condition $\text{Re}(\alpha + \beta - \gamma) < 0$ [5.18]. According to (5.51), $\alpha + \beta - \gamma = 2p$. This means that the series F_1 and F_2 converge absolutely if the value of p is negative, which means, in its turn, if the condition $q^2 < 0$ is satisfied. This simple case will be considered below.

Introducing the definition $2p = -l (l > 0), l = \sqrt{-q^2}$, one can represent the explicit solution of wave equation (2.9) for the barrier (5.40) in the form

$$\Psi = A(\text{ch}\varsigma)^{l - \frac{1}{2}}[F_1(v) + QF_2(1 - v)]. \tag{5.53}$$

Substitution of the generating function Ψ (5.53) into definition (2.8) brings the expression for both the electric component of EM field $E_x = i\omega c^{-1}\Psi$ and the magnetic component, determined by the derivative of the function Ψ with respect to z. Calculation of this derivative by means of the equalities

$$\frac{d\varsigma}{dz} = \frac{\text{ch}\varsigma}{L}, \quad \frac{dv}{dz} = -\frac{1}{2L\text{ch}\varsigma}, \tag{5.54}$$

yields the expression for H_y:

$$H_y = \frac{A(\text{ch}\varsigma)^{l - \frac{1}{2}}}{2L}\left[(2l - 1)(F_1 + QF_2)\text{sh}\varsigma - \frac{F_1' - QF_2'}{\text{ch}\varsigma}\right]. \tag{5.55}$$

$$F_1' = \frac{dF_1}{dv}, \quad F_2' = \frac{dF_2}{dh}, \quad h = 1 - v. \tag{5.56}$$

To find the reflection/transmission coefficients for barrier (5.40) by means of standard procedure, based on the continuity of the fields E_x and H_y across the boundaries $z = 0$ and $z = d$ of the barrier, the values of the variables ς and v on these boundaries are needed:

$$z = 0: \quad \varsigma = -\ln(m_+), \quad v = v_+ = \frac{1}{2}\left(1 + \frac{M}{\sqrt{1+M^2}}\right),$$

$$h = v_- = \frac{1}{2}\left(1 - \frac{M}{\sqrt{1+M^2}}\right). \quad (5.57)$$

$$z = d: \quad \varsigma = \ln(m_+), \quad v = v_-, \quad h = v_+. \quad (5.58)$$

It is remarkable that the concave and convex arcs (5.40) can be viewed as the constitutive blocks of a smoothly shaped gradient superlattice. The periodic sequence of such alternating arcs provides the profile of a superlattice, continuous across the boundaries of each arc: $U^2(z = 0) = U^2(z = d) = 1$. The demand for a smooth tangent of adjacent n-th and $(n+1)$-th arcs at the contact points $U = 1$ results in a condition linking the parameters M, g and L for these arcs,

$$\frac{M_n}{g_n L_n} = -\frac{M_{n+1}}{g_{n+1} L_{n+1}}. \quad (5.59)$$

Different profiles of superlattices, created from barriers with height $H = U_{\max}^2 - 1$ and wells with depth $D = 1 - U_{\min}^2$, measured from the level $U = 1$, are shown in Fig. 5.4. The flexibility of the parameters of these structures, combining smoothly the blocks with $H = D$ (Fig. 5.4(a)), as well as $H < D$ (Fig. 5.4(b)) and $H > D$ (Fig. 5.4(c)), opens the way to the design of new types of sophisticated superlattices and multilayer coatings.

Side by side with these non-magnetic dielectric structures it makes sense to pay attention to another type of gradient superlattices, consisting of arrays of fine wires and split-ring resonators [5.25], embedded in a plastic matrix. This composite medium is characterized by spatial distributions of its dielectric permittivity; assuming these distributions to be described by even functions, one can write the dielectric permittivity $\varepsilon = \varepsilon(\omega)U^2(z)$ and magnetic permeability

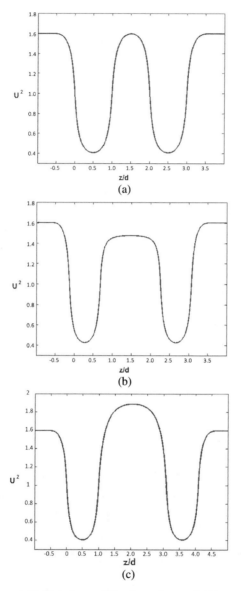

Fig. 5.4. Constituent blocks of smoothly shaped superlattices, consisting of profiles, shown in Fig. 5.3. All the concave arcs in Figs. 5.4(a)–5.4(c) are characterized by parameters: $M_1 = 2.02$, $g_1 = 1.35$. Convex arcs correspond to the values $M_2 = 2.02$, $g_2 = -1.35$ (a), $M_2 = 4.739$, $g_2 = -2.036$ (b); $M_2 = 2.843$, $g_2 = -0.971$ (c), respectively. All the profiles and their gradients are continuous at the tangent points, located at the level $U = 1$.

$\mu(z) = \mu(\omega)\Phi^2(z)$ (5.5). In the case of normal incidence of radiation on such a metamaterial layer the wave field inside the layer is governed by Eq. (5.35).

Consider the model of a left-handed medium, characterized by negative refractive index ($n_0 < 0$) and coinciding distributions $U = \Phi$ (5.5), the functions $\varepsilon(\omega)$ and $\mu(\omega)$ being arbitrary. Transformation of (5.35) to the new variable τ (3.39) leads to a simple equation

$$\frac{d^2\Psi}{d\tau^2} + \frac{\omega^2 n_0^2}{c^2}\Psi = 0, \quad n_0 = -\sqrt{\varepsilon(\omega)\mu(\omega)}. \qquad (5.60)$$

The solution of (5.60), given by harmonic waves in τ-space,

$$\Psi = \exp\left(\pm\frac{iq\tau}{c}\right), \quad q = \frac{\omega n_0}{c}, \qquad (5.61)$$

describes both propagating ($n_0^2 > 0$) and tunneling ($n_0^2 < 0$) regimes in right- and left-handed metamaterials. The dispersion in this barrier is determined by the coordinate-independent function n_0, while heterogeneity-induced dispersion in this heterogeneous medium does not arise. Note, that due to the condition $n_0 < 0$ the waves, traveling in a left-handed material in the direction $z > 0$ ($z < 0$), are characterized, in contrast to a right-handed material, by wave numbers $-q(q)$.

It is remarkable that solution (5.61) of Eq. (5.35) is rather general: it is valid for arbitrary functions $\varepsilon(\omega)$ and $\mu(\omega)$ and arbitrary distributions $U(z)$ under the condition $U(z) = \Phi(z)$. Moreover, the solutions (5.61) describe the waves in homogeneous layers in τ-space; thus, the well developed algorithms for the design of superlattices, built from homogeneous layers in conventional z-space, can be used for the design of gradient superlattices containing left-handed materials.

Note, that the generating function Ψ can be introduced, side by side with (5.34), by the familiar representation (3.37)

$$E_x = -\frac{1}{\varepsilon}\frac{\partial\Psi}{\partial z}, \quad H_y = \frac{1}{c}\frac{\partial\Psi}{\partial t}. \qquad (5.62)$$

The equation, governing this newly introduced function coincides with (5.35) after the exchanges $\varepsilon \to \mu$, $\mu \to \varepsilon$ in (3.38). This feature of the representations (3.37) and (5.34) can be considered as a manifestation of the invariance of the Maxwell equations under the replacements $\vec{B} \to -\vec{D}$, $\vec{E} \to \vec{H}$ (duality principle) [5.26].

Comments and Conclusions to Chapter 5

1. When the gradient photonic barrier $U(z)$, fabricated from a gyrotropic material, is placed in an external magnetic field $\vec{H}_0$ normal to the barrier interface, the conventional magnetooptical effects are influenced by the heterogeneity-induced dispersion of the barrier. Maxwell's equations, describing the wave field inside this gradient magnetooptical medium, are

$$\operatorname{rot}\vec{E} = -\frac{1}{c}\frac{\partial \vec{H}}{\partial t}, \quad \operatorname{rot}\vec{H} = \frac{1}{c}\frac{\partial \vec{D}}{\partial t}. \tag{5.63}$$

The electric displacement $\vec{D}$ in the wave discussed can be represented by means of the Verdet constant γ as [5.27]:

$$\vec{D} = n_0^2 U^2(z)(\vec{E} + i\gamma[\vec{E}\vec{H}_0]), \tag{5.64}$$

where $\vec{j}$ is the unit vector, oriented in the direction of the external magnetic field $\vec{H}_0$. Consider a wave incident normally on the barrier $U(z)$ (2.16). On introducing the generating functions Ψ_1 and Ψ_2 by,

$$E_x = -\frac{1}{c}\frac{\partial \Psi_1}{\partial t}, \quad H_y = \frac{\partial \Psi_1}{\partial t}, \quad E_y = \frac{1}{c}\frac{\partial \Psi_2}{\partial t},$$

$$H_x = \frac{\partial \Psi_2}{\partial z}, \quad \Psi_{1,2} = \frac{F_{1,2}}{\sqrt{U(z)}}, \tag{5.65}$$

and using the familiar variable η (2.10), we find $F_{1,2} = A_{1,2}\exp(iq_\pm \eta)$, where

$$q_\pm^2 = \left(\frac{\omega n_0}{c}\right)^2 (1 - u^2) \pm \gamma \frac{\omega n_0}{c}. \tag{5.66}$$

The splitting of the wave number q into two values, corresponding to the $\pm$ signs in (5.66), is known to describe the Faraday rotation of the polarization state [5.27]; here the factor $1 - u^2$ arises from the gradient profile of the refractive index; the variable η is defined in (2.32).

2. Solutions of Eq. (5.24), related to half-integer values of the parameter s, can be used to Illustrate the influence of the parameters g and L of gradient structure (5.20) and (5.21) on the possible enhancement of the electric field $e(x)$ in the transition layer. Thus, side by side with the distribution (5.29), corresponding to the case $s = 0.5$, one can consider the distribution $e(x)$ for $s = 1.5$:

$$e(x) = \frac{\wp}{\ell}, \quad \ell = \frac{\sin(q)}{q} - \cos(q), \tag{5.67}$$

$$\wp = \exp\left(\frac{x}{2}\right)\left\{\frac{\exp(x)\sin[q\exp(-x)]}{q} - \cos[q\exp(-x)]\right\}.$$

The expression $\sin(q)$ indicates the sine function whose argument is equal to q radians. An example of distribution (5.67) for $q = 3$ is shown by curve 2 in Fig. 5.2(b). Let us recall that curve 1 in Fig. 5.2(b) corresponds to the frequency $\omega = 3.8 \times 10^{13}$ rad/s, incident on the transition layer in a solid plasma, characterized by the spatial scale $L = 8\,\mu$m. Considering, e.g. the transition layer with a more sloping profile of the density $L = 12\,\mu$m, other parameters of the medium being unchanged, we find from (5.28) the value $q = 3$. Thus, curve 2 in Fig. 5.2(b) presents the distribution $e(x)$ for a wave with the lower frequency $\omega = 3.5 \times 10^{13}$ rad/s.

3. Solution (5.61) describes the waves in the so-called transition metamaterials, characterized by gradual changes of $\varepsilon(z)$ and $\mu(z)$ from positive to negative values. Thus, considering the profile defined by the odd function, $U(z) = \Phi(z) = \text{th}(z/L)$ [5.28], we find from (2.10) that $\eta = L\ln[\text{ch}(z/L)]$. The wave traveling in this layer in the z-direction, is written as

$$\Psi = \left[\text{ch}\left(\frac{z}{L}\right)\right]^{iqL}. \tag{5.68}$$

The wave number q was defined in (5.61). Note that the refractive index n_0 (5.60) conserves its value for any given frequency in both half-spaces $z < 0$ and $z > 0$. This means, that the wave travels from $z < 0$ to $z > 0$ without any reflection from the plane $z = 0$.

Bibliography

[5.1] H. A. Macleod, *Thin Film Optical Filters*, 2nd edn. (Adam Hilger, Bristol, 1986).
[5.2] C. Gomez-Reino, M. V. Perez and C. Bao, *Graded — Index Optics: Fundamentals and Applications* (Springer-Verlag, Berlin, 2002).
[5.3] I. Gomez, F. Dominguez-Adame, A. Flitti and E. Diez, *Phys. Lett. A* **248**, 431–433 (1998).
[5.4] S. Adachi, *Phys. Rev. B* **38**, 12966–12976 (1988).
[5.5] S. A. Maier, *Plasmonics: Fundamentals and Applications* (Springer, NY, 2007).
[5.6] V. M. Agranovich and V. L. Ginzburg, *Crystall Optics with Spatial Dispersion and Theory of Excitons* (Springer-Verlag, Berlin, 1984).
[5.7] J. B. Pendry, A. J. Holden, D. J. Robbins and W. J. Stewart, *IEEE Trans. Microwave Theory Tech.* **47**, 2075–2084 (1999).
[5.8] S. O'Brien and J. P. Pendry, *J. Phys. Cond. Matter* **14**, 6383–6394 (2002).
[5.9] T. J. Yen, W. J. Padilla, N. Fang, D. C. Vier, D. R. Smith, J. B. Pendry, D. N. Basov and X. Zhang, *Science* **303**, 1494–1496 (2004).
[5.10] S. Linden, C. Enkrich, M. Wegener, J. Zhou, T. Koschny and C. Soukoulis, *Science* **306**, 1351–1353 (2004).
[5.11] A. B. Shvartsburg and V. Kuzmiak, *European Phys. J. B* **72**, 77–88 (2009).
[5.12] D. R. Smith, W. J. Padilla, D. C. Vier, S. C. Nemat-Nasser and S. Schultz, *Phys. Rev. Lett.* **84**, 4184–4187 (2000).
[5.13] V. G. Veselago, *Sov. Phys. Uspekhi* **10**, 509–514 (1968).
[5.14] I. Ilic, P. P. Belicev, V. Milanovic and J. Radovanovic, *JOSA B* **25**, 1800–1804 (2008).
[5.15] A. B. Shvartsburg and G. Petite, *European Phys. J. D* **36**, 111–118 (2005).
[5.16] A. B. Shvartsburg, V. Kuzmiak and G. Petite, *Physics Reports* **452**, 33–88 (2007).
[5.17] A. B. Shvartsburg, G. Petite and P. Hecquet, *JOSA A* **17**, 2267–2271 (2000).
[5.18] M. Abramowitz and I. A. Stegun (eds.), *Handbook of Math. Functions* (367 Dover, NY, 1964).
[5.19] L. I. Deych, D. Livdan and A. A. Lisyansky, *Phys. Rev. E* **57**, 7254–7258 (1998).
[5.20] A. A.Bulgakov, S. A. Bulgakov and M. Nieto-Vesperinas, *Phys. Rev. B* **58**, 4438–4448 (1998).
[5.21] L. G. Wang, H. Chen and S.-Y. Zhy, *Phys. Rev. B* **70**(1–2), 245102 (2004).
R. E. Camley and D. L. Mills, *Phys. Rev. B* **26**, 1280–1288 (1982).
[5.22] F. G. Bass and A. A. Bulgakov, *Kinetic and Electrodynamic Phenomena in Classical and Quantum Semiconductor Superlattices* (Nova Science, NY, 1997).
[5.23] M. Li, X. Liu, J. Zhang, X. Ma, X. Sun, Y. Li and P. Gu, *JOSA A* **24**, 2328–2333 (2007).
[5.24] L. I. Schiff, *Quantum Mechanics* (McGrow-Hill, Tokyo, 1968).
[5.25] A. Grbic and G. V. Eleftheriades, *J. Appl. Phys.* **92**, 5930–5935 (2002).

[5.26] L. D. Landau and E. M. Lifshitz, *Electrodynamics of Continuous Media* (Pergamon Press, Oxford, 1979).
[5.27] M. Born and E. Wolf, *Principles of Optics*, 7th edn. (Cambridge University Press, 1997).
[5.28] M. Dalarsson, Z. Jaksic and P. Tassin, *Journal of Optoelectronics and Biomedical Materials* 1, 345–352 (2009).

CHAPTER 6

POLARIZATION PHENOMENA IN GRADIENT NANOPHOTONICS

This section is devoted to two-dimensional reflectance-transmittance problems for waves, incident on a gradient dielectric lossless nonmagnetic layer at an arbitrary angle δ. Unlike the case of normal incidence, here the waves have different polarization structures and are described by different equations. The standardized approach, based on exactly solvable models of gradient nanofilms, illustrating the diversity of wave reflectance spectra subject to their polarizations and angles of incidence, is developed in this Chapter. The analysis of these spectra is complicated due to the vector structure of polarized fields. Side by side with some distributions of dielectric susceptibility $\varepsilon(z)$, suitable for analytical consideration of gradient effects for both S- and P-polarized fields, this approach reveals profiles $\varepsilon(z)$ providing unusual reflectance spectra for S-polarized waves only.

To develop a common mathematical basis for analysis of S- and P-polarized fields in a gradient layer, different generating functions Ψ_s and Ψ_p, related to each of these polarizations, are required. Denoting the normal to the layer as the z-axis, and choosing the projection of the wave vector on the layer's interface as the y-axis, one can write the Maxwell equations, describing the polarization structure of an S-wave by means of its electric component E_x, parallel to the interface $z = 0$, and magnetic components H_y and H_z, situated in the plane of incidence (y, z):

$$\frac{\partial E_x}{\partial z} = -\frac{1}{c}\frac{\partial H_y}{\partial t}, \quad \frac{\partial E_x}{\partial y} = \frac{1}{c}\frac{\partial H_z}{\partial t}, \quad \frac{\partial H_z}{\partial y} - \frac{\partial H_y}{\partial z} = \frac{\varepsilon(z)}{c}\frac{\partial E_x}{\partial t}, \tag{6.1}$$

$$div(\varepsilon \vec{E}) = 0, \quad div(\mu \vec{H}) = 0. \tag{6.2}$$

Components of a P-wave (H_x, parallel to the interface $z = 0$, and electric components E_y and E_z, located in the plane of incidence) are also linked by Eqs. (6.2), but Eqs. (6.1) have to be replaced by:

$$\frac{\partial H_x}{\partial z} = \frac{\varepsilon(z)}{c}\frac{\partial E_y}{\partial t}, \quad \frac{\partial H_x}{\partial y} = -\frac{\varepsilon(z)}{c}\frac{\partial E_z}{\partial t}, \quad \frac{\partial E_z}{\partial y} - \frac{\partial E_y}{\partial z} = -\frac{1}{c}\frac{\partial H_x}{\partial t}. \tag{6.3}$$

It is worthwhile to express the field components in the Maxwell equations by means of the following polarization-dependent generating functions Ψ_s and Ψ_p [6.1]:

$$\text{S-polarization: } E_x = -\frac{1}{c}\frac{d\Psi_s}{dt},$$

$$H_y = \frac{d\Psi_s}{dz}, \quad H_z = -\frac{d\Psi_s}{dy}, \tag{6.4}$$

$$\text{P-polarization: } H_x = \frac{1}{c}\frac{d\Psi_p}{dt},$$

$$E_y = \frac{1}{\varepsilon(z)}\frac{d\Psi_p}{dz}, \quad E_z = -\frac{1}{\varepsilon(z)}\frac{d\Psi_p}{dy}. \tag{6.5}$$

Using such presentations, one can reduce the system (6.1)–(6.3) to two equations, governing S- and P-waves respectively. Restricting ourselves by plane waves we can write these equations as:

$$\frac{\partial^2 \Psi_s}{\partial z^2} + \left(\frac{\omega^2 n_0^2 U^2}{c^2} - k_y^2\right)\Psi_s = 0, \quad k_y = \frac{\omega n_1 \sin \delta}{c}. \tag{6.6}$$

$$\frac{\partial^2 \Psi_p}{\partial z^2} + \left(\frac{\omega^2 n_0^2 U^2}{c^2} - k_y^2\right)\Psi_p = \frac{2}{U}\frac{dU}{dz}\frac{\partial \Psi_p}{\partial z}. \tag{6.7}$$

By introducing the new variable η and new functions f_s and f_p:

$$f_s = \Psi_s \sqrt{U}, \quad f_p = \frac{\Psi_p}{\sqrt{U}}, \quad \eta = \int_0^z U(z_1) dz_1. \qquad (6.8)$$

one can present Eqs. (6.6) and (6.7) for S- and P-waves in similar forms:

$$\frac{d^2 f_s}{d\eta^2} + f_s \left(K - \frac{U_{\eta\eta}}{2U} + \frac{U_\eta^2}{4U^2} \right) = 0, \qquad (6.9)$$

$$\frac{d^2 f_p}{d\eta^2} + f_p \left(K + \frac{U_{\eta\eta}}{2U} - \frac{3U_\eta^2}{4U^2} \right) = 0, \qquad (6.10)$$

where

$$K = \left(\frac{\omega n_0}{c}\right)^2 - \frac{k_y^2}{U^2}, \quad U_\eta = \frac{dU}{d\eta}, \quad U_{\eta\eta} = \frac{d^2 U}{d\eta^2}.$$

Equations (6.9)–(6.10) are valid for arbitrary profiles of photonic barriers $U(z)$ and all angles of incidence δ. Owing to the transformations (6.8), which provide the presentation of equations for both S- and P-polarized waves in coinciding forms, the following analysis of polarization effects proves to be standardized.

The continuity conditions on the boundaries of a reflecting nanofilm are distinguished for S- and P-polarizations. Considering the plane wave $\Psi_0 = A_0\{i[\omega n_1(z\cos\delta + y\sin\delta)c^{-1} - t]\}$, incident from the homogeneous medium with refractive index n_1 under the angle δ on the film boundary $z = 0$, one can write these conditions, which are needed for a calculation of polarization — dependent reflection coefficients R_s and R_p as:

S-polarization:

$A_0(1 + R_s) = A_s \Psi_s|_{z=0},$

$\dfrac{i\omega n_1 A_0 \cos\delta(1 - R_s)}{c} = A_s \left.\dfrac{d\Psi_s}{dz}\right|_{z=0},$

$\dfrac{i\omega n_1 A_0 \sin\delta(1 + R_s)}{c} = ik_y A_s \Psi_s|_{z=0},$

P-polarization:

$A_0(1 + R_p) = A_p \Psi_p|_{z=0},$

$\dfrac{i\omega A_0 \cos\delta(1 - R_p)}{cn_1} = \dfrac{A_p}{\varepsilon(z)} \left.\dfrac{d\Psi_p}{dz}\right|_{z=0},$

$\dfrac{i\omega A_0 n_1 \sin\delta(1 + R_p)}{c} = ik_y A_p \Psi_p|_{z=0}.$

$$(6.11)$$

If the reflecting films are located on a homogeneous substrate with refractive index n, the continuity conditions, given on the boundary $z = d$, link the wave components (6.4) and (6.5) in the film with the corresponding components of the plane wave $\Psi = A\exp\{i[\omega(z\sqrt{n^2 - n_1^2\sin^2\delta} + yn_1\sin\delta)c^{-1} - t]\}$ in the substrate.

The difficulty of obtaining the simultaneous solutions of Eqs. (6.9) and (6.10), describing S- and P-polarized fields for one model $U(z)$, result in a limitation on the number of exactly solvable models $U(z)$. Section 6.1 is focused on a peculiar effect, inherent in the oblique propagation of waves through gradient nanofilms: profiles $\varepsilon(z)$, providing either traveling or tunneling regimes for any frequencies and arbitrary angles of incidence of both S- and P-polarized waves, are examined. Moreover, another example of a polarization selecting distribution $\varepsilon(z)$, ensuring the total transmittance of an S wave, incident under a specific angle, is presented; this heterogeneity — induced phenomenon can be viewed as an analogue of Brewster effect (total transmittance for P waves), well known in the optics of homogeneous media [6.2]. Section 6.2 is devoted for polarization — dependent filtration of fields in periodical nanogradient dielectric structures; examples of inclined propagation of polarized fields in the metamaterials with continuously distributed dielectric and magnetic response are exemplified too. Reflectionless tunneling of S wave through the multilayer stack of gradient nanofilms and Goos-Hanchen effect for this structure are considered in Sec. 6.3.

6.1. Wideangle Broadband Antireflection Coatings

Elaboration of antireflection coatings, effective simultaneously in broad spectral intervals and wide ranges of angles for arbitrary polarizations, is known to be an actual task in the optics of thin films. It is remarkable, that some gradient layers with distributions $\varepsilon(z)$, considered above in the framework of the 1D problem (normal incidence), can also be used as exactly solvable models for 2D problems (oblique incidence). This feasibility can pave the way to the design of gradient antireflection coatings and frequency — selective interfaces with parameters unattainable with homogeneous layers.

To compare and contrast gradient and homogeneous coatings let us recall films with simple distributions of refractive index $n(z) = n_0 U(z)$, and examine two different profiles $U(z)$

$$U_1(z) = \left(1 + \frac{z}{L}\right)^{-1}, \quad U_2(z) = \exp\left(-\frac{z}{L}\right). \qquad (6.12)$$

We start by considering profile $U_1(z)$. On substituting $U_1(z)$ into (6.8), we define the variable $\eta = L\ln(1 + z/L)$. Equations (6.9) and (6.10), governing the functions f_s and f_p, are reduced in this case to the standard Bessel equation (5.24). Keeping in mind applications of this equation to the analysis of some other problems, we rewrite it, changing some definitions:

$$\frac{d^2 f_{s,p}}{dx^2} + \frac{1}{x}\frac{df_{s,p}}{dx} + f_{s,p}\left(q_{s,p}^2 - \frac{l_{s,p}^2}{x^2}\right) = 0, \quad x = \frac{\eta}{L}, \qquad (6.13)$$

$$q^2 = -\left(\frac{\omega L \sin\delta}{c}\right)^2, \quad l_{s,p}^2 = \frac{1}{4}\left(1 - \frac{1}{u^2}\right),$$

$$u = \frac{\Omega}{\omega}, \quad \Omega = \frac{c}{2n_0 L}. \qquad (6.14)$$

Here the indices s and p correspond to S- and P-polarizations, Ω is some characteristic frequency, which cannot be viewed here, unlike (2.19), as a cut-off frequency. The linearly–independent solutions of Eq. (6.13) in the case $q^2 < 0$ are known to be given by the Bessel function of imaginary argument I_l and the Macdonald function K_l [6.3]. To examine the field in a layer of finite thickness d, it is worthwhile to use their linear combinations $\vartheta_l^{(1,2)} = I_l \pm iK_l$, similar to Hankel functions, related to the values $q^2 > 0$. Bringing together these solutions of Eq. (6.13) and the representations (6.8), we obtain the generating functions for S- and P-polarized waves:

$$\Psi_s = \left(1 + \frac{z}{L}\right)^{\frac{1}{2}} f_s, \quad \Psi_p = \left(1 + \frac{z}{L}\right)^{-\frac{1}{2}} f_p, \qquad (6.15)$$

$$f_{s,p} = [\vartheta_l^{(1)}(\varsigma) + Q_{s,p}\vartheta_l^{(2)}(\varsigma)], \quad \varsigma = \frac{\omega(L+z)\sin\delta}{c}. \qquad (6.16)$$

The phase factors $\exp[i(k_y y - \omega t)]$ are omitted for simplicity in the expressions for generating functions Ψ_s and Ψ_p in (6.15) and in what follows; the quantities Q_s and Q_p in (6.16) have to be defined from the continuity conditions at the air-film boundary $z = 0$ and the film-substrate boundary $z = d$. The wave components (6.4)–(6.5) as well as the reflectance spectra for these waves $|R(u)|^2$ can be found by means of a standard procedure, using the boundary conditions (6.11).

Examples of such spectra for polarized waves $|R(u)|^2$ are shown in Fig. 6.1. While computing $|R(u)|^2$ we have expressed the variable ς in (6.16) via the normalized frequency u and the value of the distribution U_1 (6.12) at the film-substrate boundary $z = d$:

$$\varsigma = \frac{\sin \delta}{2un_0 U_m}, \quad U_m = U_1(d) = \left(1 + \frac{d}{L}\right)^{-1}. \tag{6.17}$$

To illustrate the effects of the film's gradient structure, its spectra $|R(u)|^2$ are compared with the reflectance spectra for homogeneous films; here all the parameters of the incident waves, such as their polarization states, frequencies ω, angles δ, as well as all the parameters of the reflecting structure — the refractive indices of the film and substrate n_0 and n, and thickness d are equal in both cases. The complex reflection coefficients for S- and P-polarized waves from a homogeneous film, located on a substrate, are [6.4]:

$$R_s = \frac{M_s \cos \delta - r_1 - it(\cos \delta - r_1 M_s)}{M_s \cos \delta + r_1 - it(\cos \delta + r_1 M_s)}. \tag{6.18}$$

$$R_p = \frac{M_p n^2 \cos \delta - r_1 - it(n^2 \cos \delta - r_1 M_p)}{M_p n^2 \cos \delta + r_1 - it(n^2 \cos \delta + r_1 M_p)}. \tag{6.19}$$

$$M_s = \frac{r_1}{r_2}, \quad M_p = M_s \frac{n^2}{n_0^2}, \quad r_1 = \sqrt{n_0^2 - \sin^2 \delta},$$

$$r_2 = \sqrt{n^2 - \sin^2 \delta}, \quad t = \operatorname{tg}\left(\frac{\omega d r_1}{c}\right). \tag{6.20}$$

To compare these coefficients R_s and R_p for a homogeneous film with reflection coefficients for a gradient film, calculated at the same

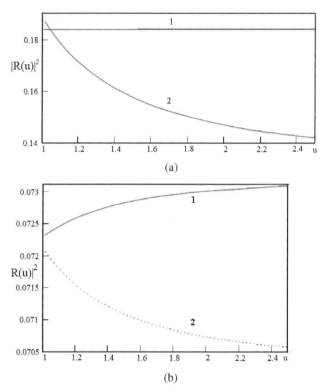

Fig. 6.1. Effects of the gradient structure of a nanofilm with the distribution of refractive index $n(z) = n_0(1 + z/L)^{-1}$ in the reflection of S- and P-polarized waves incident under the angle of illumination $\delta = 45°$ are shown in (a) and (b), respectively; $n_0 = 1.9$, the refractive index of the substrate $n = 1.5$, and u is the normalized frequency. Since the reflection coefficients for the gradient nanofilms for the same frequencies and polarizations $|R(u)|^2_{gs}$ and $|R(u)|^2_{gp}$ are much smaller than those for the homogeneous nanofilms $|R(u)|^2_{hs}$ and $|R(u)|^2_{hp}$, depicted on the curves 2, the values n_0, n and δ being equal, curves 1 on both figures are intended for calculation of $|R(u)|^2_{gs}$ and $|R(u)|^2_{gp}$ by means of values $|R(u)|^2_{1s}$ and $|R(u)|^2_{1p}$, given by the conventional curves 1: $|R(u)|^2_{gs} = 0.09|R(u)|^2_{1s}$, $|R(u)|^2_{gp} = 0.12|R(u)|^2_{1p}$. Curves 2 show the reflectance spectra for homogeneous nanofilms $|R(u)|^2_{hs}$ and $|R(u)|^2_{hp}$.

frequencies, one has to substitute into (6.18)–(6.20) the values of the factor $t = \text{tg}(\omega d r_1 c^{-1})$, expressed via the factors u and U_m:

$$t = \text{tg}\left[\left(\frac{1 - U_m}{2uU_m}\right)\sqrt{1 - \frac{\sin^2 \delta}{n_0^2}}\right]. \qquad (6.21)$$

Inspection of Fig. 6.1. shows several features of reflectance spectra of gradient films with a refractive index $U_1(z)$ profile:

1. Owing to the heterogeneity-induced dispersion the tunneling regimes arise in the Rayleigh barrier $U_1(z)$ for both S- and P- waves simultaneously; radiation transfer through this barrier is provided for both S- and P-polarized fields by evanescent modes only;
2. The reflectance of these films in a wide spectral range is about an order of magnitude smaller, than the reflectance of homogeneous films, other parameters of the films and radiation being the same;
3. The films with profile $U_1(z)$ are shown to produce a significant decrease of the reflectance despite their subwavelength thickness: thus, using the definition of the variable u (6.14) one can find the link between wavelength λ and thickness d, corresponding to any given value of the variable u:

$$\frac{d}{\lambda} = \frac{1 - U_m}{4\pi u n_0 U_m}. \qquad (6.22)$$

According to Fig. 6.1(a), the reflection coefficient of a gradient nanofilm for an S-wave, corresponding, e.g. to the value $u = 1.5$ is $|R(u)|^2_{gs} = 0.0165$; this means, that all the waves, obeying to Eq. (6.22) are characterized by this reflection coefficient. In particular, under the conditions corresponding to Fig. 6.1(a), we have from (6.22): $d = 0.007\lambda$; thus, for the infrared S- wave with $\lambda = 14.28\mu$, incident on the gradient nanofilm with thickness $d = 100$ nm under the angle $\delta = 45°$, the reflection coefficient is as small as 0.0165; the replacement of this film by a homogeneous film will result, according to Fig. 6.1(a), in an increase of the reflectance by a factor of 10: $|R(u)|^2 = 0.168$. Thus, these gradient films can be viewed as the broadband antireflection nanocoatings in a middle IR range.

Now let us consider nanofilms with an exponential profile of the refractive index $U_2(z)$ (6.12); in this case it is convenient to examine the propagation of S- and P-polarized waves separately.

Substitution of model $U_2(z)$ into Eq. (6.6) brings the equation governing the generating function Ψ_s into the form of the familiar Bessel equation (6.13); however, the variable x and parameters q^2

and l_s^2 have to be redefined:

$$x = \exp\left(-\frac{z}{L}\right), \quad q^2 = \left(\frac{\omega n_0 L}{c}\right)^2, \quad l_s^2 = \left(\frac{\omega L \sin\delta}{c}\right)^2. \quad (6.23)$$

Let us emphasize that the value of q^2 in (6.23), related to model $U_2(z)$, is always positive, in contrast to (6.14), where q^2, related to $U_1(z)$, is always negative. Due to this difference the linearly-independent solutions of Eq. (6.13) in this case are given by the Hankel functions $H_l^{(1,2)} = J_l \pm iN_l$, where J_l and N_l are Bessel and Neumann functions [6.3], $l = l_s$ (6.23); therefore generating function for S-waves in the model $U_2(z)$ can be written as

$$\Psi_s = H_l^{(1)}(\varsigma) + Q_s H_l^{(2)}(\varsigma), \quad \varsigma = \frac{\omega n_0 L}{c}\exp\left(-\frac{z}{L}\right). \quad (6.24)$$

The generating function Ψ_p, distinguished from Ψ_s, can be calculated in the same way:

$$\Psi_p = \exp\left(-\frac{z}{L}\right)\left[H_l^{(1)}(\varsigma) + Q_p H_l^{(2)}(\varsigma)\right],$$

$$l = l_p = \sqrt{1 + \left(\frac{\omega L \sin\delta}{c}\right)^2}. \quad (6.25)$$

The variable ς for both functions Ψ_s (6.24) and Ψ_p (6.25) has the same value, defined in (6.24). The derivation of the reflection coefficients, these functions being known, does not pose any mathematical problems.

Comparing the distributions $U_1(z)$ and $U_2(z)$, one can note some features of their difference and similarity:

1. In contrast to profile $U_1(z)$, both S- and P-polarized waves with any frequencies, incident under any angles, traverse the film with the profile $n(z) = n_0 U_2(z)$ as propagating modes.
2. Interaction of waves with films, possessing the refractive index profiles $n_1(z) = n_0 U_1(z)$ and $n_2(z) = n_0 U_2(z)$, illustrate the sensitivity of these interactions to the details of the distributions $U_1(z)$ and $U_2(z)$: despite the equalities $U_1(z=0) = U_2(z=0)$ and even in the case, where the scale parameters L are equal

for both distributions ($\text{grad}U_1|_{z=0} = \text{grad}U_2|_{z=0} = -L^{-1}$), the regimes of wave propagation through these media are shown to be different.

3. A simultaneous analytical solution of two Eqs. (6.9) and (6.10), describing S- and P-polarized fields, is hampered due to the necessity of finding one common model $U(z)$, suitable for solution of both these equations; some of these models are exemplified by distributions (6.12). However, restricting ourselves to the analysis of S-polarized fields only, one can develop a new insight on some well-known phenomena.

Thus, one can recall the Brewster effect — the vanishing of reflection for P-polarized waves, incident on the boundary of a homogeneous half-space with refraction index n under the angle $\delta = \text{arctg}(n)$ (Brewster angle) [6.4]. This effect is known to arise for P-waves only. One can see from (6.18), that the reflectionless interaction of S-waves with a homogeneous medium is impossible. However, in gradient media reflectionless propagation can arise for S-polarized wave as well.

To visualize this new possibility let us examine the oblique incident of an S-polarized wave from a homogeneous dielectric with index n_1 on a gradient film, characterized by the familiar profile $U(z)$ (5.40), [6.5], shown in Fig. 5.3(b). The wave structure in this film, determined by the unknown function Ψ_s, is governed by Eq. (6.6). To find Ψ_s one has to substitute the profile (5.40) into Eq. (6.6) and follow the procedure, developed in section 5.3: introduce the new function $f_s = \Psi_s\sqrt{W}$ and new variables η (5.43)–(5.44) and ς (5.45); after the replacement of $W(z)$ (5.40) by $W(\varsigma)$ (5.46) the equation, governing the function f_s, is Eq. (5.47). Here the expression for the parameter q^2 in Eq. (5.47) coincides with (5.48), while the value of the parameter Λ in Eq. (5.47) differs from the definition (5.48):

$$\Lambda = \left(\frac{\omega L}{c}\right)^2 \frac{\wp}{g} - \frac{1}{4}, \quad \wp = n_0^2(g-1) - n_1^2 g \sin^2 \delta. \qquad (6.26)$$

Now let us turn aside from an outline of the general solution of Eq. (5.47) used in Sec. 5.3, and focus the attention on the special case $\Lambda = 0$ (6.26). If the parameters of the reflecting structure n_1, n_0, g, and L are given, the following analysis is valid for any frequencies ω and angles of incidence δ, linked by the condition $\Lambda = 0$. Under this condition Eq. (5.47) is reduced to the simple form:

$$\frac{d^2 f_s}{d\varsigma^2} + q^2 f_s = 0, \quad q^2 = \frac{1}{4}\left[\frac{n_0^2}{\wp(1+M^2)} - 1\right]. \quad (6.27)$$

We will consider here the case $q^2 > 0$, where the solution of Eq. (6.27) is presented by forward and backward harmonic waves in ς-space. Taking into account the link between the functions f_s and Ψ_s introduced above ($f_s = \Psi_s \sqrt{W}$), one can write the generating function Ψ_s as [6.5]:

$$\Psi_s = \sqrt{\cos\left(\frac{z}{L}\right) + M \sin\left(\frac{z}{L}\right)} [\exp(iq\varsigma) + Q \exp(-iq\varsigma)]; \quad (6.28)$$

Expressing the wave components via Ψ_s, (6.4), and using the continuity conditions (6.11), we find the complex reflection coefficient R_s:

$$R_s = \frac{iB - M - 2iqY\sqrt{1+M^2}}{iB + M + 2iqY\sqrt{1+M^2}},$$

$$Y = \frac{\exp(-iq\varsigma_0) - Q\exp(iq\varsigma_0)}{\exp(-iq\varsigma_0) + Q\exp(iq\varsigma_0)}, \quad B = \frac{2\omega L n_1 \cos\delta}{c}. \quad (6.29)$$

The quantities q and ς_0 in (6.29) are defined in (6.27) and (5.45), respectively. The value of parameter Q, derived from the standard continuity conditions on the interface $z = d$, is:

$$Q = -\frac{(M + iB_1 - 2iq\sqrt{1+M^2})\exp(2iq\varsigma_0)}{M + iB_1 + 2iq\sqrt{1+M^2}},$$

$$B_1 = \frac{2\omega L \sqrt{n^2 - n_1^2 \sin^2\delta}}{c}. \quad (6.30)$$

Manipulations with formulae (6.29) and (6.30) yield finally the expression for the reflection coefficient R_s in a familiar form (2.31):

$$R_s = \frac{\sigma_1 + i\sigma_2}{\chi_1 + i\chi_2}, \quad t = \text{tg}(2q\varsigma_0), \qquad (6.31)$$

$$\sigma_1 = t[BB_1 + M^2 - 4q^2(1+M^2)] - 4Mq\sqrt{1+M^2},$$

$$\sigma_2 = (B - B_1)(2q\sqrt{1+M^2} - Mt),$$

$$\chi_1 = t[BB_1 - M^2 + 4q^2(1+M^2)] + 4Mq\sqrt{1+M^2},$$

$$\chi_2 = (B + B_1)(2q\sqrt{1+M^2} - Mt).$$

On setting $\sigma_1 = \sigma_2 = 0$ in (6.31), we obtain the condition for the vanishing of reflection, $R_s = 0$, for frequencies and angles, linked by the condition $\Lambda = 0$:

$$BB_1 = M^2 + 4q^2(1+M^2). \qquad (6.32)$$

After substitution of the values of B (6.29) and B_1 (6.30) into (6.32) and elimination of the parameters L and q^2 by means of (5.41) and (6.27), the Eq. (6.32) can be rewritten as

$$\frac{\omega d}{c} = (\text{arctg} M)\sqrt{\frac{n_0^2 - \wp}{\wp n_1 \cos\delta \sqrt{n^2 - n_1^2 \sin^2\delta}}}. \qquad (6.33)$$

The right side of Eq. (6.33), being independent of the frequency ω and the thickness of the gradient nanofilm d, is determined by the parameters of the gradient film (5.41) and (5.42) (n_0, M, L, g), refractive indices of the homogeneous media surrounding this film (n_1 and n), and the angle of incidence δ. Specifying these parameters, we'll determine the dimensionless factor $\omega d c^{-1}$ (6.33), with the use of this factor one can calculate the thickness of the nanofilm d, that provides the total reflection of an S-wave with frequency ω, incident under the angle δ on this reflecting nanostructure with the concave profile $U(z)$ shown in Fig. 5.3(b). Thus, in the case $n_0 = 1.9, M = 0.9632$, $g = 2.1, n_1 = 1.433, n = 2.28$ we find that for the angle $\delta = 60°$ and $d = 100$ nm the reflection vanishes for $\omega = 3.878\, 10^{15}$ rad/s (wavelength $\lambda = 485$ nm). If the S-wave is incident from an air ($n_1 = 1$)

under the angle, e.g. $\delta = 45°$, on the nanostructure with parameters $n_0 = 1.9$, $M = 0.9632$, $g = 1.3$, $n = 2$, $d = 100$ nm, the reflectionless regime arises for $\lambda = 693$ nm. Using of thicker films, other quantities in (6.33) being fixed, results in the shift of the wavelengths of reflectionless regimes to the infrared range.

Let us outline some salient features of these polarization-dependent effects:

1. The reflectionless interaction of an S-wave with the gradient nanofilm (5.40) resembles the Brewster effect, inherent in the optics of homogeneous media for P-polarized waves only; owing to heterogeneity-induced dispersion new possibilities of reflectionless penetration of radiation through the boundary between two dielectrics arise.
2. Unlike the optics of homogeneous dielectrics, the angle of illumination δ, which is needed for observation of this Brewster-like effect in the gradient nanostructure (5.40), is determined by the frequency ω according to Eq. (6.33), other parameters of structure being fixed.
3. The nanofilms discussed above can be viewed as broadband antireflection coatings, effective, in particular, for large angles δ. These coatings are based on single layer models. The reflectance of multilayer gradient structures is considered below in Sec. 6.2.

6.2. Polarization-Dependent Tunneling of Light in Gradient Optics

Conditions for the appearance of propagating or evanescent waves in nanofilms with profiles of refractive index (6.12) were shown to be independent of both the polarization of the waves and their frequency. In contrast, this section is devoted to more versatile nanostructures with heterogeneity-induced dispersion, supporting both traveling and evanescent modes subject to their polarization and frequency. These properties will be illustrated first by means of an exactly solvable single-layer model; later we will examine a periodic structure, constructed from such layers.

A simple model of a gradient layer possessing these properties can be obtained from the distribution (5.40) in the limiting case $g = 1$, $M = 0$, $U(z) = W(z)$; on this way we'll consider the concave symmetrical barrier $U(z)$ with thickness d, varying from the minimum value $U = 1$ in the center of the barrier ($z = 0$) up to the maximum value U_m at the interfaces $z = \pm d/2$ [6.6]:

$$U(z) = \left[\cos\left(\frac{z}{L}\right)\right]^{-1}, \quad U_m = m^{-1}, \quad m = \cos\left(\frac{d}{2L}\right). \tag{6.34}$$

In this case the variables η (5.44), ς (5.45), and the distribution $U(\varsigma)$ (5.46) are written as:

$$\frac{\eta}{L} = \ln\left[\frac{1 + \mathrm{tg}\left(\frac{z}{2L}\right)}{1 - \mathrm{tg}\left(\frac{z}{2L}\right)}\right], \quad \varsigma = \frac{\eta}{L}, \quad U(\varsigma) = \mathrm{ch}\,\varsigma. \tag{6.35}$$

Substitution of the distribution $U(\varsigma)$ (6.35) into the basic equations (6.39) and (6.10) yields a common equation for both functions f_s and f_p, connected with S- and P-polarized waves in the film (6.34):

$$\frac{d^2 f_{s,p}}{d\varsigma^2} + f_{s,p}\left(q^2 - \frac{\Lambda_{s,p}}{\mathrm{ch}^2\varsigma}\right) = 0, \tag{6.36}$$

$$q^2 = \frac{1}{4}\left(\frac{1}{u^2} - 1\right), \quad \Lambda_s = \left(\frac{\omega L \sin\delta}{c}\right)^2 + \frac{1}{4},$$

$$\Lambda_p = \left(\frac{\omega L \sin\delta}{c}\right)^2 - \frac{3}{4}. \tag{6.37}$$

The value of the normalized frequency u, defined in (6.14), remains valid here also.

Following the solution of the similar Eq. (5.47), we introduce the new variable ν and the new function $F(\nu)$ (5.49), governed by the hypergeometric equation (5.50) with parameters α, β and γ that have to be defined now via the quantities q^2, Λ_s and Λ_p (6.37). Note, that the sign of the parameter q^2 (6.37) changes when $\omega = \Omega$, and let us consider the low frequency spectral range $\omega < \Omega$, $q^2 < 0$. In this case the value of the exponent $-2p = \sqrt{-q^2} = l$ in the representation of both functions f_s and f_p (5.49) remains positive ($l > 0$), and the definition $\gamma = 1 - l$ (5.51) is also valid for both waves; however, the values of the parameters α and β have to be specified for each wave. Let us start the analysis with the S-wave. In this case we find:

$$\alpha_s, \beta_s = \frac{1}{2} - l \pm \frac{i\sin\delta}{2n_0 u}. \tag{6.38}$$

Since $\alpha_s + \beta_s + 1 = 2\gamma$, two linearly-independent solutions of Eq. (5.50) are given by the hypergeometric functions $F(\alpha_s, \beta_s, \gamma, v)$ and $F(\alpha_s, \beta_s, \gamma, 1 - v)$, denoted below for compactness as $F(\nu)$ and $F(1 - \nu)$. Moreover, due to the condition $\text{Re}(\alpha_s + \beta_s - \gamma) < 0$ the series representing these functions converge absolutely [6.3]. Continuing the analogy, one can be convinced that the generating function Ψ_s (6.8) coincides with (5.53), where the parameters α_s and β_s are defined in (6.38). Standard manipulations with the continuity conditions on the boundaries $z = \pm d/2$ yield the complex reflection coefficient R_s:

$$R_s = \frac{iB_1 F_1 + M_1 + Q_s(iB_1 F_2 + M_2)}{iB_1 F_1 - M_1 + Q_s(iB_1 F_2 - M_2)}, \tag{6.39}$$

$$B_1 = \frac{m\cos\delta}{n_0 u}, \quad B_2 = \frac{m\sqrt{n^2 - \sin^2\delta}}{n_0 u}, \tag{6.40}$$

$$Q_s = -\left(\frac{M_2 - iB_2 F_2}{M_1 - iB_2 F_1}\right), \quad M_{1,2} = s(2l - 1)F_{1,2} \pm m^2 F'_{1,2}. \tag{6.41}$$

Here F_1 and F_2 are the hypergeometric functions

$$F_{1,2} = F(\alpha_s, \beta_s, \gamma, \nu_{1,2}), \quad F'_{1,2} = \left.\frac{dF(\alpha_s, \beta_s, \gamma, \nu)}{d\nu}\right|_{\nu = \nu_{1,2}}, \tag{6.42}$$

$$\nu_{1,2} = \frac{1 \pm s}{2}, \quad s = \sqrt{1 - m^2}, \quad \gamma = 1 - l, \quad l = \frac{1}{2}\sqrt{1 - \frac{1}{u^2}}. \tag{6.43}$$

Proceeding in a similar fashion we obtain the generating function Ψ_p (6.8) and the reflection coefficient for P-polarized wave R_p:

$$R_p = \frac{iC_1\Phi_1 + M_3 + Q_p(iC_1\Phi_2 + M_4)}{iC_1\Phi_1 - M_3 + Q_p(iC_1\Phi_2 - M_4)}, \qquad (6.44)$$

$$C_1 = \frac{n_0 \cos\delta}{mu}, \quad C_2 = \frac{n_0\sqrt{n^2 - \sin^2\delta}}{n^2 mu}, \qquad (6.45)$$

$$Q_p = -\left(\frac{M_4 - iC_2\Phi_2}{M_3 - iC_2\Phi_1}\right), \quad M_{3,4} = s(2l+1)\Phi_{1,2} \pm m^2\Phi'_{1,2}. \qquad (6.46)$$

Here Φ_1 and Φ_2 are the hypergeometric functions, dependent, unlike (6.42), on the parameters α_p and β_p:

$$\Phi_{1,2} = \Phi(\alpha_p, \beta_p, \gamma, \nu_{1,2}), \quad \Phi'_{1,2} = \left.\frac{d\Phi(\alpha_p, \beta_p, \gamma, \nu)}{d\nu}\right|_{\nu=\nu_{1,2}}, \qquad (6.47)$$

$$\alpha_p, \beta_p = \frac{1}{2} - l \pm \sqrt{1 - \frac{\sin^2\delta}{4n_0^2 u^2}}, \quad \gamma = 1 - l. \qquad (6.48)$$

The values of the quantities $\nu_{1,2}$, s and γ, defined in (6.43) for S-waves, are also valid for P-waves.

Reflectance spectra for inclined incidence of S- and P-polarized waves on the gradient films (6.34) are depicted in Fig. 6.2. Inspection of these spectra illustrates the following polarization-related effects:

1. Reflection of S waves exceeds the reflection of P waves for both illumination angles 45° and 75°. Note, that all the spectra in Fig. 6.2. are calculated by means of the general solutions of Eq. (6.36), obtained for the case $q^2 < 0$, i.e. $u < 1$. However, subject to the parameters of reflecting nanostructure and incident wave (frequency, angle of illumination, polarization state), these solutions describe different regimes of propagation of these waves through the structure, corresponding either to propagating or to evanescent modes. The conditions for the appearance of these

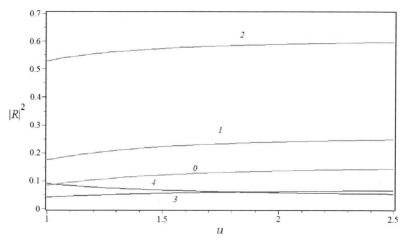

Fig. 6.2. Reflectance spectra for S- and P-polarized waves incident on a single gradient nanofilm (6.34), $m = 0.75$, are shown by curves 1,2 and 3,4, respectively, u is the normalized frequency (6.14). Curves 1(3) and 2(4) correspond to the illumination angles $45°(75°)$, curve 0 — normal incidence; $n = 2.3$, $n_0 = 1.47$.

regimes are determined by the sign of the expression in brackets in Eq. (6.36),

$$\aleph_{s,p} = q^2 - \frac{\Lambda_{s,p}}{\mathrm{ch}^2\varsigma}. \qquad (6.49)$$

The cases $\aleph_s > 0$, $\aleph_p > 0$ ($\aleph_s < 0$, $\aleph_p < 0$) correspond to the propagating (evanescent) modes for both polarizations, the mixed cases $\aleph_s > 0$, $\aleph_p < 0$ or $\aleph_s < 0$, $\aleph_p > 0$ characterize different regimes of propagation of S- and P-waves. After substitution of the quantities q^2, Λ_s and Λ_p from (6.37) into (6.49) these inequalities define the spectral ranges containing any given propagation regime. Designating the normalized frequencies u for S (P) polarization as $u_s(u_p)$, we have, e.g. the conditions of appearance of evanescent S and P modes in the gradient nanostructure (6.34):

$$u_s^2 > \frac{\mathrm{ch}^2\varsigma - n_0^{-2}\sin^2\delta}{\mathrm{ch}^2\varsigma + 1}, \quad u_p^2 > \frac{\mathrm{ch}^2\varsigma - n_0^{-2}\sin^2\delta}{\mathrm{ch}^2\varsigma - 3}. \qquad (6.50)$$

An opposite inequality for u_s or u_p indicates the formation of a propagation regime for the corresponding mode. The values of the function $\mathrm{ch}\varsigma$, describing the profile (6.34), are located in the segment $0 < \text{ch}\varsigma \leq m^{-1}$. Substituting the data, corresponding to Fig. 6.2, into inequalities (6.50), one can see that the S wave is traversing the film as an evanescent mode, unlike the P wave, which is propagating in a traveling regime. The tunneling regime in a film is known to hamper usually the penetration of wave's energy through the film interface, and this effect can cause the better reflectance of the S wave shown in Fig. 6.2.

2. The frequency dispersion of reflectance spectra in the spectral range $1 < u < 2.5$ (Fig. 6.2) is insignificant, and, therefore, these nanostructures can be viewed as broadband polarizers. Thus, rewriting the definition of frequency Ω (6.14) for the profile (6.35) in a form

$$\Omega = \frac{c \arccos(m)}{n_0 d}, \qquad (6.51)$$

and taking, e.g. the thickness of the gradient layer $d = 100\,\text{nm}$, one can consider these polarizers for the infrared radiation in the range of wavelengths $1.25\,\mu\text{m} \leq \lambda \leq 3.2\,\mu\text{m}$.

3. The difference in reflection and transmission of S- and P-polarized waves can provide a potential for broadband wide angle filtration and separation of S- and P-polarized waves: thus, comparison of curves 2 and 4, related to the large angle $\delta = 75°$, indicates the ratio of reflection coefficients $|R_s|^2/|R_p|^2 = 12$ for the frequency $u = 2.5$; the ratio of transmission coefficients for the same frequency is $|T_p|^2/|T_s|^2 = 2.3$. In this way the reflected and transmitted portions of the radiation prove to be enriched by S- and P- polarized components, respectively. Moreover, comparison of curves 1, 2, 3 and 4 with curve 0, corresponding to normal incidence, shows an important general property of these spectra: an increase of the illumination angle results in a weakening of the reflectance of P waves and in the increase of the reflectance of S waves. The monotonic nature of the spectral variations of the reflectance/transmittance spectra has to be noted also.

To generalize these results to periodic structures, containing $k > 1$ films (6.34), one can follow the analyses (6.36)–(6.48). Attributing the number $k = 1$ to the first film on the far side of structure, contiguous to the substrate, we define the parameter Q_1, describing the interference of forward and backward waves in this film. It is worthwhile to examine the propagation of S and P waves separately. In the case of S waves the parameter Q_1 coincides with the parameter Q_s (6.41). Introducing the designation $Q_{s1} = Q_s$ we have the recurrence linking the values $Q_{s,k}$ and $Q_{s,k-1}$ for k-th and (k-1)-th contiguous films ($k \geq 2$):

$$Q_{s,k} = -\left[\frac{F_1 M_2 + F_2 M_1 + 2 F_2 M_2 Q_{s,k-1}}{2 F_1 M_1 + (F_1 M_2 + F_2 M_1) Q_{s,k-1}}\right]. \tag{6.52}$$

The quantities $M_{1,2}$ and $F_{1,2}$ are defined by Eqs. (6.41) and (6.42), respectively. Substitution of the quantity $Q_{s,k}$ (6.52) instead of Q_s into Eq. (6.39) yields the expression, defining the complex reflection coefficient a for periodic nanostructure, containing k films:

$$R_{s,k} = \frac{iB_1 - \Gamma_{s,k}}{iB_1 + \Gamma_{s,k}}, \quad \Gamma_{s,k} = (2l-1)\sqrt{1-m^2} + \frac{m^2(F_1' - Q_{s,k} F_2')}{F_1 + Q_{s,k} F_2}. \tag{6.53}$$

All the quantities in Eqs. (6.53) are defined in expressions (6.40), (6.42), and (6.43), derived for a single gradient layer.

The same algorithm has to be used in a calculation of the reflection coefficient $R_{p,k}$ for P waves:

$$R_{p,k} = \frac{iC_1 + \Gamma_{p,k}}{iC_1 - \Gamma_{p,k}}, \quad \Gamma_{p,k} = (2l+1)\sqrt{1-m^2} + \frac{m^2(\Phi_1' - Q_{p,k} \Phi_2')}{\Phi_1 + Q_{p,k} \Phi_2}. \tag{6.54}$$

The recurrence relation linking $Q_{p,k}$ and $Q_{p,k-1}$ for $k \geq 2$ reads:

$$Q_{p,k} = -\left[\frac{\Phi_1 M_2 + \Phi_2 M_1 + 2 \Phi_2 M_2 Q_{p,k-1}}{2 \Phi_1 M_1 + (\Phi_1 M_2 + \Phi_2 M_1) Q_{p,k-1}}\right].$$

The quantity $Q_{p,k}$ for $k=1$ is given in (6.46); the values $C_{1,2}$, $M_{3,4}$, functions $\Phi_{1,2}$ and their derivatives, defined in (6.45), (6.46)–(6.48), coincide with the corresponding values for a single film.

The reflection coefficients for periodic nanostructures, containing 5 and 10 reflecting films, are shown in Fig. 6.3. On comparing these spectra with the spectra, depicted in Fig. 6.2 for the angle

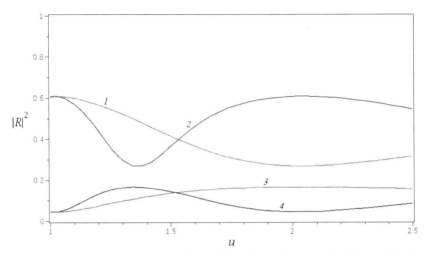

Fig. 6.3. Reflectance spectra for S- and P-polarized waves, incident under the angle 75° on a periodic multilayer nanostructure, built from the gradient layers (6.34), $m = 0.75$, are shown by curves 1(2) and 3(4), respectively, u is the normalized frequency (6.14). Curves 1(3) and 2(4) relate to the structures containing 5(10) gradient nanofilms.

$\delta = 75°$ (curves 2 and 4), one can conclude, that an increase in the number of films results in an increase of the reflectance of P waves by 1.5–2 times; a decrease of the reflectance of S waves can be noted for some spectral ranges. The interference effects, producing the non-monotonic behaviour of the spectra of both S and P waves, are well expressed for a 10-layer gradient nanostructure. Spectra for S and P polarized waves coincide for normal incidence, when the difference between S and P polarizations vanishes [6.6]. Note, that this analysis reveals the similarity of the optical properties of periodic nanostructures, discussed above, and of gradient dielectric superlattices [6.7, 6.8].

6.3. Reflectionless Tunneling and Goos–Hänchen Effect in Gradient Metamaterials

Total internal reflection of light is known to arise when the light wave impinges from a dielectric medium (index n_1) on an another dielectric medium (index n) with $n_1 > n$, for an angle of incidence $\delta > \delta_{cr} = \arcsin(n/n_1)$ [6.4]. This condition, valid for homogeneous non-dispersive media, has to be reconsidered if, at least one of the media possesses natural or heterogeneity-induced dispersion; as was shown in Sec. 5.2, the effect of total internal reflection (TIR) is possible in these cases even for normal incidence. The new opportunities for the appearance of TIR phenomena, diversified due to oblique incidence, are examined below; to illustrate the drastic changes, introduced into the classical concept of TIR by gradient effects without the use of massive mathematics, we restrict ourselves to an analysis of S-polarized waves.

Let us recall Eq. (6.27), presented in the case $q^2 > 0$, a simple example of propagating modes in a gradient layer (5.40). In contrast, our attention will be focused here on the evanescent modes, described by the same Eq. (6.27) under the same condition $\Lambda = 0$, but in the opposite case $q^2 < 0$. Introducing the parameter $q^2 = -p^2$, where $p^2 > 0$ and following the scheme of solution of Eq. (6.27), we obtain the generating function Ψ_s for the evanescent waves in the case discussed:

$$\Psi_s = \frac{A[\exp(-p\varsigma) + Q\exp(p\varsigma)]}{\sqrt{\mathrm{ch}\varsigma}}. \tag{6.55}$$

The analogy between the functions Ψ_s, written in the forms (6.55) and (6.28) for the cases $q^2 < 0$ and $q^2 > 0$, respectively, allows one to present the reflection coefficient R_s in the case $q^2 < 0$ in the form (6.31), related to $q^2 > 0$, by making in (6.31) the following replacements:

$$q^2 \to -p^2; \quad q \to ip; \quad t = \mathrm{tg}(2q\varsigma_0) \to i\mathrm{th}(2p\varsigma_0). \tag{6.56}$$

Let us examine the condition for the reflectionless tunneling of S waves ($R_s = 0$), supported by evanescent modes (6.55), through the symmetrical structure $n_1 = n, B = B_1$. After the replacements (6.56) this condition can be derived from expression (6.31)

$$\mathrm{th}(2p\varsigma_0) = \frac{4Mp\sqrt{1+M^2}}{B^2 + M^2 + 4p^2(1+M^2)}. \tag{6.57}$$

Thus, choosing the parameters of the nanostructure $n_1 = 1.415$, $n_0 = 1.8$, $M = 0.9632$, g $= 1.5$, $d = 110\,\mathrm{nm}$, we find from (6.57), that reflectionless tunneling arises for an S wave with $\lambda = 800$ nm, $\delta = 22°$; in this case the wave energy is transferred through the film by an evanescent mode.

To examine the reflectance of a periodic nanostructure containing $k \geq 2$ adjoined films (5.40), one can attribute the number $k = 1$ to the first film on the far side of the structure, contiguous to the substrate; the parameter Q_1, describing the interference of forward and backward waves in this film, can be obtained by means of making the replacement (6.56) in (6.30). Use of the continuity conditions on the interfaces of contiguous films in the consecutive order yields the recurrence relation linking the values of Q_k and Q_{k-1} for $k \geq 2$:

$$Q_k = \exp(-2p\varsigma_0)$$
$$\times \left[\frac{M\exp(p\varsigma_0) + Q_{k-1}(2p\sqrt{1+M^2}+M)\exp(-p\varsigma_0)}{(2p\sqrt{1+M^2}-M)\exp(p\varsigma_0) - MQ_{k-1}\exp(-p\varsigma_0))} \right]. \tag{6.58}$$

The reflection coefficient of a stack containing k such nanofilms, can be written as

$$R_k = \frac{iB - M + 2p\sqrt{1+M^2}\Lambda_k}{iB + M - 2p\sqrt{1+M^2}\Lambda_k}. \quad \Lambda_k = \frac{1-Q_k}{1+Q_k}. \tag{6.59}$$

Recalling the marked possibility of radiation transfer through one gradient nanofilm due to an appropriate choice of the parameters of the film and radiation, we can find by means of Eqs. (6.58) and (6.59) that the same choice of parameters yields reflectionless tunneling through the multilayer stack containing several, e.g. five or ten such layers.

Note, that the similar effect of reflectionless tunneling was discussed in Ch. 4 for normal incidence; in this case the direction of energy flow (Poynting vector), normal to the interface of gradient layer, remained unchanged inside the layer as well. However, in case of oblique incidence, the directions of the Poynting vectors $\vec{P}$ for incident and tunneling waves are different; this difference causes some lateral shift of the ray at the output of the layer (Goos-Hänchen (G-H) shift). To visualize the underlying physics of this effect it

Polarization Phenomena in Gradient Nanophotonics

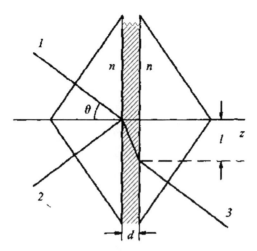

Fig. 6.4. Frustrated total internal reflection (waves 1–2) and Goos-Hänchen shift Y of transmitted wave 3, tunneling in a bi-prism configuration; the slit between prisms is filled with a gradient multilayer nanostructure.

makes sense to recall first the traditional bi-prism configuration for the demonstration of the G-H shift: two prisms of refractive index n_1 are placed with their hypotenuses in close proximity, forming a narrow gap of width d between them, filled by air (Fig. 6.4). The traditional theory of the lateral displacement of a wave, tunneling through this air gap (G-H shift), is based on the so-called frustrated total internal reflection of light (FTIR), incident from the medium with the larger value of the refractive index n_1 on the boundary of the gap ($z = 0$) with smaller value $n = 1$; subject to the value of the ratio d/λ some part of the radiation is transferred through the gap in the FTIR regime [6.9, 6.10].

For simplicity of calculations let us consider the an incident S wave; to find its G-H shift one can use the generating function for evanescent wave inside the air gap

$$\Psi = A[\exp(-p_1 z) + Q \exp(p_1 z)], \quad (6.60)$$

$$p_1 = \frac{\omega}{c} N, \quad N = \sqrt{n_1^2 \sin^2 \delta - 1}. \quad (6.61)$$

Substitution of (6.60) into Eq. (6.4) yields the wave components E_x, H_y, H_z inside the gap, and allows us to calculate the reflection coefficient

$$R = \frac{(n_1^2 - 1)\text{th}(p_1 d)}{(n_1^2 \cos^2 \delta - N^2)\text{th}(p_1 d) + 2i n_1 N \cos \delta}. \quad (6.62)$$

and the components of the Pointing vector $\vec{P}$ [6.2]

$$P_x = \frac{c}{4\pi}\text{Re}(E_x H_y^*), \quad P_y = -\frac{c}{4\pi}\text{Re}(E_x H_z^*). \quad (6.63)$$

The ray trajectory in the (yz) plane is described by a differential equation that links the displacements of photons, tunneling with group velocity $\vec{v}_g$ during the time dt, in the y- and z-directions

$$dy = v_{gy} dt, \quad dz = v_{gz} dt, \quad \vec{v}_g = \vec{P} W_{em}^{-1}. \quad (6.64)$$

W_{em} is the density of electromagnetic field energy in the medium [6.4]. Combining the equalities (6.64) we obtain the differential equation of ray trajectory,

$$dy = \frac{P_y}{P_z} dz. \quad (6.65)$$

After substitution the of expressions for P_y and P_z (6.63) into Eq. (6.65) this equation can be rewritten as

$$dy = \frac{\text{tg}\delta}{2N^2}[(n_1^2 - 1)\text{ch}(2p_1 d) - n_1^2 \cos^2 \delta + N^2] dz. \quad (6.66)$$

Integration of Eq. (6.66) brings finally the coordinate Y of the point, where the ray trajectory, passing at the layer boundary $z = 0$ through the point $y = 0$, is crossing it's another boundary $z = d$; the value Y is the G-H shift [6.11]

$$\frac{Y}{d} = \frac{\text{tg}\delta}{2N^2}\left[\frac{(n_1^2 - 1)\text{sh}(2p_1 d)}{2p_1 d} - n_1^2 \cos^2 \delta + N^2\right]. \quad (6.67)$$

Unlike the shift, inherent to a ray traveling through a transparent refractive layer, the shift (6.67) is formed by evanescent waves.

The G-H shift for a gradient layer can be found in the same way. Substitution of the generating function (6.55) into (6.4) yields the expressions for the field components E_x, H_y, H_z. Calculating the P_y and P_z components we can write the ratio P_y/P_z in Eq. (6.65) as

$$\frac{P_y}{P_z} = \frac{a\mathrm{sh}^2[p(\varsigma-\varsigma_0)] + b\mathrm{ch}^2[p(\varsigma-\varsigma_0)] + h\mathrm{sh}[2p(\varsigma-\varsigma_0)]}{8q^2\sqrt{1+M^2}\mathrm{ch}\varsigma}, \quad (6.68)$$

$$a = M^2 + B^2, \ b = 4q^2(1+M^2),$$

$$h = 4qM\sqrt{1+M^2}, \ \varsigma_0 = \ln(m_+). \quad (6.69)$$

The fraction P_y/P_z (6.68) has to be substituted into Eq. (6.65) as well as the differential dz, represented due to manipulations with (5.43) and (5.46), in the form

$$dz = \frac{d}{2(\mathrm{arctg}M)}\frac{d\varsigma}{\mathrm{ch}\varsigma}. \quad (6.70)$$

After these substitutions the G-H shift Y can be obtained due to integration of Eq. (6.65):

$$\frac{Y}{d} = \frac{I}{16q^2\sqrt{1+M^2}(\mathrm{arctg}M)}, \quad (6.71)$$

$$I = aI_1 + bI_2 + hI_3, \quad I_1 = \int_{-\varsigma_0}^{\varsigma_0} \frac{\mathrm{sh}^2[p(\varsigma-\varsigma_0)]}{\mathrm{ch}^2\varsigma}d\varsigma,$$

$$I_2 = 2\mathrm{th}\varsigma_0 + I_1, \quad I_3 = \int_{-\varsigma_0}^{\varsigma_0} \frac{\mathrm{sh}[2p(\varsigma-\varsigma_0)]}{\mathrm{ch}^2\varsigma}d\varsigma.$$

Comparing the examples of frustrated total internal reflection and the G-H shift in the bi-prism configuration (Fig. 6.4) for an air gap (case I) and a gap filled by a gradient structure (case II), we can outline some drastic differences between these cases. For convenience, we will illustrate these differences, carrying out the numerical evaluations for the bi-prism, characterized by the values of parameters $n_1 = 1.415$ and $d = 110\,\mathrm{nm}$, which were used earlier in this section for demonstration of TIR for an S wave with wavelength $\lambda = 800\,\mathrm{nm}$.

1. The critical angle in case I is $\delta_{cr} = \arcsin(n_1^{-1}) = 45°$; meanwhile, in case II, the angle of illumination, providing TIR, is

much smaller: $\delta = 22°$; therefore, the simple condition of TIR on the interface between homogeneous media $\delta > \delta_{cr}$ proves to be broken due to the heterogeneity of one of the bordering media. The appearance of TIR in case II is determined by a more complicated condition, dependent on the spatial profile $n(z)$ in the heterogeneous layer.

2. Designating the normalized G-H shifts in cases I and II as Y_I/d and Y_{II}/d and calculating Y_I/d by means of Eq. (6.67) for the illumination angle $\delta = 55° > \delta_{cr}$ we have: $Y_I/d = 1.8$; thus the G-H shift exceeds in this case the width of the gap. A decrease of the angle δ to its critical value δ_{cr} results in an increase of the G-H shift (6.67), since the slope of the tunneling ray with respect to the z-axis is increasing, approaching the y-direction (Fig. 6.4). Experimental measurements of this shift attract attention, because these data could open the way for computation of the velocity of tunneling photons in vacuum [6.9–6.11]. Inspection of Eqs. (6.67) and (6.62) shows two opposite trends in these experiments: from one viewpoint, the G-H shift is increasing exponentially when the gap width d is large enough: $p_1 d \gg 1$ (6.67). However, this trend, useful for observations, gives rise to an increase of the reflectance $|R|^2 \to 1$ (6.62) and, thus, to an exponential decrease of the transmittance of the gap $|T|^2 = 1 - |R|^2 \to 0$; this exponential weakening of radiation flow impedes the observations.

In contrast, the normalized G-H shift in the case II (6.71), calculated for the aforesaid parameters of the gradient nanostructure $n_0 = 1.8$, $M = 0.9632$, $g = 1.5$, proves to be much smaller, than Y_I/d: $Y_{II}/d = 0.25$. However, unlike the case I, the reflectance of gap, filled by nanostructure in this case was shown to vanish and, therefore, the transmittance of the gap in case II is almost complete: $|T|^2 \to 1$.

3. Note, that the G-H shift, formed by reflectionless tunneling through the bi-prism configuration (Fig. 6.4), can be increased by placing a multilayer gradient nanostructure in the gap. In this case the width of the gap increases. However, unlike in the case I, the regime of reflectionless tunneling was shown to arise in this geometry. Thus, the multilayer gradient configuration under discussion

illustrates a way to increase the G-H shift without any appreciable attenuation of the transmitted radiation.

Comments and Conclusions to Chapter 6

1. Introduction of the new variable $\eta = \eta(z)$, widely used in Ch. 6, is not a universal method for the solution of the Maxwell equations in gradient media. In dealing with some profiles $U(z)$ it is worthwhile to operate directly with the variable z. Thus, exploring the solution of Eq. (6.6) for the convex profile

$$U(z) = \left[\text{ch}\left(\frac{z}{L}\right) - M\text{sh}\left(\frac{z}{L}\right)\right]^{-1},$$

$$U(z) = (1 - M^2)^{-\frac{1}{2}}\left[\text{ch}\left(\frac{z}{L} - \zeta\right)\right]^{-1}, \qquad (6.72)$$

$$\zeta = \frac{1}{2}\ln\left(\frac{1+M}{1-M}\right), \quad 0 < M < 1,$$

and substituting (6.72) into Eq. (6.6) we can rewrite this equation in a familiar form (5.47), where the variable ς and parameters q^2 and Λ are:

$$\varsigma = \frac{z}{L} - \zeta, \quad q^2 = -\left(\frac{\omega L \sin\delta}{c}\right)^2, \quad \Lambda = -\left(\frac{\omega L}{c}\right)^2 \frac{1}{1-M^2}. \qquad (6.73)$$

Now the solution of Eq. (5.47), given in (5.49)–(5.58), can be used.

2. The models of gradient dielectric films, discussed above, can be used for analysis of the reflectance of some gradient metamaterial layers with negative refraction ($n_0 < 0$). Thus, let us consider the oblique incidence of an S wave; introducing the generating function Ψ by analogy with (5.34),

$$E_x = -\frac{1}{c}\frac{\partial \Psi}{\partial t}, \quad H_y = \frac{1}{\mu}\frac{\partial \Psi}{\partial z}, \quad H_z = -\frac{1}{\mu}\frac{d\Psi}{dz}.$$

and assuming that the spatial distributions of the dielectric permittivity $\varepsilon(z)$ and magnetic permeability $\mu(z)$ are given by the same function $\Phi(z)$ [6.12], so that

$$\varepsilon(z) = \varepsilon(\omega)\Phi(z), \quad \mu(z) = \mu(\omega)\Phi(z), \qquad (6.74)$$

we obtain the equation governing the generating function Ψ,

$$\frac{d^2\Psi}{dz^2} + \frac{\omega^2}{c^2}[n_0^2\Phi^2(z) - \sin^2\delta]\Psi = \frac{1}{\Phi(z)}\frac{d\Phi(z)}{dz}\frac{d\Psi}{dz}. \qquad (6.75)$$

Transforming Eq. (6.75) to a new variable η,

$$\eta = \int_0^z \Phi(z_1)dz_1. \qquad (6.76)$$

one can eliminate its right side to obtain

$$\frac{d^2\Psi}{d\eta^2} + \left(\frac{\omega}{c}\right)^2\left[n_0^2 - \frac{\sin^2\delta}{\Phi^2(z)}\right]\Psi = 0. \qquad (6.77)$$

Note, that the function $\Phi(z)$ until now remains unknown and can be chosen freely. To use the solution of Eq. (6.36), let us choose $\Phi(z)$ in the form

$$\Phi(z) = \text{ch}\varsigma, \quad \varsigma = \frac{\eta}{L}, \qquad (6.78)$$

where L is some unknown spatial scale. Bringing together the equality (6.76), written in the differential form $d\eta = \Phi(z)dz$, and Eq. (6.78), we can obtain the link between the variables z and η,

$$\text{tg}\left(\frac{z}{2L}\right) = \text{th}\left(\frac{\eta}{2l}\right). \qquad (6.79)$$

Substitution of (6.79) into Eq. (6.78) brings the explicit form of the profile $\Phi(z)$

$$\Phi(z) = \left[\cos\left(\frac{z}{L}\right)\right]^{-1}. \qquad (6.80)$$

Since profile (6.80) coincides with Eq. (6.34), we can express the unknown scale L via the width of the layer d by means of (6.34):

$$L = \frac{d}{2\text{arc}\cos(m)}. \qquad (6.81)$$

Now Eq. (6.77) coincides with Eq. (6.36), and we can use its solution (6.37)–(6.43).

3. Reflectance spectra for oblique incidence of polarized waves on gradient structures are presented usually by more complicated expressions than the expressions for such spectra for normal incidence; therefore it is useful to check these expressions by means of limiting cases, which have to coincide with formulae, well known from the optics of homogeneous media. Thus, in the limit $d \to 0$, $n_0 \to 1$ expressions (6.18)–(6.20) for layered media are reduced to the usual Fresnel formulae, describing the reflection of waves from a homogeneous half-space.

Let us consider now a more intricate problem, connected with the verification of reflection coefficients for gradient layers (6.34), given by Eqs. (6.39) and (6.44). It is worthwhile to rewrite Eq. (6.39), replacing the quantity Q_s by its expression (6.41):

$$R_s = \frac{i(F_1 M_1 - F_2 M_2)(B_1 - B_2) + M_1^2 - M_2^2 + B_1 B_2(F_1^2 - F_2^2)}{i(F_1 M_1 - F_2 M_2)(B_1 + B_2) - M_1^2 + M_2^2 + B_1 B_2(F_1^2 - F_2^2)}. \quad (6.82)$$

When the heterogeneity vanishes ($L \to \infty$), we have from (6.34) and (6.43): $m \to 1, s \to 0$; in this case it follows from (6.41)–(6.43) that $v_1 = v_2 = 0.5; F_1 = F_2; F_1' = F_2'; M_1 = -M_2$. Substitution of these values into Eq. (6.82) yields the result

$$R_s = \frac{B_1 - B_2}{B_1 + B_2}. \quad (6.83)$$

By using the definitions of B_1 and B_2 (6.40), we obtain from (6.83) the well-known formula, describing the reflection of an S wave from a homogeneous half-space:

$$R_s = \frac{\cos \delta - \sqrt{n^2 - \sin^2 \delta}}{\cos \delta + \sqrt{n^2 - \sin^2 \delta}}. \quad (6.84)$$

The same scheme of calculations, applied to Eqs. (6.44)–(6.47), yields another classical expression for the reflection of a P wave:

$$R_p = \frac{n^2 \cos\delta - \sqrt{n^2 - \sin^2\delta}}{n^2 \cos\delta + \sqrt{n^2 - \sin^2\delta}}; \qquad (6.85)$$

Bibliography

[6.1] A. B. Shvartsburg, V. Kuzmiak and G. Petite, *Phys. Rep.* **452**(2–3), 33 (2007).
[6.2] L. D. Landau and E. M. Lifshitz, *Electrodynamics of Continuous Media* (Pergamon Press, Oxford, 1986).
[6.3] M. Abramowitz and I. Stegun, *Handbook of Math. Functions* (Dover Publications, NY, 1968).
[6.4] J. D. Griffits, *Introduction to Electrodynamics* (Prentice-Hill, NJ, 1999).
[6.5] A. B. Shvartsburg and V. Kuzmiak, *European Phys. J. B* **72**, 77 (2009).
[6.6] A. B. Shvartsburg, V. Kuzmiak and G. Petite, *Phys. Rev. E* **76**, 01603 (2007).
[6.7] A. A. Bulgakov, A.S. Bulgakov and M. Nieto-Vesperinas, *Phys. Rev. B* **58**, 4438 (1998).
[6.8] K. V. Vytovtov, *JOSA A* **22**(4), 689 (2005).
[6.9] C.-F. Li and Q. Wang, *JOSA B* **18**(8), 1174 (2001).
[6.10] R. M. Azzam, *JOSA A* **23**(4), 960 (2006).
[6.11] A. B. Shvartsburg, *Physics — Uspekhi*, **50**(1), 39 (2007).
[6.12] N. M. Litchinitser, A. I. Maimistov, I. R. Gabitov, R. Z. Sagdeev and V. M. Shalaev, *Opt. Lett.*, **33**, 2350 (2008).

CHAPTER 7

GRADIENT OPTICS OF GUIDED AND SURFACE ELECTROMAGNETIC WAVES

A planar interface between a semi-infinite homogeneous dielectric medium and a semi-infinite homogeneous metal cannot support a surface electromagnetic wave of S polarization [7.1]. This is because the solutions of Maxwell's equations in both media that vanish at infinite distances from the interface are descending exponentials. While the continuity of these fields across the interface can be achieved, the continuity of their normal derivatives across the interface cannot be achieved. For the same reason an S-polarized surface electromagnetic wave cannot be supported by a planar interface between two different semi-infinite homogeneous dielectric media. What is needed, therefore, to produce an S-polarized electromagnetic wave that is localized to the interface in either of these situations is to relax the assumption of the homogeneity of at least one of the two media in contact across a planar interface in such a way that an S-polarized electromagnetic field in that medium still decreases exponentially at a distance far from the interface, but now has an oscillatory dependence on the coordinate normal to the interface in the neighborhood of the interface, so that the continuity of its normal derivative can be achieved. The resulting electromagnetic field, strictly speaking, is not that of a surface wave, since it does not display purely exponential decay into each medium. Rather, it is a guided wave.

Guided waves in asymmetric planar waveguides have been studied theoretically and experimentally for many years. Perhaps the simplest purely dielectric structure that acts as such a waveguide is a film whose dielectric constant is ϵ_2 sandwiched between semi-infinite dielectric constants ϵ_1 ad ϵ_3 [7.2]. Solutions of Maxwell's equations for S-polarized fields that decay exponentially in each of the media surrounding the film can be found. If ϵ_2 is larger than both ϵ_1 and ϵ_3, standing wave solutions can be found for the field within the film. Their oscillatory nature enables the satisfaction of the boundary conditions at both surfaces of the film, which gives rise to a series of discrete eigenmodes that propagate in a wavelike manner in directions parallel to the surfaces of the film, but whose fields are localized to the vicinity of the film. One of the earliest experimental investigations of optical wave propagation in such slab waveguides was carried out by Osterberg and Smith in 1964 [7.3]. Today these guided waves form the basis for much of integrated optics technology [7.4].

A different approach to the fabrication of integrated optics waveguides consists of diffusing a suitable material into a substrate that increases the refractive index near its surface [7.5]. Alternatively, the refractive index of a dielectric compound such as $LiNbO_3$ can be increased in the vicinity of its surface by out-diffusion of Li_2O from the surface [7.6]. These procedures produce asymmetric graded index waveguides in which the index of refraction decreases in a continuous fashion with increasing distance into the substrate from its surface.

The guided waves supported by such graded index waveguides have been studied by several authors who used a variety of methods in their investigations. A commonly used approach is the use of the Wentzel-Kramers-Brillouin (WKB) method [7.7] to obtain approximate results for the dispersion relation for these waves and the corresponding field profiles [7.8–7.13]. Exact dispersion relations have been obtained in analytic form for a continuous index profile consisting of several straight line segments [7.8], and one having an exponentially decreasing profile [7.14]. Numerical methods have also been used in obtaining the dispersion relations and field profiles of the modes of graded-index waveguides [7.15–7.18]. In Sec. 7.1 of this chapter we will present a new graded-index profile that yields analytic expressions for the dispersion relation for guided waves of s polarization, and the corresponding field profiles, that display features not present in the modes studied in earlier work.

To produce S-polarized guided waves at a vacuum-metal interface, which do not exist when the interface is planar, we show in Sec. 7.2, that it is sufficient for the interface to be a portion of a circularly cylindrical interface between vacuum and a homogeneous metal. When the metal is concave to the vacuum this interface supports S-polarized guided waves. The constant radius of curvature of the vacuum-metal interface is equivalent to a planar interface between vacuum and a metal that is no longer homogeneous but is characterized by a dielectric function that varies with increasing distance into it from the interface in a manner that allows it to support guided waves.

Since a portion of a circular vacuum-metal interface supports S-polarized electromagnetic guided waves, it is perhaps not surprising that a rough vacuum-metal interface, which can be regarded as composed of many such segments, can support surface electromagnetic waves of S polarization. In Sec. 7.3 we study the propagation of surface waves of this polarization on vacuum-metal interfaces that are periodically corrugated and on vacuum-metal interfaces that are randomly corrugated. These are surface electromagnetic waves that do not exist at planar vacuum-metal interfaces. They are dispersive, and possess features that can make them useful in technological applications.

7.1. Narrow-Banded Spectra of S-polarized Guided Electromagnetic Waves on the Surface of a Gradient Medium: Heterogeneity-Induced Dispersion

In this section we describe a new type of S-polarized guided electromagnetic wave in a system consisting of a semi-infinite vacuum in contact across a planar interface with a semi-infinite graded-index dielectric medium of a special type. In earlier work [7.19, 7.20] it was shown that such waves can be realized by the use of a dielectric function that has the simple free electron [7.19] or Drude [7.20] form with a plasma frequency that is a smooth continuously varying function of the coordinate normal to the interface in one of the two dielectric media in contact. Thus, the medium characterized by such dielectric functions can be considered to be an n-type semiconductor with a spatially varying conduction electron number density.

In contrast, in this section we assume that the medium in the region $z > 0$ contains no conduction electrons. It is characterized by a real, positive, dielectric constant that is frequency independent and decreases with increasing z until it saturates at a bulk value. It is given by

$$\epsilon(z) = n_0^2 \left[1 - \frac{1}{g} + \frac{1}{g(1+\frac{z}{L})^2} \right], \qquad (7.1)$$

where n_0 is the (real) index of refraction at $z = 0$. The parameter g is assumed to be greater than unity. Therefore, as $z \to \infty$ the dielectric constant (7.1) saturates at the value $n_0^2[1 - (1/g)] \equiv n_v^2$, which satisfies the inequality $n_v^2 < n_0^2$. It is further assumed that $n_v > 1$. The region $z < 0$ consists of vacuum. We note that this system is invariant in the y direction.

This structure yields an exact dispersion relation and electric field profiles of the S-polarized guided waves it supports in analytic forms. This was noted briefly in the review article [7.21], but no results were presented there. In this section, following Ref. [7.22], we expand on the work reported in Ref. [7.21], and present plots of the dispersion curves and the associated electric field profiles of these guided waves.

The electric and magnetic vectors of an S-polarized field propagating in the x direction have the form

$$\mathbf{E}(x,z;t) = (0, E_y(x,z;t), 0) \qquad (7.2a)$$

$$\mathbf{H}(x,z;t) = (H_x(x,z;t), 0, H_z(x,z;t)). \qquad (7.2b)$$

Maxwell's equations for the electromagnetic field in the dielectric medium $z > 0$ become

$$\frac{\partial E_y^>}{\partial x} = -\frac{1}{c}\frac{\partial H_z^>}{\partial t} \qquad (7.3a)$$

$$\frac{\partial E_y^>}{\partial z} = \frac{1}{c}\frac{\partial H_x^>}{\partial t} \qquad (7.3b)$$

$$\frac{\partial H_x^>}{\partial z} - \frac{\partial H_z^>}{\partial x} = \frac{1}{c}\frac{\partial D_y^>}{\partial t} = \frac{1}{c}\epsilon(z)\frac{\partial E_y^>}{\partial t}. \qquad (7.3c)$$

Since we are seeking a solution that describes a wave propagating in the x direction, we write the field components in the following forms,

$$E_y^>(x, z; t) = E_y^>(z|\omega) \exp[ikx - i\omega t] \qquad (7.4a)$$

$$H_x^>(x, z; t) = H_x^>(z|\omega) \exp[ikx - i\omega t] \qquad (7.4b)$$

$$H_z^>(x, z; t) = H_z^>(z|\omega) \exp[ikx - i\omega t]. \qquad (7.4c)$$

When we substitute Eqs. (7.4) into Eqs. (7.3) we find that the amplitude $E_y^>(z|\omega)$ satisfies the equation

$$\left\{ \frac{d^2}{dz^2} - \left[k^2 - \epsilon(z) \left(\frac{\omega}{c}\right)^2 \right] \right\} E_y^>(z|\omega) = 0. \qquad (7.5)$$

We seek a solution of this equation that decays to zero as $z \to \infty$. To obtain such a solution we assume that

$$E_y^>(z|\omega) = u^{\frac{1}{2}} f(u), \qquad (7.6)$$

where

$$u = 1 + \frac{z}{L}. \qquad (7.7)$$

On substituting Eq. (7.6) into Eq. (7.5) and making use of Eqs. (7.1) and (7.7), we find that the function $f(u)$ satisfies the equation

$$\frac{d^2 f(u)}{du^2} + \frac{1}{u} \frac{df(u)}{du} - \left[L^2 p^2 + \frac{s^2}{u^2} \right] f(u) = 0, \qquad (7.8)$$

where

$$p^2 = \frac{\omega^2}{c^2}(b^2 - n_v^2) \qquad (7.9a)$$

$$s^2 = \frac{1}{4}\left(1 - \frac{\omega^2}{\Omega_c^2}\right), \qquad (7.9b)$$

with

$$b = \frac{c}{\omega} k \qquad (7.10a)$$

$$\Omega_c = \frac{c}{2L} \frac{1}{\sqrt{n_0^2 - n_v^2}}. \qquad (7.10b)$$

Equation (7.8) is the equation satisfied by the modified Bessel functions. In order to have a solution that decays to zero exponentially as u (and hence z) tends to infinity, the coefficient p^2 should be positive. We can assume that p is real and positive. From Eqs. (7.9b) and (7.10b) we see that this restricts the allowed values of the wavenumber k to lie in the region $k > n_v(\omega/c)$. Then the solution of Eq. (7.8) is given by the modified Bessel function of the second kind and order s,

$$f(u) = AK_s(Lpu), \tag{7.11}$$

where A is an arbitrary amplitude. The coefficient s^2 in Eq. (7.8), however, can be either positive or negative. Either choice produces a solution $f(u)$ that decreases to zero exponentially as $u \to \infty$ [7.23].

When s is real and positive ($0 < \omega < \Omega_c$), the function $K_s(z)$ has the integral representation [7.24]

$$K_s(z) = \int_0^\infty dx e^{-z \cosh x} \cosh sx, \tag{7.12}$$

is real for z real and positive, is non-negative, and is an even function of s. For fixed s and $z \to 0+$ [7.25]

$$K_s(z) \sim \frac{1}{2}\Gamma(s)\left(\frac{z}{2}\right)^{-s}, \tag{7.13}$$

where $\Gamma(z)$ is the gamma function. For fixed s and large z it has the asymptotic form [7.26]

$$K_s(z) \sim \left(\frac{\pi}{2z}\right)^{\frac{1}{2}} e^{-z}\left\{1 + \frac{4s^2 - 1}{8z} + \frac{(4s^2 - 1)(4s^2 - 9)}{2!(8z)^2} + \cdots\right\}. \tag{7.14}$$

If we set $s^2 = -s_1^2$ ($\omega > \Omega_c$), the solution for $f(u)$ becomes

$$f(u) = AK_{is_1}(Lpu). \tag{7.15}$$

The function $K_{is_1}(z)$ has the integral representation [7.27]

$$K_{is_1}(z) = \int_0^\infty dx e^{-z \cosh x} \cos s_1 x, \tag{7.16}$$

is real for s_1 real and z real and positive, and is an even function of s_1. For fixed s_1 and z approaching zero through real positive values, $K_{is_1}(z)$ is an oscillatory function of z [7.23],

$$K_{is_1}(z) = -\left(\frac{\pi}{s_1 \sinh(s_1\pi)}\right)^{\frac{1}{2}} \left[\sin\left(s_1 \ell n \frac{z}{2}\right) - \phi_{s_1,0}\right] + O(z^2) \tag{7.17}$$

with

$$\phi_{s_1,0} = \arg\{\Gamma(1+is_1)\}. \tag{7.18}$$

For fixed s_1 and large z it has the asymptotic form

$$K_{is_1}(z) \sim \left(\frac{\pi}{2z}\right)^{\frac{1}{2}} e^{-z} \left\{1 - \frac{4s_1^2+1}{8z} + \frac{(4s_1^2+1)(4s_1^2+9)}{2!(8z)^2} + \cdots\right\}. \tag{7.19}$$

We turn now to the electromagnetic field in the vacuum region $z < 0$. The three nonzero components of this field satisfy the equations

$$\frac{\partial E_y^<}{\partial x} = -\frac{1}{c}\frac{\partial H_z^<}{\partial t} \tag{7.20a}$$

$$\frac{\partial E_y^<}{\partial z} = \frac{1}{c}\frac{\partial H_x^<}{\partial t} \tag{7.20b}$$

$$\frac{\partial H_x^<}{\partial z} - \frac{\partial H_z^<}{\partial x} = \frac{1}{c}\frac{\partial D_y^<}{\partial t} = \frac{1}{c}\frac{\partial E_y^<}{\partial t}. \tag{7.20c}$$

We will seek these field components in the form of a wave of frequency ω propagating in the x direction,

$$E_y^<(x,z;t) = E_y^<(z|\omega) \exp[ikx - i\omega t] \tag{7.21a}$$

$$H_x^<(x,z;t) = H_x^<(z|\omega) \exp[ikx - i\omega t] \tag{7.21b}$$

$$H_z^<(x,z;t) = H_z^<(z|\omega) \exp[ikx - i\omega t]. \tag{7.21c}$$

The substitution of Eqs. (7.21) into Eqs. (7.20) yields the result that the amplitude $E_y^<(z|\omega)$ satisfies the equation

$$\left\{\frac{d^2}{dz^2} - \left[k^2 - \left(\frac{\omega}{c}\right)^2\right]\right\} E_y^<(z|\omega) = 0. \tag{7.22}$$

The solution of this equation that tends to zero as $z \to -\infty$ can be written as

$$E_y^<(z|\omega) = B\exp[\beta_0(k,\omega)z], \tag{7.23}$$

where

$$\beta_0(k,\omega) = \left[k^2 - \left(\frac{\omega}{c}\right)^2\right]^{\frac{1}{2}} \quad Re\,\beta_0(k,\omega) > 0,$$
$$Im\,\beta_0(k,\omega) < 0. \tag{7.24}$$

For the expansion given by Eqs. (7.23) and (7.24) to represent an electric field that decays to zero exponentially with increasing distance into the vacuum from the plane $z = 0$ requires that $k > (\omega/c)$. However, since we will be interested in structures for which $n_v > 1$, the condition on k obtained above, namely that $k > n_v(\omega/c)$, is the more restrictive one, and it is in this range of values of k that the dispersion relation for the surface-localized guided waves will be sought.

We are now in a position to satisfy the boundary conditions at the interface $z = 0$. These require the continuity of the tangential components of the electric and magnetic fields across this interface, and can be written as

$$E_y^<(z|\omega)|_{z=0} = E_y^>(z|\omega)|_{z=0} \tag{7.25a}$$

$$\left.\frac{dE_y^<(z|\omega)}{dz}\right|_{z=0} = \left.\frac{dE_y^>(z|\omega)}{dz}\right|_{z=0}. \tag{7.25b}$$

With the results given by Eqs. (7.6) and (7.23) these equations take the forms

$$B = u^{\frac{1}{2}}f(u)|_{u=1} \tag{7.26a}$$

$$\beta_0(k,\omega)B = \frac{1}{L}\left[\frac{1}{2u^{\frac{1}{2}}}f(u) + u^{\frac{1}{2}}f'(u)\right]\bigg|_{u=1}, \tag{7.26b}$$

where the prime denotes differentiation with respect to argument. There are now two cases to consider.

7.1.1. $0 < \omega < \Omega_c$

When the frequency ω lies in the interval $0 < \omega < \Omega_c$, the function $f(u)$ is given by Eq. (7.11). The dispersion relation for guided electromagnetic waves in the region $0 < \omega < \Omega_c$, $k > n_v(\omega/c)$ of the (ω, k) plane is therefore

$$L\beta_0(k,\omega) = \frac{1}{2} + Lp\frac{K'_s(Lp)}{K_s(Lp)}, \qquad (7.27a)$$

where

$$s = \frac{1}{2}\left(1 - \frac{\omega^2}{\Omega_c^2}\right)^{\frac{1}{2}}. \qquad (7.27b)$$

Equation (7.27) has to be solved numerically. The resulting dispersion curve is plotted in Fig. 7.1, and consists of a single branch. The values of the material parameters assumed in obtaining Fig. 7.1 are $n_v = 1.05$, $n_0 = 2$, and $L = 30$ nm.

An approximate analytic solution of Eq. (7.27) can be found in the case where $(Lp)^2 \ll 1$. In this limit $K_s(Lp)$ can be approximated by Eq. (7.13). Substitution of this approximation into Eq. (7.27a)

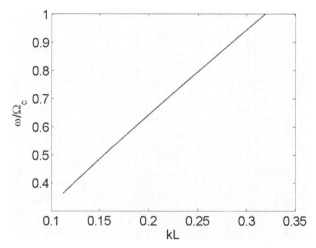

Fig. 7.1. The dispersion curve for an s-polarized guided electromagnetic wave in the frequency range $0 < \omega < \Omega_c$ propagating on a graded index dielectric medium characterized by the values $n_v = 1.05, n_0 = 2$, and $L = 30$ nm [7.22].

yields the equation

$$2L\beta_0(k,\omega) = 1 - 2s. \tag{7.28}$$

It is convenient now to introduce the dimensionless frequency $\nu = \Omega_c/\omega \geq 1$, where Ω_c is defined by Eq. (7.10b). With the substitution of Eqs. (7.9b) and (7.24) into Eq. (7.28), we obtain an explicit expression for the dependence of the wave number k on the frequency ν,

$$k^2 = \frac{\omega^2}{c^2} + \frac{1}{4L^2\nu^2[\nu + \sqrt{\nu^2 - 1}]^2}. \tag{7.29}$$

To determine the spectral range of existence of the wave with this dispersion relation, we recall the inequality $n_\nu^2 < b^2$ that follows from the necessity of p^2 being positive. With the expression for b given by Eq. (7.10a), the expression for k^2 given by Eq. (7.29), and the definition of Ω_c given by Eq. (7.10b), this inequality can be rewritten as

$$1 > \frac{(n_\nu^2 - 1)(\nu + \sqrt{\nu^2 - 1})^2}{n_0^2 - n_\nu^2}, \tag{7.30a}$$

or

$$\left(\frac{n_0^2 - n_\nu^2}{n_\nu^2 - 1}\right)^{\frac{1}{2}} - \nu > \sqrt{\nu^2 - 1}. \tag{7.30b}$$

This inequality is satisfied for

$$1 < \nu < \nu_c = \frac{n_0^2 - 1}{2[(n_\nu^2 - 1)(n_0^2 - n_\nu^2)]^{\frac{1}{2}}}. \tag{7.31}$$

Thus, the S-polarized guided wave whose dispersion relation is given by Eq. (7.29) exists in the frequency range defined by the inequality (7.31).

The dielectric constant defined in Eq. (7.1) contains two free parameters, the dimensionless factor g and the gradient scale L. The factor g defines the relation between the value of the refractive index n_0 on the surface of the medium and its value deep inside the medium, $n_0 = n_\nu/(1 - g^{-1})^{\frac{1}{2}} > n_\nu$. The cutoff frequency Ω_c defined by Eq. (7.10b) depends on both g and L. By varying the values of these two parameters it is possible to place the domain of existence of the surface wave whose dispersion relation is given by Eq. (7.29) in different spectral ranges.

The spectral range of existence is defined in terms of its limiting frequencies, namely $\omega_1 = \Omega_c(\nu = 1)$, and $\omega_2 = \Omega_c/\nu_c$. Thus, for example, let us choose a dielectric medium with $n_v = 1.42$, $n_0 = 2$, and $L = 50$ nm. In this case we find that $\nu_c = 1.05642$, $\omega_1 = 2.13 \times 10^{15}$ rad/s ($\lambda = 884.5$ nm) and $\omega_2 = 1.99 \times 10^{15}$ rad/s ($\lambda = 945.5$ nm). The S-polarized guided electromagnetic wave exists in a spectral range of width $\Delta\lambda = 61$ nm in the near infrared. However, if we reduce the gradient scale L to 30 nm, keeping the values of the remaining material parameters unchanged, we obtain $\omega_1 = 3.55 \times 10^{15}$ rad/s ($\lambda = 530.95$ nm) and $\omega_2 = 3.35 \times 10^{15}$ rad/s ($\lambda = 560.89$ nm). Thus, in this case the surface electromagnetic surface wave exists in a narrower spectral range $\Delta\lambda = 29.95$ nm in the visible region of the optical spectrum.

The preceding analytic treatment of S-polarized guided electromagnetic waves is based on the assumption that $(pL)^2 \ll 1$. On combining Eqs. (7.9a), (7.29), and (7.10b), we obtain

$$(pL)^2 = \frac{1}{4\nu^2} \left[\frac{1}{(\nu + \sqrt{\nu^2 - 1})^2} - \frac{n_v^2 - 1}{n_0^2 - n_v^2} \right]. \tag{7.32}$$

At $\nu = 1$, this expression becomes

$$(pL)^2|_{\nu=1} = \frac{1}{4}\left(1 - \frac{n_v^2 - 1}{n_0^2 - n_v^2}\right); \tag{7.33a}$$

At $\nu = \nu_c$ we find that

$$(pL)^2|_{\nu=\nu_c} = 0. \tag{7.33b}$$

If we assume the values $n_v = 1.05$ and $n_0 = 2$, Eq. (7.33a) becomes

$$(pL)^2|_{\nu=1} = 0.2412. \tag{7.34}$$

The function $(pL)^2$ decreases monotonically from this value to zero as ν increases from $\nu = 1$ to $\nu = \nu_c$, where $\nu_c = 2.75244$ from Eq. (7.31). Thus, the condition $(pL)^2 \ll 1$ is satisfied better the closer ν is to $\nu_c = 2.75244$. For example, if we assume the value $\nu = 2$, we find

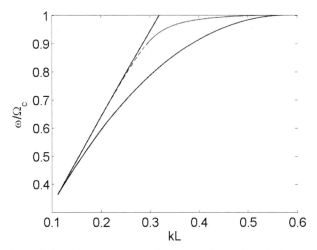

Fig. 7.2. Plots of the dispersion curve for an s-polarized guided electromagnetic wave in the frequency-range $0 < \omega < \Omega_c$ propagating on a graded index dielectric medium characterized by the values $n_v = 1.05, n_0 = 2$ and $L = 30\,\text{nm}$. The numerical solution of Eq. (7.27a) is depicted by a solid curve; the solution of Eq. (7.29) is plotted as a dash-dotted curve; and the solution of Eq. (7.37) is depicted by a dashed curve [7.22].

from Eq. (7.32) that

$$(pL)^2|_{\nu=2} = 0.002276. \tag{7.35}$$

In this case the assumption that $(pL)^2 \ll 1$ is well justified.

In Fig. 7.2 we plot the dispersion curve given by Eq. (7.29) as a dash-dotted curve for the case that $n_v = 1.05$, $n_0 = 2$, and $L = 30\,\text{nm}$, for comparison with the exact result given by a numerical solution of Eq. (7.27a), which is plotted as a solid curve. The approximate dispersion curve is seen to be a good approximation to the exact curve only when $\omega/\Omega_c = 1/\nu$ is close to $1/\nu_c = 0.3633$, as is expected, but deviates from it significantly as kL increases.

A more accurate approximate analytic dispersion relation is obtained from Eq. (7.27a) by retaining the next term in the expansion of $K_s(z)$ is powers of z beyond the expression given by Eq. (7.13). In the case that $0 < s < 1/2$ this results in

$$\frac{zK'_s(z)}{K_s(z)} = -s - \frac{2s\Gamma(1-s)}{\Gamma(1+s)}\left(\frac{z}{2}\right)^{2s}. \tag{7.36}$$

With this result the dispersion relation given by Eq. (7.27a) becomes

$$L\beta_0(k,\omega) = \frac{1}{2} - s - \frac{2s\Gamma(1-s)}{\Gamma(1+s)}\left(\frac{Lp}{2}\right)^{2s}. \qquad (7.37)$$

The solution of this equation is plotted as a dashed curve in Fig. 7.2. It is seen that this refined approximate dispersion curve is in very good agreement with the exact dispersion curve given by the numerical solution of Eq. (7.27a) over nearly the entire frequency range within which this surface wave exists.

The electric field amplitude $E_y(z|\omega)$ corresponding to a typical point on the dispersion curve plotted in Fig. 7.1 is presented in Fig. 7.3. The field in the vacuum is seen to be well localized to the interface $z = 0$, but the field in the graded index dielectric penetrates more deeply into it. However, for a value of $L = 30$ nm, as in the result plotted in this figure, a $1/e$ decay length in the dielectric medium of $100L = 3\,\mu$m is approximately six times the vacuum wavelength of the guided wave. Thus, it is still well localized to the interface $z = 0$.

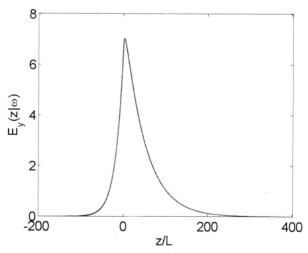

Fig. 7.3. The electric field amplitude $E_y(z|\omega)$ as a function of z for the guided wave corresponding to the point on the dispersion curve plotted in Fig. 7.1 defined by $kL = 0.18$, $\omega/\Omega_c = 0.57978$ [7.22].

7.1.2. $\omega > \Omega_c$

When the frequency ω lies in the range $\omega > \Omega_c$, the function $f(u)$ is given by Eq. (7.15). The dispersion relation for guided electromagnetic waves in the region $\omega > \Omega_c$, $k > n_v(\omega/c)$ of the (ω, k) plane is then

$$L\beta_0(k,\omega) = \frac{1}{2} + Lp\frac{K'_{is_1}(Lp)}{K_{is_1}(Lp)}, \qquad (7.38a)$$

where

$$s_1 = \frac{1}{2}\left(\frac{\omega^2}{\Omega_c^2} - 1\right)^{\frac{1}{2}}. \qquad (7.38b)$$

Equation (7.38) also has to be solved numerically. The calculations of $K_{i\mu}(z)$ and its derivative were carried out by the use of the algorithms due to Gil et al. [7.28, 7.29]. Due to the oscillatory nature of $K_{i\mu}(z)$ as z approaches zero through real positive values, a multiplicity of solutions can be found. The resulting dispersion curves are plotted in Fig. 7.4.

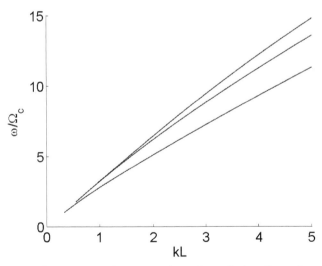

Fig. 7.4. The three lowest frequency branches of the dispersion curve for s-polarized guided electromagnetic waves in the frequency range $\omega > \Omega_c$ propagating on a graded index dielectric medium characterized by the values $n_v = 1.05$, $n_0 = 2$, and $L = 30$ nm [7.22].

The electric field amplitudes $E_y(z|\omega)$ corresponding to typical points on the dispersion curves plotted in Fig. 7.4 are presented in Fig. 7.5(a)–(c). They have the forms of the fields of waveguide modes with oscillations in the vicinity of the plane $z = 0$ and exponential decay as $z \to \pm\infty$. The number of nodes in the fields equals their branch numbers if the lowest frequency branch in this frequency range is labeled the zero branch.

The results obtained in this section can be summarized as follows.

(1) The surface localized waves supported by the dielectric structure considered here have different natures in the two frequency regions $0 < \omega < \Omega_c$ and $\omega > \Omega_c$.
(2) In the low frequency region ($0 < \omega < \Omega_c$) only a single guided wave exists, and the spectral range in which it exists is a narrow one that is defined by the technologically controlled heterogeneity scale L.
(3) In the high frequency region ($\omega > \Omega_c$) a multiplicity of guided waves exist that are characterized by an oscillatory dependence of their electric fields on z in the gradient medium.
(4) The existence of the critical frequency Ω_c and the different natures of the modes with frequencies below and above it does not appear to occur in the slab optical waveguides or in the continuously graded-index optical waveguides studied until now.
(5) Large values of Ω_c, caused by the negative gradient of the dielectric permittivity of the gradient medium afford the possibility of extending the domain of existence of these S-polarized guided electromagnetic waves to the near infrared, and even to the visible region of the optical spectrum.

7.2. Surface Electromagnetic Waves on a Curvilinear Interface: Geometrical Dispersion

We have noted earlier that a planar interface between a semi-infinite dielectric, e.g. vacuum, and a semi-infinite metal cannot support an S-polarized surface electromagnetic wave. The situation is quite different if the dielectric-metal interface is not planar but curved. In this section we consider the propagation of an S-polarized electromagnetic

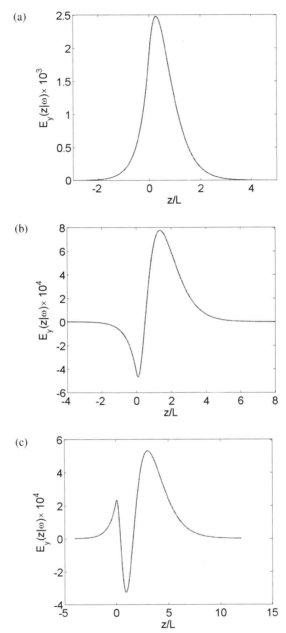

Fig. 7.5. The electric field amplitudes $E_y(z|\omega)$ as functions of z for the guided waves corresponding to the points on the three lowest frequency branches of the dispersion curve plotted in Fig. 7.4 defined by (a) $kL = 3.5$, $\omega/\Omega_c = 8.2758$; (b) $kL = 3.5, \omega/\Omega_c = 10.0514$; (c) $kL = 3.5, \omega/\Omega_c = 10.8332$.

wave propagating circumferentially around a portion of the circular boundary between a metal and vacuum. The electromagnetic fields in this case are not required to be single valued. We show that under suitable conditions the curvature of the interface can localize the wave to its vicinity, producing thereby an S-polarized surface plasmon polariton.

Azimuthal surface electromagnetic waves on cylinders have not been studied extensively in the literature. When they have been studied the structures studied have been either corrugated conducting cylinders [7.30], or dielectric-clad conducting cylinders [7.30, 7.31]. Moreover, it was assumed in these studies [7.30] that the surface waves propagated completely around the cylinder, so that the corresponding electromagnetic fields had to satisfy a single-valuedness condition.

The propagation of a P-polarized electromagnetic wave around a portion of a cylindrical boundary between a vacuum and a metal was studied many years ago by Berry [7.32]. Among the several results obtained in this work was the result that in the case that the metal is convex toward the vacuum the wave is not perfectly bound to the interface, but is attenuated as it propagates around it, the lost energy being radiated to infinity in the vacuum. In contrast, when the metal is concave toward the vacuum no attenuation of the wave occurs, and the wave is a true surface wave bound to the interface.

Because P-polarized surface electromagnetic waves exist at a dielectric-metal interface in the absence of any curvature of it, and in light of Berry's thorough treatment of P-polarized azimuthal surface waves on a portion of a cylindrical surface, we have opted to focus our attention here on the case of S-polarized azimuthal surface electromagnetic waves propagating circumferentially around a portion of a cylindrical vacuum-metal interface, where the curvature of the interface can induce the existence of a wave that does not exist in its absence [7.33].

In the propagation of an S-polarized electromagnetic wave around a portion of the circular boundary between a metal and vacuum, the only non-zero components of the electric and magnetic fields in the system in cylindrical coordinates (r, ϕ, z) are $E_z(r, \phi|\omega)$, $H_r(r, \phi|\omega)$, and $H_\phi(r,\phi|\omega)$. The Maxwell equations satisfied by these components are

$$\frac{1}{r}\frac{\partial E_z}{\partial \phi} = \frac{i\omega}{c}H_r, \qquad (7.39\text{a})$$

$$-\frac{\partial E_z}{\partial r} = \frac{i\omega}{c}H_\phi, \qquad (7.39\text{b})$$

$$\frac{1}{r}\left[\frac{\partial}{\partial r}(rH_\phi) - \frac{\partial H_r}{\partial \phi}\right] = -i\epsilon\frac{\omega}{c}E_z, \qquad (7.39\text{c})$$

where ϵ is the dielectric function of the medium in which the fields are being calculated. When Eqs. (7.39a) and (7.39b) are used to eliminate H_r and H_ϕ from Eq. (7.39c), we find that the equation satisfied by $E_z(r,\phi|\omega)$ is

$$\frac{\partial^2 E_z}{\partial r^2} + \frac{1}{r}\frac{\partial E_z}{\partial r} + \frac{1}{r^2}\frac{\partial^2 E_z}{\partial \phi^2} + \epsilon\frac{\omega^2}{c^2}E_z = 0. \qquad (7.40)$$

We solve Eq. (7.40) by separating the variables. We write

$$E_z(r,\phi|\omega) = R(r)\Phi(\phi), \qquad (7.41)$$

and find that $\Phi(\phi)$ and $R(\rho)$ satisfy the equations

$$\frac{d^2\Phi}{d\phi^2} + \mu^2\Phi = 0, \qquad (7.42)$$

$$\frac{d^2R}{dr^2} + \frac{1}{r}\frac{dR}{dr} + \left(\epsilon\frac{\omega^2}{c^2} - \frac{\mu^2}{r^2}\right)R = 0, \qquad (7.43)$$

respectively, where μ^2 is the separation constant. The sign of μ^2 was chosen so that a wavelike solution of Eq. (7.42) is possible.

We now consider two cases.

(i) Metal convex to vacuum

We consider first the case where the region $0 < r < R$ is occupied by the metal, which is characterized by the dielectric function $\epsilon(\omega)$, while the region $r > R$ is vacuum (Fig. 7.6). In this case we choose

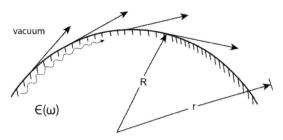

Fig. 7.6. A portion of a cylindrical interface between vacuum and a metal when the metal is convex toward the vacuum.

the solution of Eq. (7.42) to have the form

$$\Phi(\phi) = \exp(i\mu\phi), \qquad (7.44)$$

so that it represents a wave propagating in a clockwise sense around the cylinder. Thus, we initially assume that μ is real and positive.

Since we are considering propagation over only a portion of a cylindrical boundary, it is not necessary to impose a single-valuedness requirement on the electric field component $E_z(r, \phi|\omega)$, and hence on $\Phi(\phi)$. Thus, μ need not be an integer. If we rewrite Eq. (7.44) in the form

$$\Phi(\phi) = \exp[i(\mu/R)(R\phi)], \qquad (7.45)$$

and recall that $R\phi$ is the path length measured along the cylindrical surface, we see that

$$\frac{\mu}{R} \equiv k \qquad (7.46)$$

can be regarded as the wave number characterizing the propagation of this cylindrical wave.

We turn now to Eq. (7.43). We recognize it as Bessel's equation. If we wish to obtain a solution that decreases exponentially into the cylinder with increasing distance from the surface $r = R$, we have to work in the frequency range in which the metal's dielectric function $\epsilon(\omega)$ is negative. For in this frequency range Eq. (7.44) becomes

$$R'' + \frac{1}{r}R' - \left(|\epsilon|\frac{\omega^2}{c^2} + \frac{\mu^2}{r^2}\right)R = 0. \qquad (7.47)$$

The solutions of this equation are the modified Bessel functions $I_\mu(\sqrt{|\epsilon(\omega)|}(\omega/c)r)$ and $K_\mu(\sqrt{|\epsilon(\omega)|}(\omega/c)r)$. The former of these functions decays exponentially away from the surface $r = R$ toward the origin $r = 0$, as the field of a surface wave should. The field $E_z(r,\phi|\omega)$ inside the metal can therefore be written as

$$E_z^<(r,\phi|\omega) = A e^{i\mu\phi} I_\mu(\sqrt{|\epsilon(\omega)|}(\omega/c)r) \quad 0 < r < R, \quad (7.48)$$

where A is an arbitrary constant.

Turning now to the field outside the cylinder, $r > R$, where $\epsilon \equiv 1$, the solution of Eq. (7.42) is still given by Eq. (7.44).

With this result the equation for the function $R(r)$ is obtained from Eq. (7.43) in the form

$$\frac{d^2 R}{dr^2} + \frac{1}{r}\frac{dR}{dr} + \left(\frac{\omega^2}{c^2} - \frac{\mu^2}{r^2}\right) R = 0. \quad (7.49)$$

This is Bessel's equation, and its solutions are the Bessel functions of the first and second kinds, $J_\mu((\omega/c)r)$ and $Y_\mu((\omega/c)r)$, or the Hankel functions of the first and second kinds $H_\mu^{(1)}((\omega/c)r)$, and $H_\mu^{(2)}((\omega/c)r)$. Each of these functions is an oscillatory rather than an exponentially decreasing function of $(\omega/c)r$, as is any linear combination of them. This tells us that surface waves whose amplitudes decay purely exponentially into each medium with increasing distance from the surface $r = R$ cannot exist.

We have chosen for the solution of Eq. (7.49) the Hankel function of the first kind and order μ, $R^>(r) = H_\mu^{(1)}((\omega/c)r)$. In choosing this solution we have used the fact that it is the only one that describes an outgoing wave as $r \to \infty$. Thus, we find that S-polarized surface electromagnetic waves that decay with increasing r cannot exist in the situation under consideration: they must radiate. In this case the field $E_z(r,\phi|\omega)$ in the vacuum region $r > R$ becomes

$$E_z^>(r,\phi|\omega) = B e^{i\mu\phi} H_\mu^{(1)}((\omega/c)r) \quad r > R. \quad (7.50)$$

The boundary conditions satisfied by $E_z^<(r,\phi|\omega)$ and $E_z^>(r,\phi|\omega)$ on the surface $r = R$ require the continuity E_z and its normal derivative $\partial E_z/\partial r$ across it,

$$AI_\mu(|\epsilon|^{\frac{1}{2}}(\omega/c)R) = BH_\mu^{(1)}((\omega/c)R), \tag{7.51a}$$

$$A|\epsilon|^{\frac{1}{2}}I'_\mu(|\epsilon|^{\frac{1}{2}}(\omega/c)R) = BH_\mu^{(1)'}((\omega/c)R), \tag{7.51b}$$

where the prime denotes differentiation with respect to argument. The solvability condition for this pair of homogeneous equations is

$$\frac{1}{\sqrt{|\epsilon(\omega)|}} \frac{I_\mu(\sqrt{|\epsilon(\omega)|}(\omega/c)R)}{I'_\mu(\sqrt{|\epsilon(\omega)|}(\omega/c)R)} \frac{H_\mu^{(1)'}((\omega/c)R)}{H_\mu^{(1)}((\omega/c)R)} = 1. \tag{7.52}$$

Because $H_\mu^{(1)}((\omega/c)R)$ is a complex function of its argument, this equation has no real solution. The surface wavenumber $k = \mu/R$ is now a complex function of (real) ω,

$$k(\omega) = k_R(\omega) + ik_I(\omega), \tag{7.53}$$

where $k_R(\omega) > 0$ and $k_I(\omega) > 0$. This means that the wave decays as it propagates around the cylinder, with the lost energy radiated to infinity. The last statement follows from the outgoing wave nature of $H_\mu^{(1)}(z)$ for large values of its argument.

However, no solution of this equation has been found up to now. It is conjectured [7.33] that no physically acceptable solution exists, namely one for which $k_R(\omega) > 0$, $k_I(\omega) > 0$, and $k_I(\omega) \ll k_R(\omega)$, just as no S-polarized surface plasmon polariton exists at a planar dielectric-metal interface.

(ii) Metal concave to vacuum

In the case that vacuum occupies the region $0 < r < R$ (Fig. 7.7), the electric field component $E_z(r,\phi|\omega)$ satisfies the equation

$$\left(\frac{\partial^2}{\partial r^2} + \frac{1}{r}\frac{\partial}{\partial r} + \frac{1}{r^2}\frac{\partial^2}{\partial \phi^2} + \frac{\omega^2}{c^2}\right)E_z = 0, \quad 0 < r < R, \tag{7.54}$$

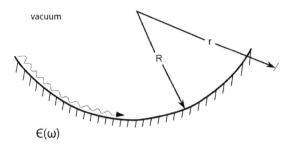

Fig. 7.7. A portion of a cylindrical interface between vacuum and a metal when the metal is concave toward the vacuum.

in this region, and the equation

$$\left(\frac{\partial^2}{\partial r^2} + \frac{1}{r}\frac{\partial}{\partial r} + \frac{1}{r^2}\frac{\partial^2}{\partial \phi^2} + \epsilon\frac{\omega^2}{c^2}\right) E_z = 0, \quad r > R, \qquad (7.55)$$

in the region $r > R$. We consider them in turn.

When $0 < r < R$, the method of separation of variables leads to the following pair of equations:

$$\frac{d^2 R}{dr^2} + \frac{1}{r}\frac{dR}{dr} + \left(\frac{\omega^2}{c^2} - \frac{\mu^2}{r^2}\right) R(r) = 0, \qquad (7.56)$$

$$\frac{d^2\Phi}{d\phi^2} + \mu^2 \Phi = 0. \qquad (7.57)$$

We again choose the solution of Eq. (7.57) to be $\Phi(\phi) = \exp(i\mu\phi)$. The solution of Eq. (7.56) will be taken to be $J_\mu((\omega/c)r)$. This choice for the solution is dictated by the following consideration. The Bessel function $J_\mu(x)$ for a fixed value of (real, nonzero) μ increases exponentially with increasing x until a value of $x \sim \mu$ is reached, at which it acquires an oscillatory dependence on x that continues for $x > \mu$. Such an oscillatory dependence of $J_\mu((\omega/c)r)$ for $r \sim R$ makes it possible to satisfy the boundary condition at $r = R$, as we will see. Since we are concerned with a solution for r in the range of $0 < r < R$ that is localized for r in the vicinity of R, i.e. is small for $r \to 0$, $J_\mu((\omega/c)r)$ has this behavior provided μ is of the order of $(\omega/c)R$. The field component $E_z(r, \phi|\omega)$ in the region $0 < r < R$ is therefore given by

$$E_z^<(r, \phi|\omega) = Ae^{i\mu\phi} J_\mu((\omega/c)r), \quad 0 < r < R. \qquad (7.58)$$

When $r > R$ the method of separation of variables leads to the following pair of equations:

$$\frac{d^2R}{d^2r} + \frac{1}{r}\frac{dR}{dr} + \left(\epsilon\frac{\omega^2}{c^2} - \frac{\mu^2}{r^2}\right)R = 0, \qquad (7.59)$$

$$\frac{d^2\Phi}{d\phi^2} + \mu^2\Phi = 0. \qquad (7.60)$$

The solution of Eq. (7.60) is again chosen to be $\Phi(\phi) = \exp(i\mu\phi)$.

In order to obtain a solution of Eq. (7.59) that decays to zero exponentially as r tends to infinity, we have to work in the frequency region in which $\epsilon(\omega)$ is negative. The solution of Eq. (7.59) in this frequency range that we choose is $K_\mu(\sqrt{|\epsilon(\omega)|}(\omega/c)r)$, where $K_\mu(x)$ is the modified Bessel function of the second kind of order μ. The field component $E_z(r,\phi|\omega)$ in the region $r > R$ is therefore given by

$$E_z^>(r,\phi|\omega) = Be^{i\mu\phi}K_\mu(\sqrt{|\epsilon(\omega)|}(\omega/c)r), \quad r > R. \qquad (7.61)$$

The boundary conditions at $r = R$ take the forms

$$AJ_\mu((\omega/c)R) = BK_\mu(\sqrt{|\epsilon(\omega)|}(\omega/c)R), \qquad (7.62a)$$

$$AJ'_\mu((\omega/c)R) = B\sqrt{|\epsilon(\omega)|}K'_\mu(\sqrt{|\epsilon(\omega)|}(\omega/c)R). \qquad (7.62b)$$

The dispersion relation for the surface electromagnetic waves in the case that the metal is concave to the vacuum becomes

$$\frac{1}{\sqrt{|\epsilon(\omega)|}}\frac{K_\mu(\sqrt{|\epsilon(\omega)|}(\omega/c)R)}{K'_\mu(\sqrt{|\epsilon(\omega)|}(\omega/c)R)}\frac{J'_\mu((\omega/c)R)}{J_\mu((\omega/c)R)} = 1. \qquad (7.63)$$

This is a real equation with real solutions that relate $\mu = kR$ to ω. If we assume that the dielectric function of the metal has the simple free electron form

$$\epsilon(\omega) = 1 - \frac{\omega_p^2}{\omega^2}, \qquad (7.64)$$

where ω_p is the plasma frequency of the electrons in the bulk of the metal, these solutions must be sought in the region $0 < \omega < \omega_p$. In the numerical calculations whose results are reported here the value $\omega_p = 12.708 \times 10^{15}$ s^{-1} was used.

Equation (7.63) was solved numerically by assuming a value of $k = \mu/R$, increasing ω in steps of equal size in the interval $0 \leq \omega \leq \omega_p$, and looking for sign changes in the difference between the left- and right-hand sides of this equation. A new value of k was selected and the calculation was repeated. In this way a plot of ω/ω_p as a function of kR was constructed.

The resulting dispersion curve possesses several branches in the frequency range $0 < \omega < \omega_p$. Typical results are plotted in Fig. 7.8. It is seen that with an increase of the radius R the number of branches increases and their separation in frequency decreases. The modes are seen to be dispersive. This is due in part to the frequency dependence of $\epsilon(\omega)$, and in part to the presence of a characteristic length in the present problem, namely the radius R.

The radial dependencies of $E_z(r, \phi|\omega)$, namely

$$R_s(r) = J_\nu((\omega/c)r) \qquad 0 < r < R \qquad (7.65a)$$

$$= \frac{J_\nu((\omega/c)R)}{K_\nu(|\epsilon(\omega)|^{\frac{1}{2}}(\omega/c)R)} K_\nu(|\epsilon(\omega)|^{\frac{1}{2}}(\omega/c)r), \qquad r > R, \qquad (7.65b)$$

corresponding to typical points on the three lowest frequency branches of the dispersion curve plotted in Fig. 7.8 are presented in Fig. 7.9. It is seen that the field is corresponding to each branch possesses as many nodes in the region of the metal as the number of the branch, if the lowest frequency branch is denoted the zero branch, and decays exponentially with increasing radial distance into each medium from the concave surface of the cylinder. Thus, these modes are more appropriately described as waveguide modes, with the lowest frequency mode — the fundamental mode — possessing several properties of a surface wave. In fact, the cylindrical surface is equivalent to a planar asymmetric graded-index waveguide, as can be seen by carrying out a coordinate transformation that maps the circular boundary into a planar one. The resulting spatial dependence of the dielectric constant in each medium is such that it is larger in the vicinity of the interface than it is far from it in each medium. We will see a simple example of this in Sec. 10.2.1.

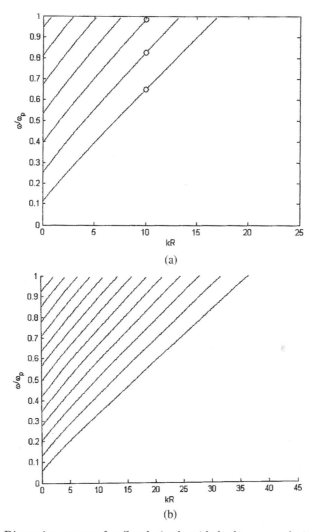

Fig. 7.8. Dispersion curves for S-polarized guided plasmon polaritons propagating circumferentially around a portion of a cylindrical interface between vacuum and silver when the metal is concave toward the vacuum. (a) $R = 0.5\,\mu\text{m}$, $\omega_p = 12.708 \times 10^{15}\,\text{rad s}^{-1}$ (b) $R = 1.0\,\mu\text{m}$, $\omega_p = 12.708 \times 10^{15}\,\text{rad s}^{-1}$.

Thus, in this section we have shown that guided electromagnetic waves of S polarization can propagate circumferentially around a portion of a cylindrical vacuum-metal interface when the homogeneous metal is concave to the vacuum. A surface electromagnetic wave of this polarization does not exist at the planar interface between

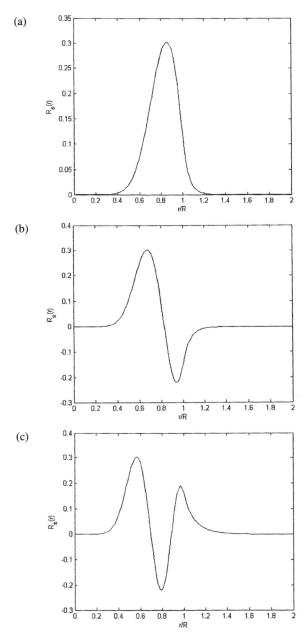

Fig. 7.9. The radial dependencies of the electric field amplitude $R_s(r)$ corresponding to the points indicated by open circles on the three lowest frequency branches of the dispersion curve depicted in Fig. 7.8(a). (a) $kR = 10, \omega/\omega_p = 0.6514$, (b) $kR = 10, \omega/\omega_p = 0.8271$; (c) $kR = 10, \omega/\omega_p = 0.9841$.

vacuum and a homogeneous metal. The dispersion relation for such waves possesses several branches within the frequency region in which dielectric function of the metal is negative.

7.3. Surface Electromagnetic Waves on Rough Surfaces: Roughness-Induced Dispersion

The dispersion relation of a surface plasmon polariton propagating along the planar interface between vacuum and a metal characterized by the simple free electron dielectric function (7.64) is [7.34]

$$\epsilon(\omega)\beta_0(k,\omega) + \beta(k,\omega) = 0, \qquad (7.66)$$

where

$$\beta_0(k,\omega) = [k^2 - (\omega/c)^2]^{\frac{1}{2}}, \quad \text{Re}\beta_0(k,\omega) > 0$$
$$\text{Im}\beta_0(k,\omega) < 0 \qquad (7.67a)$$

$$\beta(k,\omega) = [k^2 - \epsilon(\omega)(\omega/c)^2]^{\frac{1}{2}}, \quad \text{Re}\beta(k,\omega) > 0$$
$$\text{Im}\beta(k,\omega) < 0. \qquad (7.67b)$$

Its solution [7.35]

$$\omega_0(k) = \left\{ \frac{1}{2}\omega_p^2 + c^2k^2 - [\frac{1}{4}\omega_p^4 + c^4k^4]^{\frac{1}{2}} \right\}^{\frac{1}{2}}, \qquad (7.68)$$

$$k \to 0 \xrightarrow{\quad} ck, \qquad (7.69a)$$

$$k \to \infty \xrightarrow{\quad} \omega_p/\sqrt{2}, \qquad (7.69b)$$

is dispersive, i.e. the phase and group velocities of this surface electromagnetic wave are functions of the wavenumber k. This is due to the presence of a hidden characteristic length in this system, namely the wavelength corresponding to the plasma frequency ω_p, $\lambda_p = (2\pi c)/\omega_p$, which in turn corresponds to a wavenumber $k_p = \omega_p/c$. For values of k larger than k_p the effects of retardation are unimportant: light propagates over the short distances that correspond to such values of k essentially instantaneously. For values of k smaller than k_p the finite speed of light has to be taken into account in considering its propagation over such larger distances. The surface plasmon polariton dispersion relation has a qualitatively different form in each of these limits, which produces a dispersive dispersion curve.

This dispersion can be enhanced by structuring the vacuum-metal interface, either periodically or randomly. The additional dispersion induced in this manner can be termed roughness-induced dispersion. In addition, in the case of a periodically corrugated interface, the surface plasmon polariton dispersion curve can acquire additional, higher frequency, branches, whose number depends on the period of the grating and the plasma frequency of the metal.

In this section we study the roughness-induced dispersion produced by a periodic interface as well as by a randomly corrugated interface, and see how these features arise. Although the propagation of surface plasmon polaritons on doubly periodic [7.36] and on two-dimensional randomly rough [7.37–7.40] metallic surfaces has been studied, we have chosen to study this propagation on one-dimensional periodically and randomly rough surfaces when the sagittal plane is perpendicular to the generators of these surfaces. The dispersion curves obtained for these surfaces display the significant features possessed by the corresponding curves obtained for the two-dimensional surfaces, and are derived more simply.

We begin by considering a general system consisting of vacuum in the region $z > \zeta(x)$, and a metal, characterized by an isotropic, frequency-dependent dielectric function $\epsilon(\omega)$ in the region $z < \zeta(x)$. In common with most determinations of the dispersion relations of surface plasmon polaritons we will assume that the dielectric function $\epsilon(\omega)$ is real. The surface profile function $\zeta(x)$ is assumed to be a single-valued function of x that is differentiable. It is assumed that a surface plasmon polariton propagates in the x direction along the interface $z = \zeta(x)$. It is convenient to work with the single nonzero component of the magnetic field in this system, namely $H_y(\mathbf{x};t) = H_y(x,z|\omega)\exp(-i\omega t)$, since it satisfies a scalar wave equation in each medium,

$$\left(\frac{\partial^2}{\partial x^2} + \frac{\partial^2}{\partial z^2} + \frac{\omega^2}{c^2}\right) H_y^>(x,z|\omega) = 0, \quad z > \zeta(x), \quad (7.70a)$$

$$\left(\frac{\partial^2}{\partial x^2} + \frac{\partial^2}{\partial z^2} + \epsilon(\omega)\frac{\omega^2}{c^2}\right) H_y^<(x,z|\omega) = 0, \quad z < \zeta(x). \quad (7.70b)$$

The boundary conditions at the interface $z = \zeta(x)$ require the continuity of the tangential components of the magnetic and dielectric fields in the system across it:

$$H_y^>(x,z|\omega)|_{z=\zeta(x)} = H_y^<(x,z|\omega)|_{z=\zeta(x)}, \qquad (7.71a)$$

$$\frac{\partial}{\partial n} H_y^>(x,z|\omega)\bigg|_{z=\zeta(x)} = \frac{1}{\epsilon(\omega)} \frac{\partial}{\partial n} H_y^<(x,z|\omega)\bigg|_{z=\zeta(x)}. \qquad (7.71b)$$

In these equations

$$\frac{\partial}{\partial n} = [1 + \zeta'(x)^2]^{-\frac{1}{2}} \left[-\zeta'(x) \frac{\partial}{\partial x} + \frac{\partial}{\partial z} \right], \qquad (7.72)$$

is the derivative along the normal to the surface at each point of it, directed from the metal into the vacuum. We also assume vanishing boundary conditions at infinity.

The solutions of Eqs. (7.70) that satisfy the boundary conditions at $|z| = \infty$ can be written as

$$H_y^>(x,z|\omega) = \int_{-\infty}^{\infty} \frac{dq}{2\pi} A(q,\omega) \exp[iqx - \beta_0(q,\omega)z] z > \zeta_{\max}, \qquad (7.73a)$$

and

$$H_y^<(x,z|\omega) = \int_{-\infty}^{\infty} \frac{dq}{2\pi} B(q,\omega) \exp[iqx + \beta(q,\omega)z] z < \zeta_{\min}, \qquad (7.73b)$$

where

$$\beta_0(q,\omega) = [q^2 - (\omega/c)^2]^{\frac{1}{2}}, \text{Re}\beta_0(q,\omega) > 0, \qquad \text{Im}\beta_0(q,\omega) < 0, \qquad (7.74a)$$

$$\beta(q,\omega) = [q^2 - \epsilon(\omega)(\omega/c)^2]^{\frac{1}{2}}, \text{Re}\beta(q,\omega) > 0, \quad \text{Im}\beta(q,\omega) < 0. \qquad (7.74b)$$

Strictly speaking, the expressions for the fields given by Eq. (7.73) are exact only outside the grooves and ridges of the surface. We will invoke the Rayleigh hypothesis [7.41], which is the assumption that these fields can nonetheless be continued in to the interface itself and used in satisfying the boundary conditions (7.71). When this is done we obtain the following pair of homogeneous integral equations for determining the amplitude functions $A(q,\omega)$ and $B(q,\omega)$:

$$\int_{-\infty}^{\infty} \frac{dq}{2\pi} \Big\{ [A(q,\omega)\exp[iqx - \beta_0(q,\omega)\zeta(x)]$$
$$- B(q,\omega)\exp[iqx + \beta(q,\omega)\zeta(x)] \Big\} = 0, \qquad (7.75a)$$

$$\int_{-\infty}^{\infty} \frac{dq}{2\pi} \Big\{ [-iq\zeta'(x) - \beta_0(q,\omega)]A(q,\omega)\exp[iqx - \beta_0(q,\omega)\zeta(x)]$$
$$- \frac{1}{\epsilon(\omega)}[-iq\zeta'(x) + \beta(q,\omega)]B(q,\omega)\exp[iqx + \beta(q,\omega)\zeta(x)] \Big\} = 0.$$
$$(7.75b)$$

It is much simpler to work with a single integral equation than with two coupled integral equations. We can eliminate the amplitude functions $B(q,\omega)$ from the pair of equations (7.75) to obtain a single homogeneous integral equation satisfied by the amplitude function $A(q,\omega)$ alone. We do this by first multiplying Eq. (7.75a) by $[ip\zeta'(x) + \beta(p,\omega)]\exp[-ipx + \beta(p,\omega)\zeta(x)]$ and then integrating the result with respect to x. We then multiply Eq. (7.75b) by $-\epsilon(\omega)\exp[-ipx + \beta(p,\omega)\zeta(x)]$ and integrate the result with respect to x. We finally add the resulting pair of equations to obtain

$$\int_{-\infty}^{\infty} \frac{dq}{2\pi} \frac{I(\beta(p,\omega) - \beta_0(q,\omega)|p-q)}{\beta(p,\omega) - \beta_0(q,\omega)} [pq - \beta(p,\omega)\beta_0(q,\omega)]A(q) = 0,$$
$$(7.76)$$

where we have introduced the function $I(\gamma|Q)$ through

$$\exp[\gamma\zeta(x)] = \int_{-\infty}^{\infty} \frac{dQ}{2\pi} I(\gamma|Q)\exp(iQx). \qquad (7.77)$$

It follows from this result that

$$\zeta'(x)\exp[\gamma\zeta(x)] = \int_{-\infty}^{\infty} \frac{dQ}{2\pi} \frac{iQ}{\gamma} I(\gamma|Q) \exp(iQx). \qquad (7.78)$$

Both of these relations were used in obtaining Eq. (7.76). For the evaluation of $I(\gamma|Q)$ we need the inverse relation

$$I(\gamma|Q) = \int_{-\infty}^{\infty} dx \exp(-iQx) \exp[\gamma\zeta(x)]. \qquad (7.79)$$

Equation (7.76) was first obtained in Ref. [7.42] by a different approach.

As we will see later, it is sometimes convenient to remove a delta function from the function $I(\gamma|Q)$ by rewriting Eq. (7.79) as

$$\begin{aligned} I(\gamma|Q) &= \int_{-\infty}^{\infty} dx \exp(-iQx)\{1 + \exp[\gamma\zeta(x)] - 1\} \\ &= 2\pi\delta(Q) + \gamma J(\gamma|Q), \end{aligned} \qquad (7.80)$$

where

$$J(\gamma|Q) = \int_{-\infty}^{\infty} dx \exp(-iQx) \frac{\exp[\gamma\zeta(x)] - 1}{\gamma}. \qquad (7.81)$$

On substituting Eq. (7.80) into Eq. (7.76) we obtain the equation satisfied by $A(q,\omega)$ in the form

$$\frac{\epsilon(\omega)\beta_0(p,\omega) + \beta(p,\omega)}{\epsilon(\omega) - 1} A(p,\omega)$$
$$= \int_{-\infty}^{\infty} \frac{dq}{2\pi} J(\beta(p,\omega) - \beta_0(q,\omega)|p-q)[pq - \beta(p,\omega)\beta_0(q,\omega)]A(q,\omega).$$
$$(7.82)$$

The identity

$$\frac{p^2 - \beta(p,\omega)\beta_0(p,\omega)}{\beta(p,\omega) - \beta_0(p,\omega)} = -\frac{\epsilon(\omega)\beta_0(p,\omega) + \beta(p,\omega)}{\epsilon(\omega) - 1}, \quad (7.83)$$

was used in obtaining Eq. (7.82).

We now turn to an application of the results obtained in this section to the determination of the dispersion relation for surface plasmon polaritons on a periodically corrugated surface and on a randomly rough surface.

7.3.1. *Periodically corrugated surfaces*

We assume that the surface profile function $\zeta(x)$ is a periodic function of x with a period a, $\zeta(x+a) = \zeta(x)$. In this case the function $I(\gamma|Q)$ defined by Eq. (7.79) becomes

$$\begin{aligned}
I(\gamma|Q) &= \sum_{n=-\infty}^{\infty} \int_{(n-\frac{1}{2}a)}^{(n+\frac{1}{2}a)} dx \exp(-iQx) \exp[\gamma\zeta(x)] \\
&= \sum_{n=-\infty}^{\infty} \int_{-\frac{1}{2}a}^{\frac{1}{2}a} dx \exp[-iQ(x+na)] \exp[\gamma\zeta(x+na)] \\
&= \sum_{n=-\infty}^{\infty} \exp(-iQna) \int_{-\frac{1}{2}a}^{\frac{1}{2}a} dx \exp(-iQx) \exp[\gamma\zeta(x)] \\
&= \sum_{m=-\infty}^{\infty} 2\pi\delta(Q - (2\pi m/a))\mathcal{I}_m(\gamma), \quad (7.84)
\end{aligned}$$

where

$$\mathcal{I}_m(\gamma) = \frac{1}{a} \int_{-\frac{1}{2}a}^{\frac{1}{2}a} dx \exp\left(-i\frac{2\pi m}{a}x\right) \exp[\gamma\zeta(x)]. \quad (7.85)$$

The result that

$$\sum_{n=-\infty}^{\infty} \exp(-iQna) = \sum_{m=-\infty}^{\infty} \frac{2\pi}{a} \delta\left(Q - \frac{2\pi m}{a}\right) \quad (7.86)$$

was used in obtaining Eq. (7.84).

We must also express the amplitude function $A(q,\omega)$ in the form

$$A(q,\omega) = 2\pi \sum_{n=-\infty}^{\infty} A_n(k)\delta\left(q - k - \frac{2\pi n}{a}\right), \qquad (7.87)$$

in order that the expression for the magnetic field $H_y^>(x,z|\omega)$, Eq. (7.73a), that results,

$$H_y^>(x,z|\omega) = \sum_{n=-\infty}^{\infty} A_n(k)\exp\left[i\left(k + \frac{2\pi n}{a}\right)x\right]$$
$$\times \exp\left[-\beta_0\left(k + \frac{2\pi n}{a},\omega\right)z\right], \qquad (7.88)$$

satisfy the Bloch-Floquet theorem [7.43], $H_y^>(x+a,z|\omega) = \exp(ika) H_y^>(x,z|\omega)$. The wavenumber k entering Eq. (7.88) is thus the wave number of the surface plasmon polariton.

When the expansions (7.84) and (7.87) are substituted into Eq. (7.76), it becomes

$$\sum_{m=-\infty}^{\infty} 2\pi\delta(p - k_m) \sum_{n=-\infty}^{\infty} \frac{k_m k_n - \beta(k_m,\omega)\beta_0(k_n,\omega)}{\beta(k_m,\omega) - \beta_0(k_n,\omega)}$$
$$\times \mathcal{I}_{m-n}(\beta(k_m,\omega) - \beta_0(k_n,\omega))A_n(k) = 0, \qquad (7.89)$$

where we have introduced the notation $k_m = k + (2\pi m/a)$. This equation can be satisfied only if the coefficient of $2\pi\delta(p - k_m)$ vanishes for each m. In this way we obtain the equation satisfied by the $\{A_n(k)\}$ [7.44]:

$$\sum_{n=-\infty}^{\infty} \frac{k_m k_n - \beta(k_m,\omega)\beta_0(k_n,\omega)}{\beta(k_m,\omega) - \beta_0(k_n,\omega)}$$
$$\times \mathcal{I}_{m-n}(\beta(k_m,\omega) - \beta_0(k_n,\omega))A_n(k) = 0 \quad m = 0,\pm 1,\pm 2\ldots$$
$$(7.90)$$

The dispersion relation for surface plasmon polaritons on a grating is obtained by equating to zero the determinant of the matrix of coefficients in Eq. (7.90).

The solutions $\omega(k)$ of this dispersion relation have two general properties. The first is that $\omega(k)$ is a periodic function of the wavenumber k with a period $2\pi/a$. The second property of $\omega(k)$ is that it is an even function of k. These two properties of $\omega(k)$, which are independent of the surface profile function, have the consequence that all the distinct solutions of the dispersion relation are obtained if the wave number k is restricted to the interval $0 \leq k \leq \pi/a$.

In addition, it is not difficult to determine from Eq. (7.74a) that if k is real, is in the interval $(0, \pi/a)$, and is larger than ω/c, $\beta_0(k_n, \omega)$ is real and positive for all n when ω is also real. (Since we are working in a frequency range where $\epsilon(\omega)$ is negative, because it is in such a range that a surface plasmon polariton exists, $\beta(k_n, \omega)$ is real for all n. From Eq. (7.88) we see that the magnetic field in the vacuum region tends to zero as $z \to \infty$. Thus, the Bloch-like surface plasmon polaritons that are true eignemodes of the corrugated structure exist only in the triangular region of the (ω, k) plane bounded from the left by the vacuum light line $\omega = ck$, and from the right by the boundary of the first Brillouin zone $k = \pi/a$. This region is called the *non-radiative* region of the (ω, k) plane. The region of the (ω, k) plane in which k is smaller than ω/c, and which is bounded on the left by the line $k = 0$ and on the right by $k = \pi/a$ is called the *radiative region*. In this region $\beta_0(k_n, \omega)$ can become purely imaginary with a negative sign(Eq. (7.74a)) for some values of n, and the expansion (7.88) describes a wave that radiates energy into the vacuum as it propagates along the surface, and is attenuated thereby. In this case the solutions of the dispersion relation for real k become complex, $\omega(k) = \omega_1(k) - i\omega_2(k)$, and the lifetime of the energy of the wave, $\tau(k)$, is given by $\tau(k) = [2\omega_2(k)]^{-1}$. Solutions of the dispersion relation will be sought only in the non-radiative region.

These general properties of the dispersion curve are present in the results of a numerical solution of the dispersion relation. We assume a sinusoidal surface profile function

$$\zeta(x) = \zeta_0 \cos(2\pi x/a), \qquad (7.91)$$

for which the function $\mathcal{I}_m(\gamma)$ defined by Eq. (7.86) is given by

$$\mathcal{I}_m(\gamma) = I_m(\gamma \zeta_0), \qquad (7.92)$$

where $I_m(z)$ is a modified Bessel function of the first kind and order m. For the dielectric function $\epsilon(\omega)$ we assume the simple free electron form (7.64). We assume for the plasma frequency the value $\omega_p = 12.708 \times 10^{15} \text{s}^{-1}$. The grating profile was characterized by the values $a = 5000\,\text{Å}$ and $\zeta_0 = 0$ and $500\,\text{Å}$.

In the numerical calculations of the dispersion curves the infinite determinant formed from the matrix of coefficients in Eq. (7.90) was replaced by the determinant of the $(2N+1) \times (2N+1)$ matrix obtained by restricting m and n to run from $-N$ to N. The zeros of this truncated determinant were found numerically by fixing k, increasing ω from 0 to ck in small increments $\Delta\omega$, and looking for changes in the sign of the determinant. The convergence of the solutions found in this way was tested by increasing N and seeing if they approached stable limiting values.

In Fig. 7.10 we plot the dispersion curves for surface plasmon polaritons propagating on this metallic grating. It is seen to consist of three branches. The two lowest frequency branches are separated by a gap at $k = \pi/a$. Any other branches in this case

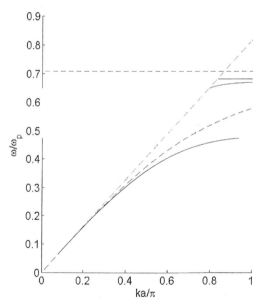

Fig. 7.10. The dispersion curves for surface plasmon polaritons on a silver grating defined by the profile function $\zeta(x) = \zeta_0 \cos(2\pi x_1/a)$, for $\zeta_0 = 500\,\text{Å}$ and $a = 5000\,\text{Å}$; $\omega_p = 12.708 \times 10^{15}\,\text{s}^{-1}$ (———). The dispersion curve for a surface plasmon polariton on a planar silver surface is also presented (- - - -), together with the dispersion curve $\omega = ck$ (- - - - -).

have higher frequencies and lie outside the non-radiative region. The three branches inside the non-radiative region in first approximation are obtained by plotting the dispersion curve for surface plasmon polaritons on a planar vacuum-metal interface in the reduced zone scheme, i.e. by folding the portions of this curve that lie in the second, third, ..., Brillouin zones of the grating into the first Brillouin zone by translating them to the left and to the right by suitable integer multiples of $2\pi/a$. The higher frequency branch terminates when it crosses the vacuum light line $\omega = ck$, and the corresponding wave becomes a radiative mode. The gap at the zone boundary opens up when the grating is turned on because the dispersion curve of the unperturbed structure is degenerate at the wavenumbers $k = \pm\pi/a$ that are separated by a translation vector of the reciprocal lattice of the grating, $2\pi/a$. Since the Fourier coefficient $\frac{1}{2}\zeta_0$ of the surface profile function corresponding to this translation vector is nonzero, degenerate perturbation theory tells us that the degeneracy is lifted when the grating is turned on. A gap therefore opens up in the dispersion curve at the zone boundary.

In the limit $k \to 0$ the lower frequency branch is tangent to the dispersion curve for a surface plasmon polariton on a planar surface for any value of ζ_0/a. In addition, we see that the lowest frequency branch displays the phenomenon of wave slowing, namely the phase and group velocities of the wave are smaller than those of the surface plasmon polariton at a planar vacuum-metal interface. The slowing down of a surface plasmon polariton wave packet by this mechanism was recently observed experimentally [7.45]. This is a consequence of the periodicity of $\omega(k)$, $\omega(k + (2\pi/a)) = \omega(k)$, and its evenness, $\omega(-k) = \omega(k)$, which force each branch of $\omega(k)$ to come into the Brillouin zone boundary $k = \pi/a$ with zero slope. This, in turn, forces the lowest frequency branch of the dispersion curve to bend away from the planar surface dispersion curve, and this bending of the dispersion curve reduces the group and phase velocities of this wave.

As the value of ζ_0 is increased the lower frequency branch moves to lower frequencies, the width of the gap increases, and at some critical value of ζ_0 the highest frequency branch moves into the radiative region, and ceases to be a true surface wave.

7.3.2. A randomly rough surface

A convenient starting point for the determination of the dispersion relation for a surface plasmon polariton on a one-dimensional randomly rough vacuum-metal interface is Eq. (7.82) [7.46]. This is because when the surface profile function $\zeta(x)$ is identically zero, the right-hand side of this equation vanishes, and the equation becomes

$$\frac{\epsilon(\omega)\beta_0(p,\omega) + \beta(p,\omega)}{\epsilon(\omega) - 1} A(p,\omega) = 0. \tag{7.93}$$

A nontrivial solution of this equation requires that

$$\epsilon(\omega)\beta_0(p,\omega) + \beta(p,\omega) = 0. \tag{7.94}$$

This is recognized as the dispersion relation for surface plasmon polaritons at a planar vacuum-metal interface, Eq. (7.66). Thus, the right-hand side of Eq. (7.82) represents a correction to this dispersion relation in the presence of surface roughness.

The surface profile function $\zeta(x_1)$ of a randomly rough surface is unknown in general. This forces us to characterize it by certain statistical properties. Underlying this characterization is the assumption that there is not a single function $\zeta(x)$. Instead there is an ensemble of realizations of this function. Physical properties associated with a statistically rough surface are to be averaged over this ensemble, and it is assumed that this ensemble average does not differ from the spatial average over a single realization of the surface in the limit of a large surface.

In common with most theoretical treatments of random surface roughness we assume that the surface profile function $\zeta(x)$ is a single-valued function of x that is differentiable and constitutes a zero-mean, stationary Gaussian random process defined by

$$\langle \zeta(x) \rangle = 0, \tag{7.95a}$$

$$\langle \zeta(x)\zeta(x') \rangle = \delta^2 W(|x - x'|). \tag{7.95b}$$

In Eqs. (7.95) the angle brackets denote an average over the ensemble of realizations of the function $\zeta(x)$. The quantity δ appearing in

Eq. (7.95b) is the rms height of the random surface

$$\delta = \langle \zeta^2(x) \rangle^{\frac{1}{2}}. \tag{7.96}$$

The normalized surface height autocorrelation function $W(|x|)$ possesses some important general properties. It follows from Eqs. (7.95b) and (7.96) that

$$W(0) = 1. \tag{7.97}$$

$W(|x|)$ is clearly an even function of x_1, because $\langle \zeta(x)\zeta(x') \rangle = \langle \zeta(x') \zeta(x) \rangle$. It is also easy to show that $W(|x|)$ satisfies the inequalities

$$-1 \leq W(|x|) \leq 1. \tag{7.98}$$

In addition, from the fact that on a statistically rough surface the heights of the surface at two widely separated points are uncorrelated, $W(|x|)$ tends to zero as $|x| \to \infty$.

It is necessary to introduce the Fourier representation of the surface profile function:

$$\zeta(x) = \int_{-\infty}^{\infty} \frac{dQ}{2\pi} \hat{\zeta}(Q) \exp(iQx). \tag{7.99}$$

The Fourier coefficient $\hat{\zeta}(Q)$ is now a random process. Since $\zeta(x)$ is a real function of x, $\hat{\zeta}(Q)$ has the property $\hat{\zeta}(-Q) = \hat{\zeta}^*(Q)$. With the aid of the Fourier inversion theorem and the results given by Eqs. (7.95a) and (7.95b) we find that

$$\langle \hat{\zeta}(Q) \rangle = 0, \tag{7.100a}$$

$$\langle \hat{\zeta}(Q)\hat{\zeta}(Q') \rangle = 2\pi\delta(Q + Q')\delta^2 g(Q). \tag{7.100b}$$

The function $g(Q)$ appearing in Eq. (7.100b) is called the *power spectrum* of the surface roughness, and is defined by

$$g(Q) = \int_{-\infty}^{\infty} dx W(|x|) \exp(-iQx).$$

$$= 2 \int_{0}^{\infty} dx\, W(|x|) \cos Qx. \tag{7.101}$$

This result shows that in fact $g(Q)$ is a real and even function of Q. It can also be shown to be a non-negative function of Q.

From the inversion formula

$$W(|x|) = \int_{-\infty}^{\infty} \frac{dQ}{2\pi} g(Q) \exp(iQx), \tag{7.102}$$

and Eq. (7.97) we see that $g(Q)$ is normalized according to

$$\int_{-\infty}^{\infty} \frac{dQ}{2\pi} g(Q) = 1. \tag{7.103}$$

In the numerical calculations whose results will be presented later in this section, we will adopt a Gaussian form for $W(|x|)$,

$$W(|x|) = \exp(-x^2/a^2), \tag{7.104}$$

where the characteristic length a is called the *transverse correlation length* of the surface roughness. It is a measure of the average distance between consecutive peaks and valleys on the surface [7.47]. The power spectrum of the surface roughness corresponding to the autocorrelation function (7.104) is

$$g(Q) = \sqrt{\pi} a \exp(-a^2 Q^2/4). \tag{7.105}$$

To obtain an analytic result for the dispersion relation of a surface plasmon polariton at a one-dimensional randomly rough vacuum-metal interface, we will make the small roughness approximation [7.40]. In this approximation the function $J(\gamma|Q)$ in Eq. (7.82), which is defined by Eq. (7.81), is expanded in powers of the surface profile function, and only the leading nonzero term in this expansion is retained:

$$J(\gamma|Q) = \hat{\zeta}^{(1)}(Q) + \frac{1}{2}\gamma\hat{\zeta}^{(2)}(Q) + \cdots, \tag{7.106}$$

where

$$\hat{\zeta}^{(1)}(Q) \equiv \hat{\zeta}(Q) \tag{7.107a}$$

$$\hat{\zeta}^{(n)}(Q) = \int_{-\infty}^{\infty} dx \exp(-iQx)\zeta^n(x), \quad n > 1. \tag{7.107b}$$

In this approximation Eq. (7.82) takes the form

$$\frac{\epsilon(\omega)\beta_0(p,\omega) + \beta(p,\omega)}{\epsilon(\omega) - 1} A(p,\omega)$$
$$= \int_{-\infty}^{\infty} \frac{dq}{2\pi} \hat{\zeta}(p-q)[pq - \beta(p,\omega)\beta_0(q,\omega)]A(q,\omega). \quad (7.108)$$

Equation (7.108) is a stochastic integral equation because of the presence of the stochastic function $\hat{\zeta}(p-q)$ in its kernel. The solution $A(q,\omega)$ therefore is also a stochastic function. Instead of seeking the probability density function of $A(q,\omega)$ we will seek its first moment $\langle A(q,\omega) \rangle$, which describes the propagation of the mean wave across the random surface.

To extract the equation satisfied by $\langle A(q,\omega) \rangle$ from Eq. (7.108) satisfied by $A(q,\omega)$ we introduce the smoothing operator P that averages everything on which it acts over the ensemble of realizations of the surface profile function: $Pf \equiv \langle f \rangle$ [7.48]. We also introduce the complementary operator Q that produces the fluctuating part of everything on which it acts: $Qf = f - \langle f \rangle$. We now apply these operators in turn to both sides of Eq. (7.108):

$$\frac{\epsilon(\omega)\beta_0(p,\omega) + \beta(p,\omega)}{\epsilon(\omega) - 1} PA(p,\omega)$$
$$= \int_{-\infty}^{\infty} \frac{dq}{2\pi} P\hat{\zeta}(p-q)[pq - \beta(p,\omega)\beta_0(q,\omega)]$$
$$\times [PA(q,\omega) + QA(q,\omega)] \quad (7.109a)$$
$$\frac{\epsilon(\omega)\beta_0(p,\omega) + \beta(p,\omega)}{\epsilon(\omega) - 1} QA(p,\omega)$$
$$= \int_{-\infty}^{\infty} \frac{dq}{2\pi} Q\hat{\zeta}(p-q)[pq - \beta(p,\omega)\beta_0(q,\omega)]$$
$$\times [PA(q,\omega) + QA(q,\omega)]. \quad (7.109b)$$

Since $P\hat{\zeta}(p-q) = 0$, Eq. (7.100a), Eq. (7.109a) can be rewritten as

$$\frac{\epsilon(\omega)\beta_0(p,\omega) + \beta(p,\omega)}{\epsilon(\omega) - 1} PA(p,\omega)$$
$$= \int_{-\infty}^{\infty} \frac{dq}{2\pi} P\hat{\zeta}(p-q)[pq - \beta(p,\omega)\beta_0(q,\omega)]QA(q,\omega). \quad (7.110)$$

We seek the right-hand side of this equation only to second order in $\hat{\zeta}(Q)$. Therefore we need to solve Eq. (7.109b) for $QA(q,\omega)$ only to first order in $\hat{\zeta}(Q)$. This solution is given by

$$\frac{\epsilon(\omega)\beta_0(q,\omega) + \beta(q,\omega)}{\epsilon(\omega) - 1} QA(q,\omega)$$
$$= \int_{-\infty}^{\infty} \frac{dr}{2\pi} \hat{\zeta}(q-r)[qr - \beta(q,\omega)\beta_0(r,\omega)]PA(r,\omega). \quad (7.111)$$

On substituting Eq. (7.111) into Eq. (7.110) we obtain the equation for $\langle A(q,\omega)\rangle$ in the form

$$[\epsilon(\omega)\beta_0(p,\omega) + \beta(p,\omega)]\langle A(p,\omega)\rangle$$
$$= [\epsilon(\omega) - 1]^2 \int_{-\infty}^{\infty} \frac{dq}{2\pi} \int_{-\infty}^{\infty} \frac{dr}{2\pi} \langle \hat{\zeta}(p-q)\hat{\zeta}(q-r)\rangle$$
$$\times \frac{[pq - \beta(p,\omega)\beta_0(q,\omega)][qr - \beta(q,\omega)\beta_0(r,\omega)]}{\epsilon(\omega)\beta_0(q,\omega) + \beta(q,\omega)} \langle A(r,\omega)\rangle$$
$$= \delta^2 [\epsilon(\omega) - 1]^2 \int_{-\infty}^{\infty} \frac{dq}{2\pi} g(p-q)$$
$$\times \frac{[pq - \beta(p,\omega)\beta_0(q,\omega)][qp - \beta(q,\omega)\beta_0(p,\omega)]}{\epsilon(\omega)\beta_0(q,\omega) + \beta(q,\omega)} \langle A(p,\omega)\rangle.$$
$$(7.112)$$

The dispersion relation for a surface plasmon polariton on a one-dimensional randomly rough vacuum-metal interface is thus found to be

$$\epsilon(\omega)\beta_0(p,\omega) + \beta(p,\omega) = \delta^2[\epsilon(\omega)-1]^2$$
$$\times \int_{-\infty}^{\infty} \frac{dq}{2\pi} g(p-q) \frac{[pq-\beta(p,\omega)\beta_0(q,\omega)][qp-\beta(q,\omega)\beta_0(p,\omega)]}{\epsilon(\omega)\beta_0(q,\omega)+\beta(q,\omega)}.$$
(7.113)

A surface plasmon polariton propagating on a planar vacuum-metal interface is attenuated by the dissipative processes in the bulk of the metal, i.e. by the processes that give rise to the imaginary part of its dielectric function. An expression for the attenuation length of the surface plasmon polariton in this case can be obtained by inserting the complex dielectric function $\epsilon(\omega) = \epsilon_1(\omega) + i\epsilon_2(\omega)$ of the metal into the dispersion relation for the surface plasmon polariton on a planar surface, $\epsilon(\omega)\beta_0(p,\omega) + \beta(p,\omega) = 0$, and obtaining the wavenumber $p(\omega)$ in the form $p(\omega) = p_1(\omega) + ip_2(\omega)$. The attenuation length for energy flow is then $\ell_{sp}(\omega) = (2p_2(\omega))^{-1}$.

In contrast, in the presence of surface roughness a surface plasmon polariton is attenuated even if the dielectric function $\epsilon(\omega)$ is real. This attenuation is due to the roughness-induced scattering of the surface plasmon polariton into volume electromagnetic waves in the vacuum traveling away from the surface, and into other surface plasmon polariton modes. Both scattering processes remove energy from the incident beam and attenuate the surface plasmon polariton thereby. This is the situation on which we wish to focus our attention. Therefore, to separate the attenuation of a surface plasmon polariton that has its origin in surface roughness from that due to dissipation we have assumed that $\epsilon(\omega)$ is real everywhere in Eq. (7.113), except in the denominator $\epsilon(\omega)\beta_0(q,\omega) + \beta(q,\omega)$ in the integrand on the right-hand side of this equation. The inclusion of an infinitesimal positive imaginary part of the dielectric function here is only for the purpose of defining the way in which the poles of the integrand that occur at the values of q at which $\epsilon(\omega)\beta_0(q,\omega) + \beta(q,\omega) = 0$ are to be treated in the evaluation of the integral over q.

If we introduce the definitions

$$F(p,\omega) = \epsilon(\omega)\beta_0(p,\omega) + \beta(p,\omega) \qquad (7.114)$$

$$G(p,\omega) = \int_{-\infty}^{\infty} \frac{dq}{2\pi} g(p-q)$$
$$\times \frac{[pq - \beta(p,\omega)\beta_0(q,\omega)][qp - \beta(q,\omega)\beta_0(p,\omega)]}{\epsilon(\omega)\beta_0(q,\omega) + \beta(q,\omega)}$$
$$= G^{(1)}(p,\omega) - iG^{(2)}(p,\omega), \qquad (7.115)$$

and denote the solution of the equation $F(p,\omega) = 0$ by $\omega_0(p)$, the frequency of a surface plasmon polariton on a planar vacuum-metal interface, then the solution of Eq. (7.113) for the frequency of a surface plasmon polariton on a randomly rough interface can be written in the form

$$\omega(p) = \omega_0(p) + \Delta(p) - i\Gamma(p), \qquad (7.116)$$

where to lowest nonzero order in δ

$$\Delta(p) = \delta^2[\epsilon(\omega_0(p)) - 1]^2 \frac{G^{(1)}(p,\omega_0(p))}{\frac{\partial}{\partial \omega}F(p,\omega)\big|_{\omega=\omega_0(p)}}, \qquad (7.117a)$$

$$\Gamma(p) = \delta^2[\epsilon(\omega_0(p)) - 1]^2 \frac{G^{(2)}(p,\omega_0(p))}{\frac{\partial}{\partial \omega}F(p,\omega)\big|_{\omega=\omega_0(p)}}. \qquad (7.117b)$$

The attenuation length of the surface plasmon polariton, $\ell_{sp}(p)$, is the distance over which the energy of the surface wave decays to $1/e$ of its initial value. It is given by

$$\ell_{sp}(p) = \frac{v_E(p)}{2\Gamma(p)}, \qquad (7.118)$$

where $v_E(p)$ is the energy transport velocity of the surface plasmon polariton. In the absence of dissipation it is equal to the group velocity of the surface plasmon polariton:

$$v_E(p) = c^2 \frac{p}{\omega_0(p)} \frac{(\frac{1}{4}\omega_p^4 + c^4p^4)^{\frac{1}{2}} - c^2p^2}{(\frac{1}{4}w_p^4 + c^4p^4)^{\frac{1}{2}}}. \qquad (7.119)$$

The inverse decay length of the electromagnetic field of the surface plasmon polariton in the vacuum region is given by the real part of

$$\beta_0(p,\omega(p)) = \left\{ p^2 - \frac{1}{c^2}[\omega_c(p) + \Delta(p)]^2 + \frac{1}{c^2}\Gamma^2(p)] \right.$$
$$\left. + i\frac{2}{c^2}[\omega_0(p) + \Delta(p)]\Gamma(p) \right\}^{\frac{1}{2}}. \qquad (7.120)$$

Numerical results for $\omega_0(p)$ and the roughness induced shift of it, $\Delta(p)$, are plotted in Fig. 7.11. The damping constant $\Gamma(p)$ and the contributions to it from the conversion of the surface plasmon polariton into volume waves in the vacuum and from its scattering into other surface plasmon polaritons are plotted in Fig. 7.12. The energy mean free path $\ell_{sp}(p)$ is presented in Fig. 7.13. The inverse decay length of the electromagnetic field of the surface plasmon polariton

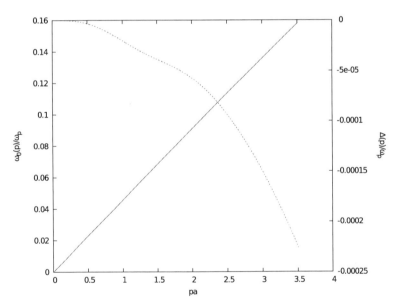

Fig. 7.11. The frequency $\omega_p(p)$ of a surface plasmon polariton, on a planar silver surface (———), and the roughness induced shift of it, $\Delta(p)$ (- - - - - -). The roughness is characterized by a Gaussian power spectrum $g(Q) = \sqrt{\pi}a\exp(-a^2Q^2/4)$ with $a = 5000$ Å, while $\delta = 500$ Å. The plasma frequency is $\omega_p = 13.12 \times 10^{15}$ s^{-1} [7.46].

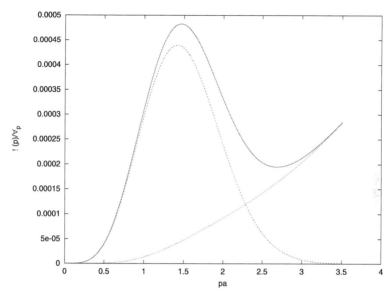

Fig. 7.12. The roughness induced damping constant, $\Gamma(p)$, of a surface plasmon polariton on a one-dimensional randomly rough silver surface (———) together with the contributions to it from the conversion of the surface plasmon into volume electromagnetic waves in the vacuum (·········) and from its scattering into other surface plasmon polaritons (- - - - -). The material and roughness parameters are those of Fig. 7.11 [7.46].

in the vacuum region in the absence and in the presence of surface roughness are shown in Fig. 7.14.

In carrying out these calculations the metal was assumed to be silver. It was characterized by a dielectric function with the free electron form, Eq. (7.64). The plasma frequency assumed was $\omega_p = 13.12 \times 10^{15}\,\text{s}^{-1}$. The values of δ and a used in the calculations were $\delta = 500\,\text{Å}$ and $a = 5000\,\text{Å}$.

The frequency shift $\Delta(p)$ is seen to be negative for all values of p, i.e. surface roughness depresses the frequency of a surface plasmon polariton below its value for a planar surface. This is similar to the depression of the frequency of the lowest branch of the dispersion curve of a surface plasmon polariton propagating on a grating by the periodic corrugation of its surface.

The damping function $\Gamma(p)$ is positive for all values of p. This means that the surface plasmon polariton is attenuated as it

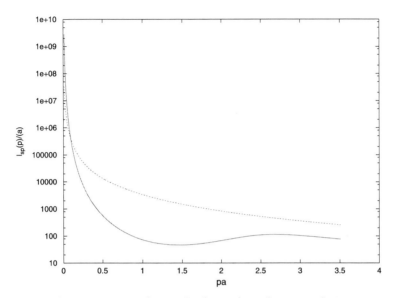

Fig. 7.13. The energy mean free path of a surface plasmon polariton on a one-dimensional randomly rough silver surface. The material and roughness parameters are those of Fig. 7.11 [7.46].

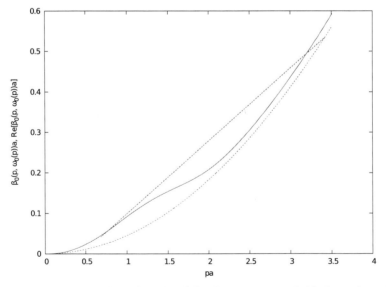

Fig. 7.14. The inverse decay lengths of the electromagnetic field of a surface plasmon polariton on a silver surface in the vacuum region in the absence ($-----$) and in the presence (———) of surface roughness. The material and roughness parameters are those of Fig. 7.11 [7.46].

propagates for all values of p. For values of p in the range $0 < pa < 2.3$ the dominant contribution to $\Gamma(p)$ is from the roughness-induced scattering of the surface plasmon polariton into other surface plasmon polaritons. For $pa > 2.3$ the dominant contribution to $\Gamma(p)$ comes from the conversion of the surface plasmon polariton into volume electromagnetic waves in the vacuum. The energy mean free path of the surface plasmon polariton due to surface roughness, $\ell_{sp}(p)$, is a decreasing function of the wavenumber p. For comparison we have also plotted in Fig. 11.13 the mean free path of a surface plasmon polariton on a planar but lossy silver surface characterized by a Drude dielectric function

$$\epsilon(\omega) = 1 - \frac{\omega_p^2}{\omega(\omega + i\gamma)}. \tag{7.121}$$

The expression for the damping function $\Gamma(p)$ in this case is

$$\Gamma(p) = \frac{\gamma}{4}\left[1 - \frac{\frac{1}{2}\omega_p^2 + c^2 p^2}{\left(\frac{1}{4}\omega_p^4 + c^4 p^4\right)^{\frac{1}{2}}}\right], \tag{7.122}$$

to lowest order in γ. The values of ω_p and γ used in this calculation were obtained by fitting the value of the dielectric function of silver at a wavelength $\lambda = 612.7$ nm, $\epsilon(\omega) = -17.2 + i0.498$ [7.49], by the expression (7.121). The values obtained in this way are $\omega_p = 13.12 \times 10^{15}\,\text{s}^{-1}$ and $\gamma = 0.8412 \times 10^{14}\,\text{s}^{-1} = 0.06411\omega_p$. It is seen that the mean free path associated with random surface roughness is smaller than that due to ohmic losses in the metal for the roughness parameters assumed.

Finally, $Re\beta_0(p,\omega(p))$ is larger than $\beta_0(p,\omega_0(p))$. Thus the surface plasmon polariton is more strongly bound to a randomly rough metal surface than it is to a planar metal surface.

Comments and Conclusions to Chapter 7

The dielectric structure supporting the electromagnetic waves studied in Sec. 7.1 is a graded-index waveguide. In the frequency range $0 < \omega < \Omega_c$ the dispersion relation (7.27a) has only a single solution that exists in a narrow spectral range whose lower edge is defined by

Eq. (7.31) and whose upper edge is given by Ω_c. Its electromagnetic field as a function of z possesses a single maximum and no nodes in the region of the waveguide, and decays to zero exponentially as $z \to \pm\infty$. There are no higher frequency branches of the dispersion curve in this frequency range. This mode thus resembles a surface plasmon polariton in its localization to the surface $z = 0$, even though the maximum of its electromagnetic field occurs inside the graded-index medium. It differs from a surface plasmon polariton in that it exists in a narrow spectral domain, which can range from the near infrared to the visible region of the optical spectrum. The technologically controlled parameter — the heterogeneity scale L — defines the narrow spectral range within which the low frequency wave exists, which is bounded from above by the critical frequency Ω_c, while the waveguide acts as a high-pass filter. Since it occurs in a dielectric structure, which has small ohmic losses, its energy mean free path can be longer than that of a surface plasmon polariton. Finally, it can be used in applications in situations, such as in an oxidizing atmosphere, where a metallic surface cannot be used.

In the frequency range $\omega > \Omega_c$ the corresponding dispersion relation, Eq. (7.38a), possesses a multiplicity of solutions, of which we have considered only the three lowest frequency modes. The electromagnetic field of the lowest frequency branch has a single maximum and no nodes in the region of the waveguide, and decays to zero exponentially as $z \to \pm\infty$. However, in this frequency range there exist higher frequency branches of the dispersion curve, whose electromagnetic fields possess nodes whose number equals the branch number, if the lowest frequency branch is denoted the zero branch. They also decay to zero exponentially as $z \to \pm\infty$. These features are characteristic of waveguide modes.

In Sec. 7.2 we have shown that electromagnetic waves of S polarization can propagate circumferentially around a portion of a cylindrical vacuum-metal interface, and be localized to it, when the homogeneous metal is concave to the vacuum. A surface electromagnetic wave of this polarization does not exist at the planar interface between vacuum and a homogeneous metal. The dispersion relation for such waves possesses several branches within the frequency region in which dielectric function of the metal is negative, and their electromagnetic fields have the nature of the fields of waveguide modes in a medium with a gradient index induced by the curvature of the vacuum-metal interface.

The dispersion curve of a surface plasmon polariton propagating on a periodically corrugated vacuum-metal interface can consist of several branches in the non-radiative region of the (ω, k) plane. The group and phase velocities of the lowest frequency branch of the dispersion curve are depressed by the periodic roughness, and the corresponding surface plasmon polariton displays the phenomenon of wave slowing. The dispersion curve of a surface plasmon polariton propagating on a randomly rough surface consists of a single branch. It also displays the phenomenon of wave slowing. The surface plasmon polariton is damped as it propagates on the randomly rough surface, and it has a shorter energy mean free path than that of a surface plasmon polariton on a planar but lossy metal surface. It is more strongly bound to the rough surface than to a planar surface. The ability to modify the dispersion curve of a surface plasmon polariton propagating on a vacuum-metal interface by structuring the interface, or to induce the existence of a surface electromagnetic wave that otherwise could not exist by structuring the interface or the system on which it propagates, can be useful in applications of these surface waves [7.50].

Bibliography

[7.1] E. Burstein, A. Hartstein, J. Schoenwald, A. A. Maradudin, D. L. Mills and R. F. Wallis, Surface polaritons-electromagnetic waves at interfaces, in *Polaritons*, eds. E. Burstein and F. de Martini (Pergamon, New York, 1974), pp. 89–110.

[7.2] J. Kane and H. Osterbeg, Optical characteristics of planar guided modes, *J. Opt. Soc. Am.* **54**, 347–356 (1964).

[7.3] H. Osterberg and L. W. Smith, Transmission of optical energy along surfaces: Part II, inhomogeneous media, *J. Opt. Soc. Am.* **54**, 1078–1084 (1964).

[7.4] P. K. Tien, Integrated optics and a new wave phenomena in optical waveguides, *Rev. Mod. Phys.* **49**, 361–420 (1977).

[7.5] T. Izawa and H. Nakagome, Optical waveguide formed by electrically induced migration of ions in glass plates, *Appl. Phys. Lett.* **21**, 584–586 (1972).

[7.6] I. P. Kaminow and J. R. Carruthers, Optical waveguiding layers in LiNbO$_3$ and LiTaO$_3$, *Appl. Phys. Lett.* **22**, 326–328 (1973).
[7.7] See, for example, L. D. Landau and E. M. Lifshitz, *Electrodynamics of Continuous Media* (Pergamon, New York, 1960), p. 235.
[7.8] D. Marcuse, TE modes of graded-index slab waveguides, *IEEE J. Quantum Electron.* **QE-9**, 1000–1006 (1973).
[7.9] P. K. Tien, S. Riva-Sanseverino, R. J. Martin, A. A. Ballman and H. Brown, Optical waveguide modes in single-crystalline LiNBO$_3$ solid solution films, *Appl. Phys. Lett.* **24**, 503–506 (1974).
[7.10] E. Conwell, Optical waveguiding in graded-index layers, *Appl. Phys. Lett.* **26**, 40–41 (1974).
[7.11] E. M. Conwell, WKB approximation for optical guide modes in a medium with exponentially varying index, *J. Appl. Phys.* **46**, 1407 (1975).
[7.12] E. M. Conwell, 'Buried modes' in planar media with nonmonotonically varying index of refraction, *IEEE J. Quantum Electron.* **QE-11**, 217–218 (1975).
[7.13] G. Stewart, C. A. Millar, P. J. R. Laybourn, C. D. W. Wilkinson and R. M. DeLarue, Planar optical waveguides formed by silver-ion migration in glass, *IEEE J. Quantum Electron.* **QE-13**, 192–200 (1977).
[7.14] E. M. Conwell, Modes in optical waveguides formed by diffusion, *Appl. Phys. Lett.* **23**, 328–329 (1973).
[7.15] D. H. Smithgall and F. W. Dabby, Graded-index planar dielectric waveguides, *IEEE J. Quantum Electron.* **QE-9**, 1023–1028 (1973).
[7.16] A. Gedeon, Comparison between rigorous theory and WKB-analysis of modes in graded-index waveguide, *Opt. Commun.* **12**, 329–332 (1974).
[7.17] Ch. Pichot, The inhomogeneous slab, a rigorous solution to the propagation problem, *Opt. Commun.* **23**, 285–288 (1977).
[7.18] J. Janta and J. Čtyroký, On the accuracy of WKB analysis of TE and TM modes in planar graded-index waveguides, *Opt. Commun.* **25**, 49–52 (1978).
[7.19] A. Shvartsburg, G. Petite and N. Auby, S-polarized surface electromagnetic waves in inhomogeneous media: exactly solvable models, *J. Opt. Soc. Am. B* **16**, 966–970 (1999).
[7.20] K. Kim, Excitation of s-polarized surface electromagnetic waves in inhomogenous dielectric media, *Opt. Express* **16**, 13354–13363 (2008).
[7.21] A. B. Shvartsburg, V. Kuzmiak and G. Petite, Optics of subwavelength gradient nanofilms, *Phys. Repts.* **452**, 33–88 (2007).
[7.22] R. M. Fitzgerald, A. A. Maradudin, J. Polanco and A. B. Shvartsburg, S-polarized guided electromagnetic waves at a planar interface between vacuum and a graded-index dielectric, (unpublished work).
[7.23] T. M. Dunster, Bessel functions of purely imaginary order, with an application to second-order linear differential equations having a large parameter, *SIAM J. Math. Anal.* **21**, 995–1018 (1990).
[7.24] M. Abramovitz and I. A. Stegun, eds., *Handbook of Mathematical Functions* (Dover, New York, 1964), p. 367, entry 9.6.24.

[7.25] Ref. [7.24], p. 375, entry 9.6.9.
[7.26] Ref. [7.24], p. 378, entry 9.7.2.
[7.27] F. Oberhettinger, *Tabellen zur Fourier Transformation* (Springer-Verlag, Heidelberg, 1957), p. 40.
[7.28] A. Gil, J. Segura and N. M. Temme, Algorithm 831: modified Bessel functions of imaginary order and positive argument, *ACM Trans. on Math. Software* **30**, 159–164 (2004).
[7.29] A. Gil, J. Segura and N. M. Temme, Algorithm 819: AIZ, BIZ: two Fortran 77 routines for the computation of complex Airy functions, *ACM Trans. On Math. Software* **28**, 325–336 (2002).
[7.30] R. S. Elliott, Azimuthal surface waves on circular cylinders, *J. Appl. Phys.* **26**, 368–376 (1955).
[7.31] K. Horiuchi, Surface wave propagation over a coated conductor with small cylindrical curvature in direction of travel, *J. Appl. Phys.* **24**, 961–962 (1953).
[7.32] M. V. Berry, Attenuation and focusing of electromagnetic surface waves rounding gentle bends, *J. Phys. A: Math. Gen.* **8**, 1952–1971 (1975).
[7.33] J. Polanco, R. M. Fitzgerald and A. A. Maradudin, Propagation of s-polarized surface polaritions circumferentially around a locally cylindrical surface, *Phys. Lett. A* **176**, 1573–1575 (2012).
[7.34] A. A. Maradudin, Interaction of surface polaritons and plasmons with surface roughness, in *Surface Polaritons*, eds. V. M. Agranovich and D. L. Mills (North-Holland, Amsterdam, 1982), pp. 405–510.
[7.35] Ref. [7.34], p. 416.
[7.36] M. Kretschmann and A. A. Maradudin, Band structures of two-dimensional surface plasmon polaritonic crystals, *Phys. Rev. B* **66**, 245408(1–8) (2002).
[7.37] D. L. Mills, Attenuation of surface polaritons by surface roughness, *Phys. Rev. B* **12**, 4036–4046 (1975).
[7.38] A. A. Maradudin and W. Zierau, Effects of surface roughness on the surface polariton dispersion relation, *Phys. Rev. B* **14**, 484–499 (1976).
[7.39] E. Kröger and E. Kretschmann, Surface plasmon and polariton dispersion at rough boundaries, *physica status solidi (b)* **76**, 515–523 (1976).
[7.40] F. Toigo, A. Marvin, V. Celli and N. R. Hill, Optical properties of rough surfaces: General theory and the small roughness limit, *Phys. Rev. B* **12**, 5618–5626 (1977).
[7.41] Lord Rayleigh, *The Theory of Sound*, Vol. II, 2nd ed. (MacMillan, London, 1896), pp. 89, 297–311.
[7.42] A. A. Maradudin, Electromagnetic surface excitations on rough surfaces, in *Electromagnetic Surface Excitations*, eds. R. F. Wallis and G. I. Stegeman (Springer-Verlag, New York, 1986), pp. 57–131.
[7.43] C. Kittel, *Introduction to Solid State Physics*, 6th ed. (John Wiley and Sons, New York, 1986), pp. 163–164, 169.
[7.44] Ref. [7.42], pp. 99–100.
[7.45] M. Sandtke and L. Kuipers, Slow guided surface plasmons at telecomm frequencies, *Nature Photonics* **1**, 573–576 (2007).

[7.46] S. Chakrabarti and A. A. Maradudin (unpublished work).
[7.47] A. A. Maradudin and T. Michel, The transverse correlation length for randomly rough surfaces, *J. Stat. Phys.* **52**, 485–501 (1990).
[7.48] U. Frisch, Wave propagaton in random media, in *Probabilistic Methods in Applied Mathematics*, ed. A. T. Bharucha-Reid (Academic Press, New York, 1968), Vol. 1, pp. 76–198.
[7.49] P. B. Johnson and R. W. Christy, Optical constants of the noble metals, *Phys. Rev. B* **6**, 4370–4379 (1972).
[7.50] Q. Gan, Y. J. Ding and F. J. Bartoli, Rainbow trapping and releasing at telecommunications wavelengths, *Phys. Rev. Lett.* **102**(1–4), 056801 (2009).

CHAPTER 8

NON-LOCAL ACOUSTIC DISPERSION OF GRADIENT SOLID LAYERS

This chapter is devoted to the physical fundamentals and mathematical basis of the theory of gradient acoustical barriers. Such barriers are formed by finite thickness layers of an inhomogeneous elastic medium with continuous distributions of density and elastic modulus in the medium inside the layer. The advent of artificial materials (metamaterials) [8.1–8.3] stimulated the development of qualitatively new concepts of gradient acoustical barriers, based on new exact analytical solutions of acoustical wave equations in heterogeneous media. This concept is being developed now in connection with the problems of sound reflection and transmission in layers of inhomogeneous alloys [8.4], composite materials [8.5] and porous structures [8.6]. The reflectance/transmittance spectra of such layers can differ drastically from the spectra of homogeneous media:

1. Gradient acoustical barriers have characteristic frequencies determined by the shape of the spatial distributions of the density and elastic properties of the barrier as well as its thickness. The influence of these frequencies on the propagation of sound waves results in a strong heterogeneity-induced non-local dispersion of reflectance/transmittance spectra of the barrier. This artificial dispersion, which can be made both normal and anomalous in a given spectral range, proves to be especially important for solids, whose natural acoustical dispersion in this range is insignificant.
2. Subject to the heterogeneity-induced dispersion of the barrier the interference of forward and backward waves can cause the peculiar

effects of weakly attenuated tunneling of longitudinal and shear acoustic waves through the barrier.
3. The exact analytical solutions of wave equations for gradient acoustic barriers, illustrating the mathematical analogies between acoustics and electromagnetics of gradient media, open the way to use some obtained results of gradient optics for analysis of related acoustical problems.

To illustrate the generalization of concepts of gradient optics for acoustics we consider the interaction of sound with gradient barriers in the simplest geometry. We assume, that a plane acoustic wave is incident from the side $z < 0$ normally on the boundary of an isotropic layer coinciding with the plane $z = 0$; another boundary of the layer is formed by the plane $z = d$. It is known, that in this configuration two acoustic waves, corresponding to longitudinal and transverse (shear) modes, can propagate in a homogeneous layer. The velocities v_ℓ and v_t of these modes as well as their wave numbers $k_{\ell,t}$ for each frequency ω are given by [8.7]:

$$v_\ell^2 = \frac{E(1-\mu)}{\rho(1+\mu)(1-2\mu)}, \quad v_t^2 = \frac{E}{2\rho(1+\mu)}, \quad k_{\ell,t} = \frac{\omega}{v_{\ell,t}}. \qquad (8.1)$$

Here E is the Young modulus, ρ is the density of the medium and μ is Poisson's ratio. Sound dispersion in medium (8.1) is absent.

Unlike (8.1), the density ρ and quantities E and μ depend in the gradient layer on the coordinate z across the layer. These dependencies can be conveniently represented by introducing dimensionless differentiable functions $F^2(z)$ and $W^2(z)$. For the density profile $\rho(z)$ we assume

$$\rho(z) = \rho_0 F^2(z), \quad \rho|_{z=0} = \rho_0, \quad F|_{z=0} = 1. \qquad (8.2)$$

Shear waves can be described by relating the function $W^2(z)$ to the coordinate-dependent shear modulus $G(z)$:

$$G(z) = G_0 W^2(z), \quad G_0 = \frac{E}{1+\mu}, \quad W|_{z=0} = 1. \qquad (8.3)$$

The values E, μ and G_0 in (8.3) correspond to the barrier boundary $z = 0$.

Theoretical problems of sound propagation in elastic media are considered based on the equations of motion relating the displacement $\vec{u}$ of particles of a medium to the components σ_{ik} of the stress tenso [8.7]:

$$\rho \frac{\partial^2 u_i}{\partial t^2} = \frac{\partial \sigma_{ik}}{\partial x_k}. \tag{8.4}$$

Here the density ρ and the tensor components σ_{ik} in gradient media depend continuously on the coordinates x_k. We will analyze by means of (8.4) two types of acoustic waves with frequency ω, propagating along the z direction:

1. Longitudinal wave, propagating with velocity v_l (8.1), and characterized by the displacement $u = u_z \exp(-i\omega t)$; in this case the right-hand side of Eq. (8.4) depends on only the component σ_{zz} of the stress tensor, which is determined as [8.7]:

$$\sigma_{zz} = \frac{E(1-\mu)}{(1+\mu)(1-2\mu)} \frac{\partial u}{\partial z}. \tag{8.5}$$

2. Transverse wave, propagating with velocity v_t (8.1), and characterized by the displacement $u = u_x \exp(-i\omega t)$, where $x \perp z$ (shear wave); in this geometry only one stress tensor component $\sigma_{xz}(z)$ has to be taken into account in Eq. (8.4):

$$\sigma_{xz} = \frac{E}{2(1+\mu)} \frac{\partial u_x}{\partial z}. \tag{8.6}$$

It has to be noted, that the local acoustical dispersion can arise in solids containing homogeneously distributed inclusions with elastic properties different from those of the host material [8.8]. The elastic moduli of such homogeneous structured materials are characterized by two constants, g and h, having the dimension of length and related to the potential and kinetic energies of inclusions in the wave field; for example, applied to the acoustics of solid porous biomaterials, this approach gives the estimate $g \approx h \approx 10^{-5}$ m [8.9]. The phase velocities of longitudinal V_l and shear V_t waves, defined in this approach, depend upon the corresponding wave numbers k_l

and k_t [8.10]

$$V_{l,t} = v_{l,t}\sqrt{\frac{1+g^2 k_{l,t}^2}{1+h^2 k_{l,t}^2}}.$$

Here $v_{l,t}$ are the phase velocities of longitudinal and shear waves in the absence of the inclusions (8.1).

In contrast, we will consider in this chapter the non-local acoustical dispersion in heterogeneous elastic solids, which is distinguished in principle from the local effect, described by (8.4). This non-local effect can be exemplified, e.g. in the acoustics of concentrationally graded alloys, where the concentration of components depends continuously upon the coordinates. This dependence determines the spatial distribution of the density and elastic properties inside the alloy. Thus, in the simple case of a normal stress being applied along the slab of a binary alloy, a weighted mean between the Young moduli of the two components is [8.11, 8.12]:

$$E_c = E_1 V_1 + E_2 V_2. \tag{8.7}$$

Here $E_{1,2}$ and $V_{1,2}$ are the Young modulus and volume fraction of each of two components. The alloy's layer with a technologically controlled spatial distribution of volume fractions, e.g. the layer with a one-dimensional distribution of V_1 across the layer in the z direction ($V_1 = V_1(z)$), can be exemplified as the gradient acoustic barrier $E_c(z)$. The reflectance and transmittance spectra of acoustic waves in gradient barriers can have a strong frequency dispersion produced in the required wavelength range by means of specially selected spatial distributions of the density $F^2(z)$ or of the elastic properties $W^2(z)$ across the barrier. We will consider distributions $F^2(z)$ and $W^2(z)$, for which the wave field inside the barrier is described by exact analytical solutions of Eq. (2.3). The reflectance/transmittance spectra are calculated from the continuity conditions for displacements and stresses at the barrier boundaries. For the normal incidence of waves on the boundary $z = 0$, these conditions may be written

a. as the equality of displacements u_i: $u_i|_{z=-0} = u_i|_{z=+0}$; (8.8)

b. as the equality of normal stresses σ_{iz}: $\sigma_{iz}|_{z=-0} = \sigma_{iz}|_{z=+0}$. (8.9)

This chapter is devoted to the reflectance spectra of gradient acoustical barriers in the case of normal incidence of both longitudinal and shear waves. The spectra obtained are based on the exact analytical solutions of the wave equations for gradient solid barriers, obtained without the use of any assumptions about the smallness or slowness of the variations of fields or media. For simplicity we assume below that the elastic media on the left and right of the barrier are identical. The expressions for spectra include the contributions to the reflection of sound caused not only by the difference of acoustical impedances, but also by gradients and curvatures of spatial distributions of density and elasticity inside the barrier; this analysis can be viewed as the parallel counterpart of an approach developed in gradient optics (Sec. 8.1). The reflectance spectra for the solid barrier with a heterogeneous distribution of density and homogeneous elastic properties is considered in Sec. 8.1 in the framework of this approach. The opposite situation (variable elastic properties and constant density) is analyzed in Sec. 8.2 by means of the special "auxiliary barrier" method. Although such a separation of medium properties is conventional, it allows choosing the approach to the design of gradient acoustic materials with specified reflectance/transmittance spectra. In contrast, the examples of gradient media with "consistent" spatial variations of density and elasticity are considered in Sec. 8.3.

8.1. Gradient Acoustic Barrier with Variable Density: Reflectance/Transmittance Spectra of Longitudinal Sound Waves

Propagation of a longitudinal sound wave incident along the direction z normally on a variable-density layer can be examined using the equation of motion (8.4), assuming, that $u = u_z$, $\rho(z) = \rho_0 F^2(z)$, $W = 1$. Substitution of the value σ_{zz} (8.5) to (8.4) yields the equation governing the displacement u:

$$\frac{d^2 u}{dz^2} + \frac{\omega^2}{v_0^2} F^2(z) u = 0. \tag{8.10}$$

Here $v_0 = v_l$ is the longitudinal sound velocity, defined at the plane $z = 0$ by Eq. (8.1) with $\rho = \rho_0$.

Equation (8.10) resembles the Eq. (2.9), used in gradient optics, and we can consider the function $F^2(z)$, coinciding with $U(z)^2$ (2.16), assuming the density distribution to have the form

$$\rho(z) = \rho_0 \left(1 + \frac{s_1 z}{L_1} + \frac{s_2 z^2}{L_2^2}\right)^{-2}. \qquad (8.11)$$

Here the spatial scales of heterogeneity L_1 and L_2 are the free parameters of the distribution (8.11). The minimum and maximum values of the density as well as the parameters L_1 and L_2 are determined via the density ρ_0 and barrier's width d,

$$\rho_{\min,\max} = \frac{\rho_0}{(1 \pm y^2)^2}, \quad y = \frac{L_2}{2L_1}, \quad L_2 = \frac{d}{2y}, \quad L_1 = \frac{d}{4y^2}. \qquad (8.12)$$

Recalling the exact analytical solution of Eq. (8.10), obtained in Sec. 2.2, we can write

$$u = \frac{\exp(iq\eta) + Q\exp(-iq\eta)}{\sqrt{F(z)}}, \quad \eta = \int_0^z F(z_1)dz_1, \quad q = \frac{\omega}{v_0} N_\pm. \qquad (8.13)$$

Effective refractive indices N are different for positive (N_+) and negative (N_-) dispersion, corresponding, accordingly to Sec. 2.1, to the concave and convex profiles (8.11)

$$N_\pm = \sqrt{1 \pm S^2}, \quad S = \frac{\Omega_\pm}{\omega}, \quad \Omega_\pm = \frac{v_0}{d}\theta_\pm, \quad \theta_\pm = 2y\sqrt{1 \mp y^2}. \qquad (8.14)$$

While using here the exactly solvable model (8.11), initially introduced in optics, we designate the ratio Ω/ω as S in order to avoid any confusion with optics, where this ratio is designated as u (2.19). The characteristic frequencies $\Omega_\pm$ are expressed in (8.14) via the parameter d/v_0, indicating the travel time of a wave, travelling with velocity v_0 through a distance d, and dimensionless form factors $\theta_\pm$, dependent on the shape of the gradient profile. This representation of the characteristic frequencies will be used below for different barriers, distinguished by expressions for the form factors $\theta_\pm$.

The reflection spectrum under discussion is calculated from the continuity conditions (8.8) and (8.9) at the boundaries of the layer $z = 0$ and $z = d$. Representing a longitudinal wave incident from a homogeneous medium $z \leq 0$ with density ρ_1 and wave velocity v_1 on the layer boundary $z = 0$ in the form $u = A_i \exp[i\omega(z/v_1 - t)]$, let us consider, e.g. the concave profile $(N = N_-)$; introducing the complex reflection coefficient, we can write the boundary conditions as

$$A_i(1+R) = A_r(1+Q), \qquad (8.15)$$

$$i\omega\rho_1 v_1(1-R)A_i = A_r \rho_0 v_0^2 \left[-\frac{1+Q}{2L_1} + iq(1-Q) \right]. \qquad (8.16)$$

From (8.15) and (8.16) we find:

$$R = \frac{i\alpha - \frac{\gamma}{2} - iN_-\Lambda}{i\alpha + \frac{\gamma}{2} + iN_-\Lambda}, \quad \Lambda = \frac{1-Q}{1+Q}. \qquad (8.17)$$

The parameter α in (8.17) is the ratio of the acoustic impedances $I_{1,2}$ of the adjacent media $(I = \rho v)$, γ is a dimensionless parameter

$$\alpha = \frac{\rho_1 v_1}{\rho_0 v_0}, \quad \gamma = \frac{v_0}{\omega L_1} = \frac{2Sy}{\sqrt{1+y^2}}. \qquad (8.18)$$

Assuming for simplicity that the medium in the region $z \geq d$ is the same as in the region $z \leq 0$ we can find the quantity Q from the continuity conditions on the boundary $z = d$:

$$Q = \exp(2iq\eta_0) \left[\frac{N_- + \frac{i\gamma}{2} - \alpha}{N_- - \frac{i\gamma}{2} + \alpha} \right], \quad \eta_0 = \eta(d). \qquad (8.19)$$

Substitution of Q from (8.19) into (8.17) yields the expression for complex reflection coefficient of the gradient acoustic barrier, presented in a form similar to (2.31):

$$R = \frac{\sigma_1 + i\sigma_2}{\chi_1 + i\chi_2}, \qquad (8.20)$$

$$\sigma_1 = t\left(\alpha^2 + \frac{\gamma^2}{4} - N_-^2\right) - \gamma N_-, \quad \sigma_2 = 0,$$

$$\chi_1 = t\left(\alpha^2 - \frac{\gamma^2}{4} + N_-^2\right) + \gamma N_-, \quad \chi_2 = 2\alpha\left(N_- - \frac{\gamma t}{2}\right), \quad (8.21)$$

$$t = \operatorname{tg}(q\eta_0), \quad q\eta_0 = \frac{N_-}{S}\ln\left(\frac{y_+}{y_-}\right), \quad y = \sqrt{\sqrt{\frac{\rho_0}{\rho_{\min}}} - 1}.$$

Formulae (8.20) and (8.21) solve the problem of the reflection of a longitudinal sound wave from the gradient layer (8.11) with the concave density profile; here the function $q = q(\omega)$ corresponds to the negative dispersion. The similar problem, related to the reflectance of a layer with a convex profile $\rho(z)$, described by model (8.11) with $s_1 = 1$, $s_2 = -1$ (positive dispersion), is solved by the analogous algorithm. Reflection spectra calculated in this way can be written due to the following replacements in (8.21):

$$N_- \to N_+, \quad \gamma \to -\frac{2Sy}{\sqrt{1-y^2}}, \quad (8.22)$$

$$t = \operatorname{tg}(q\eta_0), \quad q\eta_0 = \frac{2N_+}{S}\operatorname{arctg}\left(\frac{y}{\sqrt{1-y^2}}\right), \quad y = \sqrt{1 - \sqrt{\frac{\rho_0}{\rho_{\max}}}}.$$
$$(8.23)$$

Reflectance spectra for longitudinal waves $|R(S)|^2$, are shown in Fig. 8.1. for negative (Figs. 8.1(a) and 8.1(b)) and positive (Fig. 8.1(c)) dispersion. These graphs illustrate the controlled flexibility of the spectra $|R(S)|^2$ subject to the ratio of impedances and depth of density modulation in the gradient acoustical barrier, determined by the parameters α (8.18) and y (8.12), respectively.

8.2. Heterogeneous Elastic Layers: "Auxiliary Barrier" Method

The reflection of shear waves from a medium with spatially distributed density (8.2) and shear modulus (8.3) can be studied using the equation of motion (8.4). In the case of normal incidence of shear wave on the layer boundary $z = 0$ the only component of the stress tensor entering (8.4) can be represented by means of the

Non-Local Acoustic Dispersion of Gradient Solid Layers 211

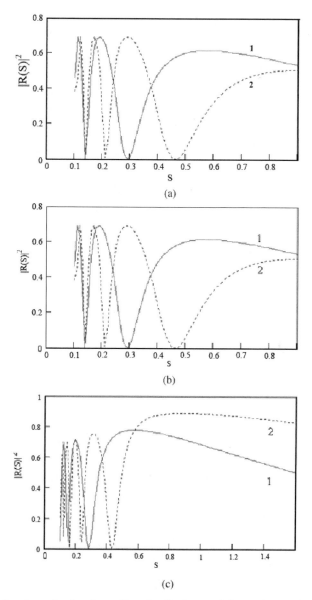

Fig. 8.1. Spectra of reflection of longitudinal sound from the gradient barrier, described by Eq. (8.11); (a) and (b) correspond to the normal non-local dispersion; the ratios of impedances of barrier and surrounding media are $\alpha = 0.3$ and $\alpha = 1.25$, respectively; (c) relates to the anomalous non-local dispersion, $\alpha = 0.3$; curves 1 and 2 on all graphs correspond to the values of the parameter $y = 0.45$ and $y = 0.7$, respectively.

dimensionless function $W^2(z)$ in the form

$$\sigma_{xz}(z) = \frac{E}{2(1+\mu)} \frac{du_x}{dz} W^2(z). \tag{8.24}$$

Substituting (8.24) into the equation of motion (8.4), taking into account (8.2), and designating $u_x = u$, we can rewrite (8.4) as

$$W^2(z)\frac{\partial^2 u}{\partial z^2} + \frac{\omega^2}{v_0^2} F^2(z)u + 2WW_z \frac{\partial u}{\partial z} = 0. \tag{8.25}$$

Here $W_z = \frac{\partial W}{\partial z}$ and $v_0 = v_t$, where v_t, defined in (8.1), is the shear wave velocity on the layer boundary. The choice of model functions $F^2(z)$ and $W^2(z)$ in Eq. (8.25) is limited so far only by conditions $F^2(0) = W^2(0) = 1$.

To separate the effects caused by the distribution $W^2(z)$, we assume that the medium density is independent of the coordinates ($\rho = \rho_0$, $F = 1$); in this case Eq. (8.25) takes the form

$$\frac{d^2 u}{dz^2} + \frac{\omega^2}{v_0^2} \frac{u}{W^2(z)} = -\frac{2W_z}{W} \frac{du}{dz}. \tag{8.26}$$

Equation (8.26) differs in its right-hand side from Eq. (8.10), used in the problem of wave propagation through the variable-density medium, and is therefore solved using a special algorithm based on the auxiliary barrier method [8.13]. This method involves the following steps:

1. Differentiation with respect to z in (8.26) is replaced by differentiation with respect to a new variable η, which is now, unlike (8.13), defined by the relation

$$d\eta = \frac{dz}{W^2(z)}. \tag{8.27}$$

Passing to the variable η removes the right-hand side in Eq. (8.26):

$$\frac{d^2 u}{d\eta^2} + \frac{\omega^2}{v_0^2} W^2(z) u = 0. \tag{8.28}$$

The function u in Eq. (8.28) depends on two variables, z and η. To solve this equation it is necessary to specify the function $W^2(z)$ and express it in terms of η. In particular, Eq. (8.28) is reduced to Eq. (8.10) solved previously by introducing an auxiliary barrier $F^2(\eta)$ in the η space:

$$W^2(z) = F^2(\eta). \tag{8.29}$$

The function $F^2(\eta)$ in (8.29) can be chosen arbitrarily. However, if it is taken in the form (2.16) with z replaced by η, we can use the ready-made solution (8.13). We can write the function $F^2(\eta)$, corresponding, for example, to the convex profile ($s_1 = -1$, $s_2 = 1$), in the form

$$F^2(x) = (1 - 2yx + x^2)^{-2}, \quad x = \frac{\eta}{L_2}. \tag{8.30}$$

The characteristic lengths L_1 and L_2 as well as the parameter y in (8.30) are unknown.

2. Substituting expressions (8.29) and (8.30) into (8.27) and using the condition $\eta|_{z=0} = 0$, following from Eqs. (8.29) and (8.30), we can find the dependence of z upon x by integrating (8.27)

$$\frac{z(x)}{L_2} = \frac{1}{2(1-y^2)^{\frac{3}{2}}} \left\{ \arctg\left(\frac{x-y}{\sqrt{1-y^2}}\right) + \arctg\left(\frac{y}{\sqrt{1-y^2}}\right) \right.$$
$$\left. + \sqrt{1-y^2}\left(y + \frac{x-y}{1-y^2 + (x-y)^2}\right) \right\} \tag{8.31}$$

To find y in (8.31), we note, that according to (8.29), the convex profile $F^2(x)$ corresponds to the convex profile $W^2(z)$ and the maximum of the convex profile $F^2_{\max} > 1$ corresponds to the maximum of profile $W^2_{\max} = F^2_{\max}$. Substituting the value $F^2_{\max} = (1-y^2)^{-2}$, we find:

$$y = \sqrt{1 - \frac{1}{W_{\max}}}. \tag{8.32}$$

The parameter x in (8.31) can be easily found by substituting (8.30) into (8.29), solving the resulting equation for x, and replacing y by means of (8.32):

$$x(W) = \sqrt{1 - \frac{1}{W_{\max}}} \pm \sqrt{\frac{1}{W} - \frac{1}{W_{\max}}}. \qquad (8.33)$$

Expressions (8.31)–(8.33) determine implicitly the coordinate dependence of the shear modulus inside the barrier $W^2(z)$; as follows from (8.30) the variable x ranges within the interval $0 \leq x \leq 2y$. In this case $z(0) = 1$, and the barrier width d, determined by the distance between the points where $W(0) = 1$ and $W(2y) = 1$, are related to the characteristic size L_2:

$$d = L_2 z(2y) = 2L_2 B_1, \qquad (8.34)$$

$$B_1 = \frac{1}{(1-y^2)^{\frac{3}{2}}} \left[y\sqrt{1-y^2} + \mathrm{arctg}\left(\frac{y}{\sqrt{1-y^2}}\right) \right]. \qquad (8.35)$$

Thus, knowing the width d and the height $W_{\max}^2$ of the barrier $W^2(z)$, specified implicitly, we can find the spatial scales L_1 and L_2 of the auxiliary barrier $F^2(\eta)$, specified explicitly in (8.30). The height and width of the auxiliary barrier, $W_{\max}^2$ and d_1, as well as the characteristic lengths L_1 and L_2, are expressed in terms of the width d and parameter y:

$$d_1 = 2yL_2 = \frac{yd}{B_1}, \quad L_1 = \frac{d}{4yB_1}, \quad L_2 = \frac{d}{2y}. \qquad (8.36)$$

The convex barrier $W^2(z)$ and corresponding auxiliary barrier $F^2(x)$ are shown in Fig. 8.2(a).

3. To calculate the reflection coefficient of the barrier $W^2(z)$, one has to find the field inside the barrier described by Eq. (8.28). Under the condition (8.29) Eq. (8.28) coincides formally with (8.10). Introducing the variable

$$\tau(\eta) = \int_0^\eta F(\eta_1) d\eta_1, \qquad (8.37)$$

we can write the solution of Eq. (8.28) in the form, similar to (8.13),

$$u = \frac{A_r [\exp(iq\tau) + Q \exp(-iq\tau)]}{\sqrt{F(\eta)}}. \qquad (8.38)$$

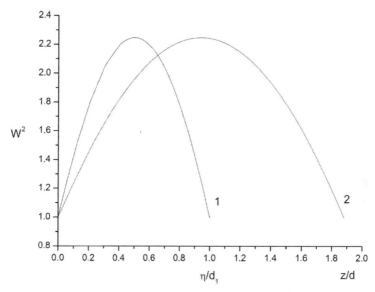

Fig. 8.2. Gradient barriers formed by parametrically specified shear modulus distributions (curves 2) and the corresponding auxiliary barriers (curves 1); (a) and (b) correspond to the convex and concave barriers.

Continuing this analogy we can calculate the reflection coefficient R for the heterogeneity of shear modulus inside the barrier (8.31). This coefficient is expressed by the same formula (8.21) as the reflection coefficient, related to the heterogeneity of density $F^2(z)$ (8.11). In this case the characteristic frequency Ω_+, entering the parameter S, differs from (8.14) by the form factor

$$\theta_+ = \frac{2}{1-y^2}\left[y\sqrt{1-y^2} + \text{arctg}\left(\frac{y}{\sqrt{1-y^2}}\right)\right]. \quad (8.39)$$

4. Reflection from the concave profile $W^2(z)$, characterized by the minimum $W_{\min}$, can be studied by choosing the concave profile of the auxiliary barrier $F^2(x)$, with $s_1 = 1$, $s_2 = -1$. Repeating the analysis in (8.29)–(8.32), we find the parameter y

$$y = \sqrt{W_{\min}^{-1} - 1}, \quad (8.40)$$

and the implicit expression for the gradient shear modulus profile inside the barrier:

$$\frac{z(x)}{L_2} = \frac{1}{(1+y^2)^{\frac{3}{2}}} \left\{ \text{arcth}\left(\frac{x-y}{\sqrt{1+y^2}}\right) + \text{arcth}\left(\frac{y}{\sqrt{1+y^2}}\right) \right.$$
$$\left. + \sqrt{1+y^2}\left[y + \frac{x-y}{1+y^2-(x-y)^2}\right] \right\}; \qquad (8.41)$$

$$x(W) = \sqrt{\frac{1}{W_{\min}} - 1} \pm \sqrt{\frac{1}{W_{\min}} - \frac{1}{W}}. \qquad (8.42)$$

The auxiliary barrier width d_2 and characteristic lengths L_1 and L_2 can be expressed, similarly to (8.36), in terms of the width d of the barrier $W^2(z)$ and parameter y (8.40):

$$d_2 = \frac{yd}{B_2}, \quad L_1 = \frac{d}{4yB_2}, \quad L_2 = \frac{d}{2y}, \qquad (8.43)$$

$$B_2 = \frac{1}{(1+y^2)^{\frac{3}{2}}}\left[y\sqrt{1+y^2}\right] + \text{arcth}\left(\frac{y}{\sqrt{1+y^2}}\right). \qquad (8.44)$$

The concave barrier $W^2(z)$ and the corresponding auxiliary barrier (8.41), characterized by negative dispersion, are shown on Fig. 8.2(b).

The reflection coefficient for the concave barrier is calculated from expressions (8.20) and (8.21), where the parameter y is defined in (8.40) and the frequency Ω_- is given by (8.14) but with a different form factor $\theta_-(y)$:

$$\theta_-(y) = \frac{2}{1+y^2}\left[y\sqrt{1+y^2} + \text{arcth}\left(\frac{y}{\sqrt{1+y^2}}\right)\right]. \qquad (8.45)$$

a. The main results in this section are the expressions for reflection coefficients for longitudinal and shear waves, reflected from gradient wave barriers formed by spatial distributions of the density and elastic properties. As the heterogeneity is weakening ($L_1 \to \infty, L_2 \to \infty$), the parameters y, γ and the characteristic frequencies Ω_+ and Ω_- tend to zero, while expression (8.20) is reduced to the well-known formula for the reflection of normally incident sound from a homogeneous layer:

$$R = \frac{\mathrm{tg}\delta(\alpha^2 - 1)}{\mathrm{tg}\delta(\alpha^2 + 1) + 2i\alpha}, \quad \delta = \frac{\omega d}{v_0}. \tag{8.46}$$

b. It is remarkable, that the analysis of sound reflection from gradient barriers involves the characteristic frequencies Ω_+ and Ω_- determined by the travel times of waves with velocity v_0 through the gradient barrier with width d and by the geometric parameters $\theta_\pm$ of the layer. These frequencies enter to the expressions for $N_\pm$ (8.14), whose structure resembles that of refractive indices in the electrodynamics of dielectrics with positive and negative dispersion.

Nonlocal artificial dispersion, formed by the geometric parameters of the barrier allows selecting the spectral range for the specified frequency band far from the absorption band of the acoustic medium.

c. Within this unified approach the auxiliary barrier method reveals the similarity and difference of reflection spectra caused by physically different gradient structures (for example, heterogeneties of the density and elastic parameters of the medium). In the framework of this approach the reflection spectra of acoustic waves reflected from barriers with negative and positive non-local dispersion are described by the general expressions (8.20) and (8.21), which are valid after the substitution of the corresponding values of the parameters y, $N_\pm$, and the form-factors $\theta_{1,2}(y)$. This generality can be extended, as is shown in Sec. 8.3, to some other classes of gradient acoustic barriers, formed by the common action of heterogeneities of density and elasticity.

8.3. Double Acoustic Barriers: Combined Effects of Gradient Elasticity and Density

Unlike the sound dispersion, caused by either the density distribution $F^2(z)$, or the elastic parameter distribution $W^2(z)$, considered in Secs. 8.1 and 8.2, the dispersion of gradient barriers discussed here depends upon the spatial distributions of both the density and elastic parameters simultaneously. Such combined dependences attract attention in the acoustics of organic materials [8.14], glasses [8.15], composite and granulated metamaterials [8.16, 8.17]. Combined action of these mechanisms leads to competing dispersion effects in the sound reflectance/transmittance spectra of gradient barriers. Since both these effects are simultaneously manifested in one barrier, we can speak about "double" barriers and their complicated spectra. Some specific features of the formation of these spectra can be distinguished by consideration of two qualitatively different problems:

a. finding the reflection spectrum of barrier in which the changes in $F^2(z)$ and $W^2(z)$ inside the barrier are described by different functions;
b. finding the spectral characteristics of a gradient barrier in which the distributions $F^2(z)$ and $W^2(z)$ are equal.

Simple examples of such spectra are examined below.

1. "Double" gradient barrier: interplay of positive and negative dispersion.

We consider a shear wave inside the gradient layer described by Eq. (8.25) and introduce a new variable η by means of formula (8.27). Then Eq. (8.25) takes the form

$$\frac{d^2u}{d\eta^2} + \frac{\omega^2}{v_0^2}F^2(z)W^2(z)u = 0. \tag{8.47}$$

By describing the distributions $F^2(z)$ and $W^2(z)$ inside the barrier of width d with the help of characteristic lengths l_1 and l_2

$$W(z) = 1 + \frac{z}{l_1}, \quad F(z) = \left(1 + \frac{z}{l_2}\right)^{-1}. \tag{8.48}$$

we can study the effects caused by the increase or decrease in the density and elastic parameters inside the double barrier in a general form, considering both positive and negative values of the lengths l_1 and l_2 independently. To distinguish these lengths, related to models of different physical quantities, from the lengths L_1 and L_2, characterizing the distribution of one quantity, e.g. the density in model (8.11), the lowercase letters are used in models (8.48).

Substitution of the function $W^2(z)$ (8.48) into (8.27) yields the explicit expression of the variable η via z:

$$\eta = z\left(1 + \frac{z}{l_1}\right)^{-1}. \tag{8.49}$$

Owing to (8.49) the product of functions $W(z)$ and $F(z)$ reads as a function of η:

$$F(z)W(z) = U(\eta) = \left(1 + \frac{\eta}{l}\right)^{-1}, \tag{8.50}$$

$$l = \frac{l_1 l_2}{l_1 - l_2}. \tag{8.51}$$

By substituting (8.50) into (8.47), we can rewrite this equation in η-space in a form similar to (8.10):

$$\frac{d^2u}{d\eta^2} + \frac{\omega^2}{v_0^2}U^2(\eta)u = 0. \tag{8.52}$$

This equation is simple to solve using the algorithm, that was already applied in Secs. 8.1 and 8.2. Introducing the new variable τ

$$\tau = \int_0^\eta U(\eta_1)d\eta_1 = l\ln\left[\frac{l_1(z + l_2)}{l_2(z + l_1)}\right], \tag{8.53}$$

we can represent the solution of Eq. (8.52) in the form of forward and backward waves, traveling along the τ-axis

$$u = \frac{A_r[\exp(iq\tau) + Q\exp(-iq\tau)]}{\sqrt{U(\eta)}}. \tag{8.54}$$

The wave number q in (8.54) corresponds to the plasma-like dispersion of the gradient layer

$$q = \frac{\omega}{v_0}\sqrt{1 - \frac{\Omega^2}{\omega^2}}, \quad \Omega = \frac{v_0}{2|l|}. \tag{8.55}$$

The characteristic frequency Ω in (8.55) depends via the parameter l on the spatial scales of variations of the density and elastic properties l_1 and l_2. Taking into account the boundary conditions (8.8) and (8.9) at the barrier boundary $\eta = 0$ ($z = 0$), we can write the expression for the reflection coefficient R in a form resembling (8.17). However, the parameter Q, describing the contribution of the backward wave to the field inside the barrier, unlike the one presented in (8.19), is asymmetric: $U(\eta = 0) \neq U(\eta_0)$ where the coordinate η_0 corresponds to the far boundary of the barrier $z = d$,

$$\eta_0 = \eta(d) = d\left(1 + \frac{d}{l_1}\right)^{-1}. \quad U_0 = U(\eta_0) = \frac{l_1(d+l_2)}{l_2(d+l_1)}. \quad (8.56)$$

Designating the coordinate τ (8.53), corresponding to the far boundary of the barrier $z = d$ as

$$\tau_0 = \tau(d) = -l \ln U_0, \quad (8.57)$$

and using the relations

$$\frac{d\eta}{dz} = \frac{1}{W^2(z)}, \quad \frac{d\tau}{dz} = \frac{F(z)}{W(z)}, \quad (8.58)$$

following from distributions (8.48), we write the continuity conditions for displacements and stresses at this boundary $\tau = \tau_0$:

$$\frac{A_r[\exp(iq\tau_0) + Q\exp(-iq\tau_0)]}{\sqrt{U_0}} = A_2, \quad (8.59)$$

$$\frac{A_r\sqrt{U_0}}{W^2(d)}\left\{\frac{\gamma}{2}[\exp(iq\tau_0) + Q\exp(-iq\tau_0)]\right.$$

$$\left. + iN_-U_0[\exp(iq\tau_0) - Q\exp(-iq\tau_0)]\right\} = i\alpha A_2. \quad (8.60)$$

Manipulations of Eqs. (8.59) and (8.60) and use of Eq. (8.56) give the value of parameter Q

$$Q = -\exp(2iq\tau_0)\left(\frac{\alpha\beta + \frac{i\gamma}{2} - N_-U_0}{\alpha\beta + \frac{i\gamma}{2} + N_-U_0}\right), \quad (8.61)$$

$$\beta = \left(1 + \frac{d}{l_1}\right)\left(1 + \frac{d}{l_2}\right). \quad (8.62)$$

Here A_2 is the amplitude of the transmitted wave, α is the ratio of impedances (8.18); it is assumed, for simplicity, that the densities and elastic parameters of the media surrounding the barrier are equal. After substitution of Q from (8.61) into expression (8.17), the formula for the complex reflection coefficient R can be presented in the standard form (8.20), where:

$$\sigma_1 = t\left(\alpha^2\beta - \frac{\gamma^2}{4} - N_-^2 U_0\right) + \frac{\gamma N_-}{2}(1 - U_0),$$

$$\sigma_2 = \alpha\left[\frac{\gamma t}{2}(1+\beta) + N_-(U_0 - \beta)\right],$$

$$\chi_1 = t\left(\alpha^2\beta + \frac{\gamma^2}{4} + N_-^2 U_0\right) - \frac{\gamma N_-}{2}(1 - U_0), \qquad (8.63)$$

$$\chi_2 = \alpha\left[\frac{\gamma t}{2}(1-\beta) + N_-(U_0 + \beta)\right],$$

$$t = \text{tg}(q\tau_0), \quad q\tau_0 = -\frac{N}{2S}\ln U_0.$$

Expressions (8.63), determining R, are written for positive values of the parameter l (8.51). It can be seen from (8.51), that the condition $l > 0$ becomes possible for three density and shear modulus profiles:

1. $l_1 > l_2$; $l_1 > 0$; $l_2 > 0$.
2. $l_1 > l_2$; $l_1 < 0$; $l_2 < 0$. (8.64)
3. $l_1 < l_2$; $l_1 < 0$; $l_2 > 0$.

Each of these combinations (1–3), determining the spatial structure of the gradient layer, corresponds, via the value of the parameter β (8.62), to its own reflection coefficient.

2. Let us consider now an another example of a "double" barrier, where the distributions of density and elastic parameters are characterized, unlike (8.48), by coinciding normalized distributions $F^2(z) = W^2(z)$. In this case Eq. (8.47) takes the form

$$\frac{d^2 u}{d\eta^2} + \frac{\omega^2}{v_0^2} F^4(z) u = 0. \qquad (8.65)$$

Equation (8.65) can be readily solved by the "auxiliary barrier" method, developed in Sec. 8.2.

We first consider a convex profile $F(z)$, containing two free parameters: the characteristic length L and the dimensionless parameter M:

$$F(z) = \cos\left(\frac{z}{L}\right) + M\sin\left(\frac{z}{L}\right), \quad 0 \leq \frac{z}{L} \leq \pi. \qquad (8.66)$$

The value of η can be found by substituting the function $W^2 = F^2$ into (8.27):

$$\eta = \frac{Lt}{1 + Mt}, \quad t = \text{tg}\left(\frac{z}{L}\right). \qquad (8.67)$$

Using (8.67), we can express $F^2(z)$ in terms of η,

$$F^2(z) = \left[1 - \frac{2M}{L}\eta + \frac{(1+M^2)}{L^2}\eta^2\right]^{-1}. \qquad (8.68)$$

It is important, that the function $F^2(z)$, written in the form (8.68), is similar to the frequently used model $F(\eta)$ (8.11), if we set in (8.11) $s_1 = -1$, $s_2 = 1$ and find the characteristic lengths L_1, L_2 and parameter y of model (8.11) by comparing it with (8.68):

$$\frac{2M}{L} = \frac{1}{L_1}, \quad \frac{(1+M^2)}{L^2} = \frac{1}{L_2^2}, \quad y = \frac{L_2}{2L_1} = \frac{M}{\sqrt{1+M^2}} < 1. \qquad (8.69)$$

The function $F^2(\eta)$ thus defined forms an auxiliary convex barrier allowing the representation of Eq. (8.65) in a form that coincides with (8.10)

$$\frac{d^2u}{d\eta^2} + \frac{\omega^2}{v_0^2}F^2(\eta)u = 0. \qquad (8.70)$$

The maxima of the "auxiliary barrier" $F(\eta)$ and of the barrier $F^2(z)$, located in η- and z-spaces, respectively, are equal: $F_{\max}^2(z) = F_{\max}(\eta) = 1 + M^2$. The width d of the symmetrical barrier in z-space, determined from the condition $F(0) = F(d) = 1$, is given by

$$d = 2L\,\text{arctg}(M). \qquad (8.71)$$

The width d_1 of the auxiliary barrier in η- space is determined from the condition (8.12)

$$d_1 = 2yL_2 = \frac{2ML}{1+M^2}. \tag{8.72}$$

By reducing Eq. (8.65) for the convex barrier to the form (8.70), coincident formally with (8.10), we can use the solution (8.13) to obtain the reflection coefficient of the "double" gradient barrier, given by distributions $F^2(z) = W^2(z)$, in a form (8.20), where the characteristic frequency $\Omega = \Omega_+$ and the phase shift of reflected wave $q\eta_0$ (2.31) are given by:

$$\Omega_+ = \frac{v_0}{d}\theta_+, \quad \theta_+ = \frac{2M}{1+M^2}, \quad q\eta_0 = \frac{2N_+ \operatorname{arctg}(M)}{S}. \tag{8.73}$$

The "auxiliary barrier" method also allows finding the reflectance spectrum of a concave profile, containing two free parameters L and M as in (8.66), under the same condition $F^2(z) = W^2(z)$:

$$F(z) = \operatorname{ch}\left(\frac{z}{L}\right) - M\operatorname{sh}\left(\frac{z}{L}\right) = W(z), \quad 0 \leq M < 1. \tag{8.74}$$

Using the algorithm developed in (8.67)–(8.70) for the convex profile and substituting (8.74) into (8.27), we introduce the new variable η

$$\eta = \frac{Lt}{1 - Mt}, \quad t = \operatorname{th}\left(\frac{z}{L}\right). \tag{8.75}$$

Expressing the function $F^2(z)$ in terms of η, we obtain the concave profile of the auxiliary barrier

$$F^2(z) = \left[1 + \frac{2M}{L}\eta - \frac{(1-M^2)}{L^2}\eta^2\right]^{-1}. \tag{8.76}$$

Profile (8.76) coincides with the model (8.11), if the characteristic lengths L_1 and L_2 and the parameter y of model (8.11) are defined by the expressions

$$\frac{2M}{L} = \frac{1}{L_1}, \quad \frac{1-M^2}{L^2} = \frac{1}{L_2^2}, \quad y = \frac{M}{\sqrt{1-M^2}}. \tag{8.77}$$

The widths d of barrier (8.74) and d_1 of auxiliary barrier (8.76) are

$$d = L \text{ arcth}\left(\frac{2M}{1+M^2}\right), \quad d_1 = \frac{2ML}{1-M^2}. \tag{8.78}$$

The minima of barrier $F^2(z)$ (8.74) and of auxiliary barrier (8.78) coincide ($F^2_{\min} = 1 - M^2$). The reflection coefficient for the concave profile (8.74) can be calculated from expression (8.20) by using the expressions

$$\Omega_- = \frac{v_0}{d}\theta_-, \quad \theta_- = \frac{2M}{1-M^2}, \quad q\eta_0 = \frac{2N_-\text{arcth}(M)}{S}, \quad S = \frac{\Omega_-}{\omega}. \tag{8.79}$$

Note, that the reflection spectra for gradient solid layers with spatial distributions of density $F^2(z)$ and elasticity $W^2(z)$, described by different "consistent" models, are given by the general expression (8.20), which depends on the characteristic frequencies of the nonlocal dispersion $\Omega_\pm$. It is remarkable that in reflection upon the concave profiles of $F^2(z)$ and $W^2(z)$ in the low frequency region $S > 1$, $N^2_- < 0$ the phase shift $q\eta_0$ (8.79) becomes imaginary. A peculiar effect of tunneling of sound through such acoustical barriers will be considered in Sec. 9.3.

Comments and Conclusions to Chapter 8

1. To examine the propagation of shear waves in a gradient material with a spatially variable density and constant shear modulus one can use Eq. (8.25), putting there $W = 1, W_z = 0$ and, respectively, $\eta = z$. The equation, obtained under this condition, coincides formally with Eq. (8.10), describing the propagation of longitudinal waves through this barrier; thus one can calculate the reflection of shear waves from this barrier, using the exact solution (8.13), keeping in mind, that the symbol v_0 designates now the velocity of shear waves $v_0 = v_t$ (8.1).
2. Analysis of the "double" barrier (8.48) with the equal characteristic scales $l_1 = l_2$ reveals a particular case when, according to Eq. (8.51), we have $l \to \infty$ and, thus, $\Omega \to 0$, $N_- = 1$. The values

of the quantities U, τ, and the phase shift $q\tau_0$ in this limit are

$$U \to 1, \quad \tau = \eta, \quad q\tau_0 = \frac{\omega d}{v_0}\left(1 + \frac{d}{l_1}\right)^{-1}. \qquad (8.80)$$

Substitution of (8.80) into (8.63) yields the value of the reflection coefficient in the limit discussed. The non-local dispersion in this case vanishes.

3. Another particular case of dispersionless propagation through the "double" barrier is described by Eq. (8.47) under the condition $F^2(z)W^2(z) = 1$. The solution of Eq. (8.47) under this condition reads

$$u = \exp\left(\frac{i\omega\eta}{v_0}\right) + Q\exp\left(-\frac{i\omega\eta}{v_0}\right). \qquad (8.81)$$

Considering, e.g. the familiar density profile $F^2(z)$ (8.66), we find the variable η, determined by expressions (8.67). Forward and backward waves (8.81) are represented in η-space by harmonic waveforms with constant wave numbers $k = \omega/v_0$, while the propagation in z-space is accompanied by the reshaping of waveforms.

Bibliography

[8.1] P. Sheng et al., *Physica B* **338**, 201–205 (2003).
[8.2] G. W. Milton, M. Briane and J. R. Willis, *New J. Physics* **8**, 248–267 (2006).
[8.3] H. Chen and C. T. Chan, *Appl. Phys. Lett.* **91**(1–3), 183518 (2007).
[8.4] S.-J. Lee et al., *Appl. Phys. Lett.* **82**, 2133–2135 (2003).
[8.5] M. Z. Ben-Amoz, *Angew. Math. Phys.* **27**, 83–99 (1976).
[8.6] A. Chakraborty, *J. Acoust. Soc. Am.* **123**, 56–67 (2008).
[8.7] L. D. Landau and E. M. Lifshitz, *Theory of Elasticity* (Pergamon Press, Oxford, 1986).
[8.8] V. I. Erofeyev, *Wave Processes in Solids with Microstructure* (World Scientific, 2003).
[8.9] T. Bennett, I. M. Gitman and H. Askes, *Int. J. Fracturs* **148**, 185–193 (2007).
[8.10] M. G. Vavva et al., *J. Acoust. Soc. Am.* **125**, 3414–3427 (2009).
[8.11] T. W. Clyne and P. J. Withers, *An Introduction to Metal Matrix Composites*, (Cambridge University Press, UK, 1993).
[8.12] F. Lasagni and H. P. Degischer, *J. Composite Materials* **44**, 739–756 (2010).
[8.13] A. B. Shvartsburg and N.S. Erokhin, *Physics–Uspekhi* **54**, 627–646 (2011).

[8.14] M. Schoenberg and P. Sen, *J. Acoust. Soc. Am.* **73**, 61–67 (1983).
[8.15] A. V. Granato, *J. Phys. Chem. Solids* **55**, 931–939 (1994).
[8.16] J. Guck et al., *Phys. Rev. Lett.* **84**, 5451–5454 (2000).
[8.17] V. Aleshin, V. Gusev and V. Tournat, *J. Acoust. Soc. Am.* **121**, 2600–2611 (2007).

CHAPTER 9

SHEAR ACOUSTIC WAVES IN GRADIENT ELASTIC SOLIDS

This chapter is devoted to the diversity of the effects of heterogeneity-induced dispersion in the complex of acoustic wave phenomena in gradient solids. At first glance the formal similarity of the equations governing the optical and acoustic wave processes in heterogeneous media, opens the way to the direct use of physical concepts and mathematical solutions, elaborated in gradient optics, in the corresponding acoustic problems (acousto-optical analogy). However, the optics of gradient dielectrics, developed above, is usually deals with only one spatially distributed parameter ($\varepsilon(z)$), so this one-to-one conformity proves to be useful for solids described by models of the spatial distribution of only one normalized parameter-either density $F^2(z)$ (8.2) or Young's modulus and connected with it the shear elastic modulus $W^2(z)$ (8.3); for simplicity the variations of Poisson coefficient are ignored here. On the other hand the more complicated dependence of heterogeneity-induced acoustic dispersion on two quantities-spatially distributed density and elasticity, represented by the functions $F^2(z)$ and $W^2(z)$, has no optical counterpart and, thus, the corresponding mathematical basis for operating with these two independent functions has to be elaborated from the very beginning (see, e.g. Sec. 8.3). Both of these groups of models can be applied to real solids, e.g. composites or binary alloys.

The first type of exactly solvable models ($\rho = \rho(z)$, Young's modulus $E = $ const) is used in Sec. 9.1 for the solution of a historical long time existing theoretical problem concerning the eigenoscillations spectra of strings with variable density [9.1]. The spectra obtained provide examples of the explicit dependence of the eigenfrequencies on the distribution $\rho = \rho(z)$ and the conditions at the string endpoints. The same model is used in this section for a calculation of the fundamental eigenfrequency of a concentrationally graded plane layer. Being developed for this specific aim, the model discussed may become useful for other problems, connected with binary alloys, in cases, where the concentration dependence of the elastic parameters of a solid solution is insignificant; this effect is inherent, e.g. for Al-Mg alloy, where the admixture of Mg results in a decrease of the alloy's density, retaining the value of Young's modulus close to it's value for Al, as long as the Mg content doesn't exceed 10 mass % [9.2].

The discrete spectrum of torsional eigenoscillations of an elastic rod, formed from an array of circular cylinders with equal radii and decreasing lengths, is considered in Sec. 9.2. This mechanical effect broadens the family of Wannier–Stark-like ladders, observed initially in solid state physics [9.3] and in gradient optics [9.4]. The Wannier–Stark ladder can be viewed as the frequency domain counterpart of the Bloch oscillations of an electron accelerated by a constant external electric field and travelling in a periodic potential, e.g. in a crystal [9.5]. The optical analogue of electronic Bloch oscillations was demonstrated by oscillations in light beam in an optical superlattice, possessing a linear variation in the optical thickness of the layers along the propagation direction; this gradient is the optical counterpart of the external electric field, used for acceleration of particles in an electronic superlattice [9.4]. The similar gradient effect in the analogue of Wannier-Stark ladder in acoustics, shown in the Sec. 9.2, is imitated by the suitable distribution of lengths of elastic cylinders, supporting the propagation of torsional eigenmodes.

The peculiar effects of acoustical tunneling through a gradient elastic layer, including reflectionless tunneling, are exemplified in Sec. 9.3. These effects, resembling the tunneling of light through a gradient dielectric nanofilm, are considered for solid layers with a variable density as well as for layers with spatial distributions of both density and Young's modulus. The exactly solvable model, developed here for the solution of the latter problem, can be applied to the analysis of acoustic waves in binary alloys, such as, e.g. Ti–Hf,

Al–Si or Cu–Ni, where the variations of density are accompanied by variations of Young's modulus [9.6].

9.1. Strings with Variable Density

The equation of elastic oscillations of a thin homogeneous string became the standard equation for many problems in optics, radiophysics, and acoustics. This equation, which follows from (8.25) with $F = W = \text{const} = 1$, coincides with the one-dimensional wave equation

$$\frac{\partial^2 u}{\partial z^2} - \frac{1}{v^2}\frac{\partial^2 u}{\partial t^2} = 0, \tag{9.1}$$

which describes a bending wave, propagating at the speed v along a thin string with a constant cross-section S and constant density per unit length ρ_0, stretched by the force T, where [9.7]

$$v^2 = \frac{T}{S\rho_0}. \tag{9.2}$$

The spectrum of eigenfrequencies Ω_n of a homogeneous string stretched between the points $z = 0$ and $z = d$ such, that the string displacement at these points is zero, is described by the classical formula [9.8]

$$\omega_n = \frac{v_0 \pi n}{d}. \quad n = 1; 2; 3\ldots \tag{9.3}$$

Rayleigh extended the applications limits of expression (9.3) for perturbations of the density and studied the spectrum of oscillations of a "string with a linear density not quite constant" [9.1]. The small corrections to the spectrum (9.3) for weak density variations, based on the perturbation theory, were found in [9.1].

To illustrate the applicability of the methods of gradient acoustics, presented in Ch. 8, it is useful to reconsider this classical problem again and to find the oscillations spectrum of a variable density string without the assumption of the smallness of density variations [9.9]. Let us analyze first the oscillations of a string of length d with a convex symmetrical density distribution $\rho = \rho_0 \, F^2(z)$, described by

(8.11) with $s_1 = -1, s_2 = 1$, and characterized by the maximum of the distribution, located at the point $z = 0.5d$. The unknown lengths L_1 and L_2 in (8.11) are expressed in terms of the length d by means of the dimensionless parameter y (8.12), and the parameter y is related to the distribution maximum $\rho_{\max}$ (8.23). Recalling the exact analytical solution of Eq. (8.10), written in the form (8.13) and (8.14),

$$u = \frac{\exp(iq\eta) + Q\exp(-iq\eta)}{\sqrt{F(z)}}; \quad \eta = \int_0^z F(z_1)dz_1; \quad q = \frac{\omega}{v_0}N_+, \tag{9.4}$$

we can present the solution of Eq. (8.10), describing the standing waves, vanishing at the string endpoints in η-space ($\eta = 0$ and $\eta = \eta(d)$) in the form

$$u = \frac{\sin(q\eta)}{\sqrt{F(z)}}. \tag{9.5}$$

The eigenfrequencies of standing waves (9.5) are given by the condition

$$q\eta(d) = \pi n. \tag{9.6}$$

The value of $\eta(d)$ is calculated from (9.4) as a product of string length d on some form-factor A, dependent upon the density distribution along the string,

$$\eta(d) = dA, \quad A = \frac{1}{y\sqrt{1-y^2}} \arctg\left(\frac{y}{\sqrt{1-y^2}}\right). \tag{9.7}$$

Substituting the values of q from (8.13) and $\eta(d)$ from (9.7) into the condition (9.6), and taking the characteristic frequency Ω_+ (8.14) into account, we find the discrete mode spectrum of a gradient string with the convex density distribution along the string specified by function (8.11) with $s_1 = -1$, $s_2 = 1$:

$$(\Omega_+)_n = \omega_n D_n. \tag{9.8}$$

Here ω_n is the eigenfrequency (9.3) of homogeneous string and D_n is the dimensionless correction coefficient

$$D_n = \sqrt{A^{-2} - \frac{4y^2(1-y^2)}{\pi^2 n^2}}. \tag{9.9}$$

The mode spectrum of the string with a concave density profile, characterized by the value $\rho_{\min}$, related to the values $s_1 = 1, s_2 = -1$ in (8.11), is determined similarly. Taking the values q, N_-, Ω_- from Eq. (8.14), and linking the parameter y with $\rho_{\min}$ via (8.21), we find the function $\eta(d)$ by analogy with (9.7):

$$\eta(d) = Bd, \quad B = (2y\sqrt{1+y^2})^{-1}\ln\left(\frac{y_+}{y_-}\right), \quad y_\pm = \sqrt{1+y^2} \pm y. \tag{9.10}$$

The mode spectrum $(\Omega_-)_n$ of the string with the concave density profile (8.11) can be written in the form (9.8) by introducing the correction coefficient H_n:

$$(\Omega_-)_n = \omega_n H_n, \quad H_n = \sqrt{B^{-2} + \frac{4y^2(1+y^2)}{\pi^2 n^2}}. \tag{9.11}$$

We note that the mode spectrum of the variable-density string (8.11) was calculated without assuming of the smallness of the density variations. Formulae (9.9) and (9.11) illustrate the influence of the maximum and minimum values of the string density, expressed via the parameter y, on the correction coefficients D_n and H_n. Plots of these coefficients are shown in Figs. 9.1(a) and 9.1(b), respectively. In the limit of the vanishing inhomogeneity ($y \to 0$) it follows from (9.8) and (9.11) that

$$\lim D_n|_{y \to 0} = 1; \quad \lim H_n|_{y \to 0} = 1;$$
$$\lim(\Omega_+)_n|_{y \to 0} = \lim(\Omega_-)_n|_{y \to 0} = \omega_n. \tag{9.12}$$

As expected, the spectra (9.8) and (9.11) of gradient strings are reduced in this limit to the classical formula (9.3) for the homogeneous string.

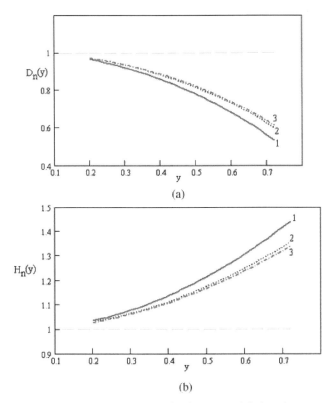

Fig. 9.1. Correction coefficients $D_n(y)$ (9.9) and $H_n(y)$ (9.11) are shown in (a) and (b), respectively. (a) and (b) relates to the convex (concave) profile of density distribution along the string (8.11), the parameter y is linked with the maximum (minimum) value of the density by Eqs. (8.23) and (8.21), respectively.

The solution of one-dimensional wave equation for the gradient medium (9.5) can be used for the calculation of the acoustic eigenfrequencies of a gradient layer with a variable density [9.10]. Unlike the aforementioned string, both of whose endpoints $z = 0$ and $z = d$ are immobile, let us consider the layer with thickness d, whose boundary $z = d$, located on a rigid substrate, is assumed to be immobile, while the other boundary $z = 0$ is free. The boundary conditions at these planes are:

$$u|_{z=d} = 0; \quad \left.\frac{\partial u}{\partial z}\right|_{z=0} = 0. \tag{9.13}$$

The solution of Eq. (9.4), satisfying to the first of the conditions (9.13), can be written as

$$u = \frac{\sin[q(\eta - \eta_0)]}{\sqrt{F(z)}}, \quad \eta_0 = \eta(d). \tag{9.14}$$

Let us examine, for definiteness, a concave profile of the density distribution $\rho = \rho_0 F^2(z)$; substitution of the solution (9.14) into the second of conditions (9.13) yields the dispersion equation, determining the eigenfrequencies of this elastic structure

$$\operatorname{tg}(q\eta_0) = 2qL_1. \tag{9.15}$$

Recalling the definition of the quantity $q\eta_0$, given in (8.21), we can present the dispersion equation (9.15) in the form

$$\frac{\operatorname{tg}(l\sqrt{S^{-2}-1})}{\sqrt{S^{-2}-1}} = \frac{\sqrt{1+y^2}}{y}, \quad l = \ln\left(\frac{\sqrt{1+y^2}+y}{\sqrt{1+y^2}-y}\right). \tag{9.16}$$

To solve Eq. (9.16) it is worthwhile to introduce the function $\wp(\vartheta)$:

$$\wp(\vartheta) = \frac{\operatorname{tg}\vartheta}{\vartheta}. \tag{9.17}$$

Using this function $\wp(\vartheta)$ we rewrite Eq. (9.16) as

$$\wp(\vartheta) = \frac{\sqrt{1+y^2}}{ly}. \tag{9.18}$$

Here $\vartheta = l\sqrt{S^{-2}-1}$; Eq. (9.18) defines the frequency-dependent quantity ϑ as a function of the parameter y: $\vartheta = \vartheta(y)$. Considering for simplicity only the fundamental eigenfrequency of oscillations of the gradient layer under discussion, ω_0, we find from (9.18):

$$\omega_0 = \frac{\Omega_-\sqrt{l^2 + [\vartheta(y)]^2}}{l}. \tag{9.19}$$

This analysis relates to the case of the concave profile of the density $F(z)$ (8.11) inside the layer; in a case of the convex profile $F(z)$ the right side of Eq. (9.15) changes its sign, and the resulting equation reads as $\operatorname{tg}(q\eta_0) = -2qL_1$; the subsequent calculations are performed in a similar fashion.

In the limit of vanishing heterogeneity ($y \to 0$) the values of the parameters l and ϑ are

$$\lim l|_{y \to 0} = 2y; \quad \lim \vartheta|_{y \to 0} = \frac{\pi}{2}. \tag{9.20}$$

Expression (9.19) is reduced in this case to the well known formula for the fundamental eigenfrequency of a homogeneous elastic layer [9.7]:

$$\omega_0 = \frac{v_0 \pi}{2d}. \tag{9.21}$$

Inspection of the obtained results shows, that:

a. the spectral intervals between the eigenfrequencies of a string with a heterogeneously distributed density are unequal; the influence of heterogeneity on the spectra of eigenoscillations decreases with the growth of the mode number n; thus, the difference between the spectra of the 2nd and 3rd eigenmodes in Fig. 9.1(a) is insignificant;
b. spectra (9.9) and (9.11), found for the shear waves, remain valid for the longitudinal waves after the replacement $v_t \to v_l$;
c. spectra (9.3) and (9.21) of the eigenoscillations of a string with the variable density (8.11) prove to be limiting cases of the more general results of gradient acoustics [9.9].

9.2. Torsional Oscillations of a Graded Elastic Rod

The graded media we have considered up to now have all had material or geometrical properties that were continuously varying functions of coordinates. In this section we consider a graded structure in which the geometrical properties are discrete functions of the spatial coordinates. Specifically, we study the torsional vibrations of an elastic rod with free ends formed from a linear array of circular cylinders with a constant radius but with a decreasing length, separated by identical very small cylinders. The mass density and shear elastic modulus of both sets of cylinders are constants independent of position along the rod. The interest in this problem has the following origins.

In the 1950's Wannier studied the motion of an electron in a periodic potential to which a constant, uniform external electric field is applied. He showed that the energy spectrum of the electron, which has a band structure in the absence of the electric field, consists of equidistant discrete energy levels in the presence of the electric field, with the separation between consecutive levels proportional to the electric field strength. These equally spaced energy levels have come to be called an *electronic Wannier–Stark ladder* [9.11].

Wannier's prediction was controversial [9.12, 9.13], but some 20 years after it was made it was confirmed theoretically by computer simulations for simple one-dimensional models [9.14, 9.15], and subsequently in experiments on high quality semiconductor superlattices [9.16].

Despite the observation of electronic Wannier–Stark ladders, experimental and theoretical searches were carried out for simpler systems, consisting of electrically neutral particles, instead of electrons, displaying this phenomenon. In early efforts of this kind a Wannier–Stark ladder was observed in a system consisting of atoms moving in an accelerating optical lattice formed by two interfering laser beams [9.17].

A major impetus to the search for such systems was the realization of analogies between the electrons in a crystal and the flow of light in photonic crystals [9.18]. This stimulated the search for macroscopic systems that can display optical analogues of Wannier–Stark ladders.

The earliest theoretical study of the existence of the optical Wannier–Stark ladder was carried out by Monsivais *et al.* [9.19], who studied the transmission of transverse electromagnetic waves through a stratified structure whose dielectric constant at a given frequency ω was the sum of a periodic function of the coordinate normal to the interfaces of the structure and a linear function of that coordinate. The transmission coefficient as a function of $[(\omega/c)\sin\theta]^2$, where θ was the angle of incidence of the electromagnetic wave, displayed a Wannier–Stark ladder structure for some values of the parameters characterizing this structure. The experimental observation of an optical Wannier–Stark ladder was reported several years later for a structure consisting of a linearly chirped Moiré grating written in the core of an optical fiber [9.20].

Analogues of Wannier–Stark ladders have also been studied in mechanical systems. These include stratified elastic media in which the square of the shear wave speed has a periodic dependence on the coordinate normal to the interfaces of the structure, supplemented by a term that increases linearly with this coordinate [9.21]; stratified piezoelectric media, in which the ratio of the mass density to the stiffened shear elastic modulus has the same dependence on the coordinate normal to the interfaces [9.22]; and the torsional waves of special rods with free ends [9.23]. Of these the last is the simplest to analyze, and it is a system that has been studied experimentally. In the remainder of this section we direct our attention to it, following the treatments of Gutiérrez et al. [9.23] and Morales et al. [9.24].

Thus, we study a special elastic rod with free ends whose torsional waves have some analogies to a Wannier–Stark ladder. This rod, which is depicted in Fig. 9.2, consists of a set of N circular cylinders of radius R and varying length ℓ_n, $n = 1, 2, \ldots, N$, separated by very small cylinders of length $\epsilon \ll \ell_n$, and a radius r such that $r < R \ll \ell_n$.

In order to design this system so that its torsional oscillations display an analogue of a Wannier–Stark ladder, we start with an independent rod model in which each cylinder oscillates independently from the rest. The frequencies $f_j^{(n)}$ of the normal torsional

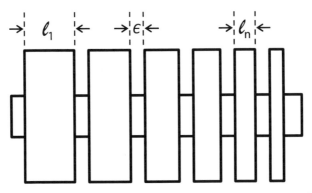

Fig. 9.2. The graded elastic rod whose torsional vibrations are studied in this section [9.23].

modes of rod n of length ℓ_n and wave velocity c_n are given by [9.25]

$$f_j^{(n)} = \frac{c_n}{2\ell_n}j, \tag{9.22}$$

where $j = 1, 2, 3, \ldots$ is the number of nodes in the amplitude of the wave. We seek a structure with equidistant frequencies. Thus we take circular rods with $\ell_n = \ell/(1 + n\gamma)$, $n = 1, 2, 3, \ldots, N$, and $c_n = (\mu/\rho)^{\frac{1}{2}}$, where μ is the shear elastic modulus of rod n, ρ is its mass density, and ℓ is a fixed arbitrary length. We note that in circular rods the wave velocity c_n is independent of their radius [9.25]. From these results and Eq. (9.22) we find that the frequencies $f_j^{(n)}$ are given by

$$f_j^{(n)} = \sqrt{\mu/\rho}(1 + n\gamma)j/2\ell, \tag{9.23}$$

so that the differences $\Delta f_j^{(n)} = f_j^{(n+1)} - f_j^{(n)}$ are given by

$$\Delta_j = \Delta f_j^{(n)} = \sqrt{\mu/\rho}\gamma j/2\ell, \tag{9.24}$$

and are independent of the index n.

When the arbitrary parameter γ is set equal to zero, we have a finite periodic rod. The torsional oscillations of this finite periodic rod possess a band spectrum [9.24]. When $\gamma \neq 0$ a completely new spectrum occurs, which resembles a Wannier-Stark ladder.

Before deriving these results it is useful to carry out a qualitative analysis to determine what kind of spectrum is to be expected for the independent rod model. At the lowest frequencies, the wavelength λ is of the order of magnitude of the length of the rod L, and the entire rod is excited. However, when λ is decreased and becomes of the order of $\ell_1 = \ell/(1 + \gamma)$, the longest rod, rod 1, is excited in a state equivalent to its lowest frequency normal mode ($j = 1$). The remaining $N - 1$ cylinders are out of resonance, so the amplitude of this mode decreases with increasing distance from rod 1. This state is therefore localized about rod 1. If we next increase the exciting frequency by Δ_1 the rod with length $\ell_2 = \ell/(1 + 2\gamma)$ will be excited, while the remaining $N - 1$ rods will be out of resonance. The amplitudes of the oscillations therefore decrease with increasing distance from rod 2. Therefore, at this higher frequency the wave amplitude is localized about rod 2. It has a shape similar to that of the wave amplitude localized about rod 1, but is slightly deformed, squeezed, and translated from rod 1 to rod 2. The same arguments can be made when the exciting frequency is increased by Δ_{n-1} when rod n of length $\ell_n = \ell/(1 + n\gamma)$ is excited.

Thus, we have produced a finite Wannier–Stark ladder, namely N localized states with a constant frequency difference given by Eq. (9.24) with $j = 1$. More ladders can exist, however, since normal modes with two or more nodes can also be excited in each rod. For example, if we take $j = 2$ in Eq. (9.23) a second ladder is obtained. From Eq. (9.24) we see that the frequency difference between consecutive steps of this ladder is twice that for the ladder corresponding to $j = 1$.

We can now put the preceding qualitative results on a firm foundation by calculating the eigenmode properties of the system depicted in Fig. 9.2 rigorously by a transfer-matrix method [9.24].

The wave amplitude in cylinder i can be written as

$$\psi_i(z) = A_i e^{ik(z-z_{i-1})} + B_i e^{-ik(z-z_{i-1})}, \tag{9.25}$$

for $z_{i-1} \leq z \leq z_i$, $i = 1, 2, \ldots, 2N + 1$. The boundary conditions satisfied by torsional modes at $z = z_i$ are

$$\psi_i|_{z=z_i} = \psi_{i+1}|_{z=z_i}, \tag{9.26}$$

$$s_i^2 \frac{\partial \psi_i}{\partial z}\bigg|_{z=z_i} = s_{i+1}^2 \frac{\partial \psi_{i+1}}{\partial z}\bigg|_{z=z_i}, \tag{9.27}$$

where $s_i = \pi r_i^2$, and r_i is the radius of cylinder i, which is defined as occupying the region $z_i \leq z \leq z_{i+1}$.

The boundary conditions (9.26) and (9.27) are only approximate, however. The wave amplitude in fact is a function of both z and the radial coordinate, and the latter dependence has not been considered here. It has been shown by Morales et al. [9.24] that if instead of the actual value $\eta = r/R$ that enters the boundary condition (9.27) an effective value η_{eff} is used, a more accurate result is obtained. The value of η_{eff} is

$$\eta_{\text{eff}} = \frac{\eta}{(1 + \alpha_T/\epsilon)^{\frac{1}{4}}}, \tag{9.28}$$

where α_T is a constant that depends on the shear modulus μ. In their work Morales et al. determined α_T from a least-squares fit of theoretical results for the frequencies of the torsional waves in the system depicted in Fig. 9.2 to experimental values for these frequencies.

When Eq. (9.25) is substituted into Eqs. (9.26) and (9.27), the amplitudes A_{i+1} and B_{i+1} are found to be related to the amplitudes A_i and B_i by the transfer matrix $\mathbf{M}_{i \to i+1}$ according to

$$\begin{pmatrix} A_{i+1} \\ B_{i+1} \end{pmatrix} = \mathbf{M}_{i \to i+1} \begin{pmatrix} A_i \\ B_i \end{pmatrix},$$

where

$$\mathbf{M}_{i \to i+1} = \frac{1}{2} \begin{pmatrix} \left[1 + \left(\frac{r_i}{r_{i+1}}\right)^4\right] e^{ik(z_i - z_{i-1})} & \left[1 - \left(\frac{r_i}{r_{i+1}}\right)^4\right] e^{-ik(z_i - z_{i-1})} \\ \left[1 - \left(\frac{r_i}{r_{i+1}}\right)^4\right] e^{ik(z_i - z_{i-1})} & \left[1 + \left(\frac{r_i}{r_{i+1}}\right)^4\right] e^{-ik(z_i - z_{i-1})} \end{pmatrix}.$$

(9.29)

From this result we find that the amplitudes A_{2N+1} and B_{2N+1} are related to the amplitudes A_1 and B_1 by

$$\begin{pmatrix} A_{2N+1} \\ B_{2N+1} \end{pmatrix} = \mathbf{M}_{2N \to 2N+1} \mathbf{M}_{2N-1 \to 2N} \ldots \mathbf{M}_{2 \to 3} \mathbf{M}_{1 \to 2} \begin{pmatrix} A_1 \\ B_1 \end{pmatrix}$$

(9.30a)

$$= \mathbf{M} \begin{pmatrix} A_1 \\ B_1 \end{pmatrix}$$

(9.30b)

$$= \begin{pmatrix} M_{11} & M_{12} \\ M_{21} & M_{22} \end{pmatrix} \begin{pmatrix} A_1 \\ B_1 \end{pmatrix}.$$

(9.30c)

We assume that the ends of the rod are stress free. This assumption is expressed by the pair of boundary conditions

$$\left. \frac{\partial \psi_1(z)}{\partial z} \right|_{z=z_0=0} = 0$$

(9.31)

$$\left. \frac{\partial \psi_{2N+1}}{\partial z} \right|_{z=z_{2N+1}=L} = 0.$$

(9.32)

The first condition yields the relation

$$A_1 - B_1 = 0. \tag{9.33}$$

The second condition yields

$$A_{2N+1}e^{ik(L-z_{2N})} - B_{2N+1}e^{-ik(L-z_{2N})} = 0. \tag{9.34}$$

When Eq. (9.30c) is used to express A_{2N+1} and B_{2N+1} in Eq. (9.34) in terms of A_1 and B_1, the result together with Eq. (9.33) leads to a pair of coupled homogeneous equations for A_1 and B_1,

$$\mathbf{P}\begin{pmatrix} A_1 \\ B_1 \end{pmatrix} = 0, \tag{9.35}$$

where

$$\mathbf{P} = \begin{pmatrix} 1 & -1 \\ M_{11}e^{ik(L-z_{2N})} & M_{12}e^{ik(L-z_{2N})} \\ -M_{21}e^{-ik(L-z_{2N})} & -M_{22}e^{-ik(L-z_{2N})} \end{pmatrix}. \tag{9.36}$$

This result corrects an error in Eq. (17) of Ref. [9.24]. The solvability condition for Eq. (9.35), $\det(\mathbf{P}) = 0$, gives the allowed values of the wavenumber k for a given value of γ. They are obtained by fixing the value of γ and scanning the determinant as a function of k, and searching for changes in its sign. When a change of sign is found, an accurate value of k at which $\det(\mathbf{P})$ vanishes can be obtained by a standard root-finding routine [9.26]. When the roots have been found the wave amplitudes can be calculated by the use of Eq. (9.25).

The frequencies and amplitudes of the torsional normal modes of a notched metallic rod with free ends of the kind depicted in Fig. 9.2 have been measured by the use of an electromagnetic acoustic transducer (EMAT) [9.27].

This EMAT consists of a simple coil and a permanent magnet (Fig. 9.3). In the configuration depicted in Fig. 9.3, it can be used to excite and detect torsional elastic waves. When it is used to excite these waves the time-varying magnetic field of the coil induces a time-varying eddy current in the metal. The field of the permanent magnet then generates a Lorentz force on the eddy current accelerating

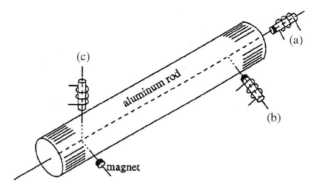

Fig. 9.3. A diagram showing an EMAT configuration for detecting or exciting: (a) compressional oscillations, (b) bending oscillations, and (c) torsional oscillations [9.27].

the rod. When it is used as a detector the motion of the rod induces a variable magnetic flux in the rod due to the field of the permanent magnet. This variable magnetic flux, according to Faraday's law, generates an emf, which in turn produces eddy currents. The variable magnetic field caused by these eddy currents is then detected by the coil of the EMAT. The emf produced by the EMAT detector is proportional to the acceleration of the surface of the rod because the eddy current is proportional to the time derivative of the magnetic flux and, in first approximation, this is proportional to the speed of the surface of the metal. Because the emf induced in the coil is proportional to the time derivative of the field produced by the eddy current, this emf, again in first approximation, is proportional to the acceleration of the surface of the rod. To excite or detect torsional waves the magnet axis must be perpendicular to the coil axis, and both must be perpendicular to the axis of the rod. In this geometry the eddy current is perpendicular to the static magnetic field, so that this device will detect the torsional waves, and in the excitation mode will apply torque to the rod. The EMAT developed in Ref. [9.27] operates at frequencies from a few hertz up to hundreds of kilohertz.

In Fig. 9.4 theoretical and experimental results for the normal mode frequencies of the rod shown in Fig. 9.2 are depicted for $\gamma = 0.091$ for $j = 1, 2, 3$. For each value of j the left-hand column

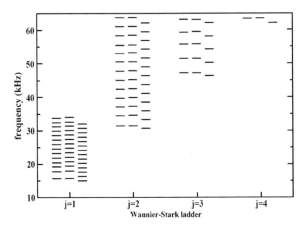

Fig. 9.4. Normal mode frequencies of the system depicted in Fig. 9.2 that display a finite Wannier–Stark ladder. For each value of j the left-hand column corresponds to the experimental values, the middle column presents the numerical results obtained by the transfer-matrix method, and the right-hand column shows the approximate results obtained from the independent rod model [9.23].

presents the experimental values, the middle column presents the numerical results obtained by the transfer matrix method, and the right-hand column presents the approximate results obtained by the independent rod model. The metal from which the system is fabricated is aluminium. The number of cells in it is $N = 14$, $\ell = 10.8$ cm, $\epsilon = 2.52$ mm, and $\sqrt{\mu/\rho} = 3104.7$ m/s. The radii of the small and large cylinders are $r = 2.415$ mm and $R = 6.425$ mm, respectively. An effective value of $\eta = r/R$ given by Eq. (9.28) with $\alpha_T = 0.88$ mm, namely $\eta_{\text{eff}} = 0.3488$, was used in these calculations. It is seen that the theoretical results agree very well with the experimental ones. Moreover, the approximate results obtained by the independent rod model represent quite a good first approximation to the experimental ones. It is seen from Fig. 9.4 that the frequencies of the torsional oscillations form a set of Wannier–Stark ladders for each value of j. The frequencies at the ends of each ladder do not have the same difference in frequency as those in the middle of the ladder. This is due to edge effects on the amplitude of the waves localized near the free ends of the cylinder.

9.3. Tunneling of Acoustic Waves Through a Gradient Solid Layer

Tunneling of sound in gradient elastic media is stipulated by heterogeneity-induced acoustic dispersion of these media. The simple example of acoustical tunneling can be examined in the framework of a model of concentrationally graded solid, whose density variation is given by the familiar concave distribution (8.11). Let us consider shear acoustical waves, assuming for simplicity the shear elastic modulus of material to be coordinate-independent (see Sec. 8.2). The wave field of these waves is governed by equation (8.10) with $v_0 = v_t$; solution of (8.10), describing the tunneling regime, related to low frequencies $(S > 1)$, reads by analogy with (9.4):

$$u = \frac{A[\exp(-p\eta) + Q\exp(p\eta)]}{\sqrt{F(z)}}, \quad p = \frac{\omega}{v_t}N, \quad N = \sqrt{S^2 - 1}. \tag{9.37}$$

The cut-off frequency, determining the upper boundary of the tunneling-related spectral interval, is given in (8.18): $\Omega_- = v_t d^{-1}\theta_-(y)$, where $\theta_-(y)$ is the form-factor for the structure under discussion

$$\theta_-(y) = 2y\sqrt{1 + y^2}. \tag{9.38}$$

Standard calculations, based on continuity conditions, yield the reflection coefficient R, presented in the form (8.20); the quantities $\sigma_{1,2}$ and $\chi_{1,2}$ in the case of tunneling are obtained from the corresponding terms (8.21) by the replacements

$$N_- \to iN; \quad N_-^2 \to -N^2; \quad t = \text{th}(q\eta_0). \tag{9.39}$$

The complex transmission coefficient for the evanescent sound mode can be found in the form

$$T = |T|\exp(i\phi_t); \quad |T| = \frac{2\alpha N\sqrt{1-t^2}}{\sqrt{|\chi_1|^2 + |\chi_2|^2}}; \quad \text{tg}\phi_t = \frac{\chi_1}{\chi_2}. \tag{9.40}$$

Transmittance spectra $|T(S)|^2$ for shear waves, tunneling through the concentrationally graded layer (8.11) are depicted in Fig. 9.5.

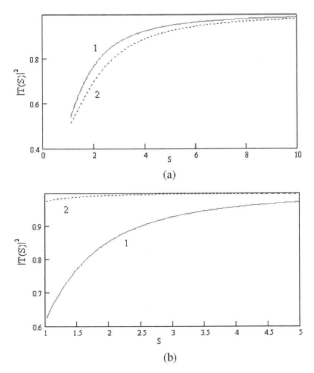

Fig. 9.5. Transmittance spectra for shear waves tunneling through the gradient acoustical barrier, formed by density distribution (8.11) subject to the density-related and impedances-related parameters y (8.21) and α (8.18). (a): spectra 1 and 2 correspond to the values of $y = 0.45$ and $y = 0.7$, respectively, $\alpha = 0.3$. (b): spectra 1 and 2 relate to the values $\alpha = 0.3$ and $\alpha = 1.25$, respectively, $y = 0.3$.

Inspection of these graphs shows the variation of $|T(S)|^2$ induced by changes of the parameters y and α, connected with the density minimum (8.21) and the ratio of impedances (8.18), respectively. Curve 2 on Fig. 9.5(b) exemplifies the high transmission coefficient in the regime of acoustical tunneling, tending to unity in a wide spectral range $2 < S < 5$.

It is remarkable, that the tunneling of sound through a gradient elastic layer can be characterized by zero reflectance of this layer. Making the replacements (9.39) in the expression for σ_1 (8.20) we can write the condition for the occurrence of the reflectionless regime ($\sigma_1 = 0, R = 0$) for the tunneling wave, connected with the ratio of impedances α (8.18),

$$\alpha^2 = -\frac{\gamma^2}{4} - N^2 + \frac{\gamma N}{t}. \quad (9.41)$$

Here the quantities γ, N and $q\eta_0$ are defined in (8.18), (9.37) and (8.21) respectively. The vanishing of the reflectance under the condition (9.41) yields the complete transmittance; here the unchanged amplitude of the tunneling wave acquires the phase shift ϕ_t

$$|T| = 1, \quad \phi_t = \arctan\left(\frac{\alpha t}{N - \frac{\gamma t}{2}}\right). \quad (9.42)$$

This change ϕ_t is distinguished from the phase shift $\phi_0 = \omega d v_t^{-1}$, accumulated by the shear sound wave with the same frequency ω, travelling with velocity v_t through the same distance d; expressing the quantity ϕ_0 via the parameter y and normalized frequency S we have

$$\phi_0 = \frac{2y\sqrt{1+y^2}}{S}. \quad (9.43)$$

Comparison of phase shifts ϕ_t (9.42) and ϕ_0 (9.43) shows that this structure, containing a gradient elastic layer between two homogeneous elastic media, acts like a sound phase shifter, retaining the amplitude of the transmitted wave unchanged.

This example shows the effect of reflectionless tunneling of sound through a gradient layer with continuous spatial variation of its density and a constant shear elastic modulus. The influence of both density and elasticity distributions on the acoustical tunneling through the solid layer, can be illustrated by the generalization of the results of Sec. 8.3, describing the propagation of sound through the double acoustic layer in the framework of model (8.48). Thus, making the replacements (9.39) in (8.63), we find the complex transmission coefficient T for the double layer (8.48) in a form (9.40) with

$$\chi_1 = t\left(\alpha^2\beta + \frac{\gamma^2}{4} - N^2 U_0\right) - \frac{\gamma N}{2}(1 - U_0). \quad (9.44)$$

$$\chi_2 = \alpha \left[\frac{\gamma t}{2}(1-\beta) + N(U_0 + \beta) \right]. \tag{9.45}$$

$$t = \text{th}(q\tau_0); \quad q\tau_0 = \frac{\sqrt{1-S^{-2}}}{2} \ln \left[\frac{l_1(d+l_2)}{l_2(d+l_1)} \right]. \tag{9.46}$$

According to Eq. (8.55), the normalized frequency S in (9.37) has to be defined as $S = v_t(2l\omega)^{-1}$. Using the expression for the complex transmittance coefficient T (9.40) and substituting the values χ_1 and χ_2 from (9.44) and (9.45), we obtain the equation $|T|^2 = 1$, determining the condition of reflectionless tunneling through the double layer in the form

$$4\alpha^2 N^2(1-t^2) = |\chi_1|^2 + |\chi_2|^2. \tag{9.47}$$

Since Eq. (9.47) is transcendental, its solution, defining the frequency S, related to the regime of reflectionless tunneling, has to be calculated numerically.

Note, that the effects of reflectionless tunneling of both sound and light waves through gradient layers, defined by the same normalized distributions (8.11) and (2.16), respectively, are described by the same exact analytical solutions of the wave equation. This similarity illustrates the simplest example of acousto-optical analogy in gradient media. This analogy opens the way for the use of more complicated solutions of wave equation, corresponding to distributions of the refractive index in dielectric optical nanofilms, examined in Chs. 2 and 3, for reaching the goals of gradient acoustics of elastic solids.

Comments and Conclusions to Chapter 9

1. The cut-off frequency Ω_-, separating the travelling and tunneling spectral ranges, is proportional to the form-factor $\theta_-(y)$ (8.14). In the case of a variable density and a constant shear elastic modulus, considered in Sec. 9.3, the form factor of the gradient acoustic layer $\theta_-(y)$, defined in (9.38), is increasing monotonically with the increase of density minimum inside the layer: $\lim \theta_-(y)|_{y \to \infty} = 2y^2$. In the opposite case (spatially distributed shear elastic modulus and constant density), represented

by the model (8.41)–(8.42), the variations of the form-factor $\theta_-(y)$, defined in (8.45), are non-monotonic: the values of $\theta_-(y)$ increase only in the interval $0 \leq y \leq y_m = 1.515$; $\theta_-(y_m) = 2.4$. The subsequent growth of values of y results in the saturation of form-factor values: $\lim \theta_-(y)|_{y\to\infty} = 2$; thus, in this case the cut-off frequency does not exceed the value $\Omega_- = 2.4 v_t d^{-1}$.

2. The formal analogy between the expressions for evanescent wave fields in optics (4.16) and acoustics (9.37) of gradient media permits comparing and contrasting the tunneling effects for both these fields. Thus, to provide the reflectionless tunneling of light through a gradient dielectric layer, located on the homogeneous substrate with the refraction index n, the value n has to obey to equation, derived from condition $R = 0$, where the reflection coefficient R is defined in (4.31):

$$n^2 = -\frac{\gamma^2}{4} - n_e^2 + \frac{\gamma n_e}{t}. \qquad (9.48)$$

Considering the substrate, fabricated from a dispersionless material, we have $n > 1$; this means that the value of the right side of Eq. (9.48) has to exceed unity, and this restriction impedes the choice of parameters of the gradient layer for the given frequency. On the other hand, the condition for reflectionless tunneling of sound, having the similar form (9.41), coincides with (9.48) due to replacement $\alpha^2 \to n^2$; however, the quantity α, determined by the ratio of impedances (8.18), can be chosen, unlike the refractive index n, in the entire interval $\alpha > 0$; this flexibility broadens the possibilities for reflectionless acoustical tunneling.

3. It is worthwhile to stress the analogy between the tunneling of sound through the double acoustical barrier described by Eq. (8.25), and the transmission line (TL) with distributed parameters. The dynamics of the electric current I and voltage V in the lossless TL is governed by the well-known system of equations [9.9]:

$$\frac{\partial I}{\partial z} + C(z)\frac{\partial V}{\partial t} = 0, \quad \frac{\partial V}{\partial z} + L(z)\frac{\partial I}{\partial t} = 0. \qquad (9.49)$$

Here $C(z) = C_0 W^{-2}(z)$ and $L(z) = L_0 F^2(z)$ are the distributions of capacity and self-inductance per unit length, dependent on the coordinate z along the line. Introducing the generating function Ψ by means of the representations

$$V(z) = \frac{W^2(z)}{C_0}\frac{\partial \Psi}{\partial z}, \quad I = -\frac{\partial \Psi}{\partial t}, \qquad (9.50)$$

we find, that the first equation in the pair (9.49) is reduced to an identity, while the second one coincides with Eq. (8.25), which describes the displacement inside the gradient layer, characterized by the distributions of density $F^2(z)$ and shear elastic modulus, $W^2(z)$ respectively. This analogy can become useful for the modelling of complicated acoustical fields in gradient media by means of a transmission line.

Bibliography

[9.1] J. W. S. Rayleigh, *The Theory of Sound* (London, McMillan and Co, 1937).
[9.2] F. Lasagni and H. P. Degischer, *J. Composite Materials*, **44**(6), 739–754 (2010).
[9.3] A. G. Chynoweth, G. H. Wannier, R. A. Logan and D. E. Thomas, *Phys. Rev. Lett.* **5**, 57–60 (1960).
[9.4] R. Sapienza, P. Costantino, D. Wiersma, M. Chulinean, C. Oton and L. Pavesi, *Phys. Rev. Lett.* **91**(1–4), 26, 263902 (2003).
[9.5] G. H. Wannier, *Phys. Rev.* **100**, 1227–1233 (1955).
[9.6] Y. L. Zhou, M. Niinomi and T. Akahori, *Material Transactions* **45**(5), 1549–1554 (2004).
[9.7] L. D. Landau and E. M. Lifshitz, *Theory of Elasticity* (Pergamon Press, Oxford, 1986).
[9.8] L. M. Brekhovskikh and O. A. Godin, *Acoustics of Layered Media, I* (Berlin, Springer-Verlag, 1990).
[9.9] A. B. Shvartsburg and N. S. Erokhin, *Physics-Uspekhi*, **54**, 627–646 (2011).
[9.10] O.A. Godin and D. M. F. Chapman, *JASA* **106**, 2367–2373 (1999).
[9.11] G. H. Wannier, *Elements of Solid State Theory* (Cambridge University Press, Cambridge, UK, 1959), pp. 190–193.
[9.12] J. Zak, Stark ladder in solids? *Phys. Rev. Lett.* **20**, 1477–1481 (1968).
[9.13] A. Rabinovitch, Invalidity of translation symmetry argument for the existence of a Stark ladder in finite crystals, *Phys. Lett. A* **33**, 403–404 (1970).
[9.14] S. Nagai and J. Kondo, Electrons in infinite one-dimensional crystals in a uniform electric field, *J. Phys. Soc. Jpn.* **49**, 1255–1259 (1980).
[9.15] F. Bentosela, V. Grecchi and F. Zironi, Approximate ladder of resonances in a semi-infinite crystal, *J. Phys. C: Solid State Phys.* **15**, 7119–7131 (1982).
[9.16] E. E. Mendez, F. Agulló-Rueda and J. M. Hong, Stark localization in GaAs-GaAℓAs superlattices under an electric field, *Phys. Rev. Lett.* **60**, 2426–2429 (1988).

[9.17] S. R. Wilkinson, C. V. Barucha, K. W. Madison, Q. Niu and M. G. Raizen, Observation of atomic Wannier-Stark ladders in an accelerating optical potential, *Phys. Rev. Lett.* **76**, 4512–4515 (1996).

[9.18] E. Yablonovitch, Inhibited spontaneous emission in solid-state physics and electronics, *Phys. Rev. Lett.* **58**, 2059–2062 (1987).

[9.19] G. Monsivais, M. del Castillo-Mossot and F. Claro, Stark-ladder resonances in the propagation of electromagnetic waves, *Phys. Rev. Lett.* **64**, 1433–1436 (1990).

[9.20] C. M. de Sterke, J. N. Bright, P. A. Krug and T. E. Hammon, Observation of an optical Wannier-Stark ladder, *Phys. Rev. E* **57**, 2365–2370 (1998).

[9.21] J. L. Mateos and G. Monsivais, Stark-ladder resonances in elastic waves, *Physics A* **207**, 445–451 (1994).

[9.22] G. Monsivais, R. Rodríguez-Ramos, R. Esquivel-Sirvent and L. Fernández-Alvarez, Stark-ladder resonances in piezoelectric composites, *Phys. Rev. B* **68**, 174109 (1–11) (2003).

[9.23] L. Gutiérrez, A. Díaz-de-Anda, J. Flores, R. A. Méndez-Sánchez, G. Monsivais and A. Morales, Wannier-Stark ladders in one-dimensional elastic systems, *Phys. Rev. Lett.* **97**, 114301 (1–4) (2006).

[9.24] A. Morales, J. Flores, L. Gutiérrez and R. A. Méndez-Sánchez, Compressional and torsional wave amplitudes in rods with periodic structures, *J. Acoust. Soc. Am.* **112**, 1961–1967 (2002).

[9.25] P. M. Morse and K. Uno Ingard, *Theoretical Acoustics* (McGraw-Hill, New York, 1968), pp. 175–185.

[9.26] See http://gams.nist.gov/ and http://www.netlib.org/

[9.27] A. Morales, L. Gutiérrez and J. Flores, Improved eddy current driver-detector for elastic vibrations, *Am. J. Phys.* **69**, 517–522 (2001).

[9.28] B. Djafari-Rouhani and A. A. Maradudin, Green's function theory of acoustic surface shape resonances, *Solid State Commun.* **73**, 173–177 (1990).

CHAPTER 10

SHEAR HORIZONTAL SURFACE ACOUSTIC WAVES ON GRADED INDEX MEDIA

In 1887 Lord Rayleigh [10.1] showed that a semi-infinite, homogeneous, isotropic elastic medium, bounded by a single, stress-free, planar surface, supports a surface vibrational mode that is wavelike in directions parallel to the surface of the solid with an amplitude that decays exponentially with increasing distance into the solid from its surface with a decay length that is of the order of the wavelength of the wave along the surface. The displacement vector of this wave lies in the sagittal plane, i.e. the plane defined by the direction of propagation of the wave and the normal to the surface. This wave is an acoustic wave in that its frequency is a linear function of the magnitude of the two-dimensional wave vector characterizing its propagation along the surface. It is consequently non-dispersive, i.e. its speed of propagation, either its phase velocity or its group velocity, is independent of its wavelength parallel to the surface. This property is due to the absence of any characteristic length in the system supporting the wave. The frequency of this surface wave lies below the continuum of frequencies allowed the normal vibration modes of an infinite elastic medium for the same value of the two-dimensional wave vector. Such surface acoustic waves are now known as *Rayleigh waves*.

Surface acoustic waves of shear horizontal polarization in which the displacement vector is perpendicular to the sagittal plane, cannot exist in the system studied by Lord Rayleigh.

However, if the constraints of the planarity of the surface and the homogeneity of the elastic medium are lifted, surface acoustic waves that are non-dispersive in the presence of these constraints become dispersive, and other types of acoustic waves localized to the surface, e.g. surface acoustic waves of shear horizontal polarization, become possible.

In this chapter we study the propagation of surface acoustic waves of shear horizontal polarization on the planar surface of an inhomogeneous elastic medium whose mass density and elastic moduli are functions of the distance into the medium from its surface, and on curved surfaces of several types on homogeneous media. The restriction of this study to surface acoustic waves of shear horizontal polarization is prompted by the following consideration. Surface acoustic waves of sagittal polarization on the planar surface of a homogeneous elastic medium are nondispersive. The introduction of inhomogeneity into the material properties of the medium supporting them, or curvature of its boundary, introduces dispersion into their dispersion curves, but is not necessary for their existence. In contrast, inhomogeneity of material properties or surface curvature is essential for the existence of surface acoustic waves of shear horizontal polarization, a much more dramatic consequence of these departures from homogeneity and planarity than their effect on surface acoustic waves of sagittal polarization. References to work in which the latter effects are studied will be given at appropriate points in what follows.

10.1. Surface Acoustic Waves on the Surface of a Gradient Elastic Medium

The earliest studies of surface waves were prompted by the problem of the propagation of seismic shocks in the earth's crust. Because the radius of the earth is much larger than the wavelength of the seismic disturbance, this problem was simplified by neglecting the curvature of the earth and considering the surface of the earth to be a plane bounding a semi-infinite elastic medium. It was on the basis of this model that Lord Rayleigh predicted the existence of the surface acoustic waves, on the stress-free, planar surface of a semi-infinite, homogeneous, isotropic elastic medium that now bear his name [10.1].

Some twenty-five years after the pioneering work on surface acoustic waves by Lord Rayleigh the first calculations were carried out that took into account the fact that the earth's crust has different elastic properties from those of the underlying material. In a planar model this situation is represented by an isotropic plate bonded rigidly to an isotropic half space (or substrate) having different material properties [10.2]. The guided acoustic waves supported by this structure initially were of interest to seismologists. However, they have taken on a new importance in recent years in the context of high frequency surface acoustic wave devices for electronic signal processing [10.3].

Even more complicated structures consisting of multiple layers of different materials are often required in the context of seismological problems. Properties of surface acoustic waves in such layered media are described in the books by Brekhovskikh [10.4] and Ewing, Jardetzky, and Press [10.5].

The displacement vector of the surface acoustic waves predicted by Lord Rayleigh lies in the sagittal plane, i.e. the plane defined by the direction of propagation of the wave and the normal to the surface. Shear horizontal surface acoustic waves, whose displacement vector is perpendicular to the sagittal plane, do not exist on the surface of the homogeneous medium studied by Lord Rayleigh, due to the impossibility of satisfying the stress-free boundary condition on the surface of the semi-infinite medium with a displacement field that decays exponentially with increasing distance into the medium from the surface. The first surface acoustic waves of shear horizontal polarization were predicted by Love [10.2], who studied the vibrational modes of a plate on a semi-infinite medium. These modes have the nature of standing waves in the plate, and their amplitudes decay exponentially with increasing distance into the substrate from the plate-substrate interface. The existence of these Love waves requires that the speed of shear elastic waves in the plate be smaller than the speed of shear elastic waves in the substrate. Their dispersion curve consists of an infinite number of branches.

Although systems consisting of a slab of one material on a substrate of a second material, or of several layers of different materials on a substrate, are met frequently in applications, and provide models for systems whose material properties vary with distance from the surface, situations arise in which it is desirable to take into account a continuous, rather than a discrete, variation of these properties. For example, a sheet of metal that has been cold rolled might be expected to have a mass density and elastic moduli whose values in the vicinity of its surfaces differ from those in the interior, with an essentially continuous variation of these properties across its thickness. Properties of surface acoustic waves propagating on the surface of such a medium have not been studied intensively.

In this section we outline an approach [10.6] to obtaining the dispersion relation and associated displacement field of a surface acoustic wave of shear horizontal polarization propagating along the planar surface of a semi-infinite anisotropic elastic medium occupying the region $z > 0$, in which the mass density ρ and the elastic modulus tensor $C_{\alpha\beta\mu\nu}$ are continuous functions of the distance into the medium from the stress-free surface $z = 0$. We formulate the problem in some generality, but will quickly go to a specific example to illustrate the approach presented here.

We begin by writing the mass density and elastic moduli of the elastic medium in the forms

$$\rho(z) = \theta(z)\hat{\rho}(z) \tag{10.1a}$$

$$C_{\alpha\beta\mu\nu}(z) = \theta(z)\hat{C}_{\alpha\beta\mu\nu}(z). \tag{10.1b}$$

In Eqs. (10.1), $\theta(z)$ is the Heaviside unit step function. Its presence indicates explicitly our assumption that the medium occupies the region $z > 0$. The elastic modulus tensor $C_{\alpha\beta\mu\nu}$ is symmetric in α and β; in μ and ν; and in the interchange of the pairs $\alpha\beta$ and $\mu\nu$.

The equations of motion of the medium are [10.7]

$$\rho(z)\frac{\partial^2}{\partial t^2}u_\alpha(\mathbf{x},t) = \sum_\beta \frac{\partial}{\partial x_\beta}T_{\alpha\beta}(\mathbf{x},t), \tag{10.2}$$

where $u_\alpha(\mathbf{x},t)$ is the α Cartesian component of the elastic displacement field at the point $\mathbf{x}$ in the medium, at the time t, while $T_{\alpha\beta}(\mathbf{x},t)$ is an element of the stress tensor

$$T_{\alpha\beta}(\mathbf{x},t) = \sum_{\mu\nu} C_{\alpha\beta\mu\nu}(z)\frac{\partial}{\partial x_\nu}u_\mu(\mathbf{x},t). \tag{10.3}$$

When Eqs. (10.1) and (10.3) are substituted into Eq. (10.2), we obtain

$$\theta(z)\hat{\rho}(z)\frac{\partial^2}{\partial t^2}u_\alpha(\mathbf{x},t) = \sum_{\mu\nu}\left[\delta(z)\hat{C}_{\alpha z\mu\nu}(0) + \theta(z)\frac{d\hat{C}_{\alpha z\mu\nu}(z)}{dz}\right]\frac{\partial u_\mu(\mathbf{x},t)}{\partial x_\nu}$$

$$+ \sum_{\beta\mu\nu}\theta(z)\hat{C}_{\alpha\beta\mu\nu}(z)\frac{\partial^2 u_\mu(\mathbf{x},t)}{\partial x_\beta\partial x_\nu}, \qquad (10.4)$$

where we have used the result that $d\theta(z)/dz = \delta(z)$. The Dirac delta function in Eq. (10.4) must be interpreted as a one-sided delta function, in the sense that

$$\int_0^\infty \delta(x)dx = 1. \qquad (10.5)$$

Its presence in Eq. (10.4) indicates that the stress-free boundary conditions

$$T_{\alpha z}(\mathbf{x},t)|_{z=0} = 0, \quad \alpha = x, y, z, \qquad (10.6)$$

are incorporated into the equations of motion of the medium. In the following we will understand that $z \geq 0$, and will not write the step functions in Eq. (10.4) explicitly.

In a homogeneous medium surface acoustic waves whose displacement vector is perpendicular to the sagittal plane can exist only if the sagittal plane is a plane of reflection symmetry for the medium or is perpendicular to a two-fold rotation axis [10.8]. This is also true in the present case of an inhomogeneous medium whose material properties are functions only of z. Thus, in what follows we will study the case of a wave of shear horizontal polarization propagating in the [100] direction (x direction) on the (001) surface of a cubic crystal, a geometry in which both of these conditions are satisfied. The nonzero elements of the elastic modulus tensor of a cubic crystal whose cube axes are parallel to the coordinate axes are

$$\begin{aligned}
C_{xxxx} &= C_{yyyy} = C_{zzzz} = C_{11}, \\
C_{xxyy} &= C_{yyzz} = C_{zzxx} = C_{12}, \\
C_{xyxy} &= C_{yzyz} = C_{zxzx} = C_{44},
\end{aligned} \qquad (10.7)$$

(recall the remark following Eq. (10.1b). The last of each of these equations expresses the corresponding elastic moduli in the Voigt contracted notation.

The displacement field in this case has the form

$$\mathbf{u}(\mathbf{x},t) = (0, \hat{u}_y(k,\omega|z), 0)\exp(ikx - i\omega t), \qquad (10.8)$$

and the equations of motion (10.4) reduce to a single equation

$$-\hat{\rho}(z)\omega^2 \hat{u}_y(k,\omega|z) = \hat{C}_{44}(0)\delta(z)\frac{d\hat{u}_y(k,\omega|z)}{dz} + \frac{d\hat{C}_{44}(z)}{dz}\frac{d\hat{u}_y(k,\omega|z)}{dz}$$
$$+ \hat{C}_{44}(z)\left[-k^2\hat{u}_y(k,\omega|z) + \frac{d^2\hat{u}_y(k,\omega|z)}{dz^2}\right]. \qquad (10.9)$$

We can solve Eq. (10.9) for arbitrary dependencies of $\hat{\rho}(z)$ and $\hat{C}_{44}(z)$ on the variable z, subject to the restriction that $(\hat{C}_{44}(0)/\hat{\rho}(0))^{\frac{1}{2}} < (\hat{C}_{44}(\infty)/\hat{\rho}(\infty))^{\frac{1}{2}}$, which is a necessary condition for the existence of Love waves in this medium. We do so by expanding $\hat{u}_y(k,\omega|z)$ according to

$$\hat{u}_y(k,\omega|z) = \sum_{n=0}^{\infty} a_n(k,\omega)\phi_n(z;\alpha), \qquad (10.10)$$

where

$$\phi_n(z;\alpha) = \alpha^{\frac{1}{2}}e^{-\frac{1}{2}\alpha z}L_n(\alpha z) \equiv |n\rangle. \qquad (10.11)$$

In Eq. (10.11) $L_n(x)$ is the nth Laguerre polynomial [10.9]. The first few Laguerre polynomials are

$$L_0(x) = 1, L_1(x) = 1 - x, \quad L_2(x) = 1 - 2x + \frac{1}{2}x^2, \ldots, \qquad (10.12)$$

and higher-order polynomials can be obtained by means of the recurrence relation [10.10]

$$(n+1)L_{n+1}(x) = (2n+1-x)L_n(x) - nL_{n-1}(x), \qquad (10.13)$$

for $n \geq 1$, or from the generating function [10.11]

$$\frac{\exp\{-[s/(1-s)]x\}}{1-s} = \sum_{n=0}^{\infty} s^n L_n(x). \tag{10.14}$$

The parameter α in Eq. (10.11) is arbitrary, and can be varied to improve the rate of convergence of the expansion (10.10). The functions $\{\phi_n(z;\alpha)\}$ are complete and orthonormal in the interval $0 \leq x \leq \infty$,

$$\int_0^{\infty} dz\, \phi_m(z;\alpha)\phi_n(z;\alpha) = \delta_{mn}. \tag{10.15}$$

The $\{\phi_n(z;\alpha)\}$ are a convenient set of basis functions in which to expand $\hat{u}_y(k,\omega|z)$, as they decay exponentially for large z, which is the behavior we expect for $\hat{u}_y(k,\omega|z)$.

If we substitute Eq. (10.10) into Eq. (10.9), multiply the resulting equation from the left by $\phi_m(z;\alpha)$, and integrate the product over z from 0 to ∞, we obtain the matrix equation

$$\omega^2 \sum_{n=0}^{\infty} N_{mn} a_n(k,\omega) = \sum_{n=0}^{\infty} M_{mn}(k) a_n(k,\omega) \quad m = 0,1,2\ldots, \tag{10.16}$$

for the coefficients $\{a_n(k,\omega)\}$ in Eq. (10.10), where

$$N_{mn} = \langle m|\hat{\rho}(z)|n\rangle, \tag{10.17}$$

$$M_{mn}(k) = -\hat{C}_{44}(0)\langle m|\delta(z)\frac{d}{dz}|n\rangle - \langle m|\frac{d\hat{C}_{44}(z)}{dz}\frac{d}{dz}|n\rangle$$
$$+ k^2 \langle m|\hat{C}_{44}(z)|n\rangle - \langle m|\hat{C}_{44}(z)\frac{d^2}{dz^2}|n\rangle, \tag{10.18}$$

and we have introduced the notation

$$\langle m|0|n\rangle = \int_0^{\infty} dz\, \phi_m(z;\alpha) 0(z) \phi_n(z;\alpha). \tag{10.19}$$

The dispersion relation for surface acoustic waves of shear horizontal polarization is obtained by solving the nonstandard eigenvalue equation represented by Eqs. (10.16)–(10.18).

If the functions $\hat{\rho}(z)$ and $\hat{C}_{44}(z)$ are known analytically or numerically, the matrix elements entering Eq. (10.16) can always be evaluated numerically if not analytically. However, if $\hat{\rho}(z)$ and $\hat{C}_{44}(z)$ can be expanded in terms of the descending exponential function,

$$\hat{\rho}(z) = \sum_i \rho_i e^{-\kappa_i z} \tag{10.20a}$$

$$\hat{C}_{44}(z) = \sum_i C_{44}^{(i)} e^{-\lambda_i z}, \tag{10.20b}$$

which suffices to model the depth dependence of a large variety of inhomogeneities, all of the required matrix elements can be evaluated analytically.

To see how this can be done, we start with the generating function for the functions $\{\phi_n(z;\alpha)\}$, that is obtained in a straightforward way from Eqs. (10.11) and (10.14),

$$\alpha^{\frac{1}{2}} \frac{e^{-\frac{1}{2}\alpha z \frac{1+s}{1-s}}}{1-s} = \sum_{m=0}^{\infty} s^m \phi_m(z;\alpha). \tag{10.21}$$

Therefore, for example,

$$\int_0^\infty dz \alpha^{\frac{1}{2}} \frac{e^{-\frac{1}{2}\alpha z \frac{1+s}{1-s}}}{1-s} e^{-\beta z} \alpha^{\frac{1}{2}} \frac{e^{-\frac{1}{2}\alpha z \frac{1+t}{1-t}}}{1-t}$$

$$= \sum_{m=0}^{\infty} \sum_{n=0}^{\infty} s^m t^n \langle m | e^{-\beta z} | n \rangle$$

$$= \frac{\alpha}{(\alpha + \beta - \beta t) - (\beta + (\alpha - \beta)t)s}$$

$$= \alpha \sum_{m=0}^{\infty} s^m \frac{(\beta + (\alpha - \beta)t)^m}{(\alpha + \beta - \beta t)^{m+1}}$$

$$= \alpha \sum_{m=0}^{\infty} s^m \sum_{n=0}^{\infty} t^n \sum_{p=0}^{\min(m,n)} \frac{(m+n-p)!}{(m-p)!(n-p)!p!}$$

$$\times \frac{(\alpha - \beta)^p \beta^{m+n-2p}}{(\alpha + \beta)^{m+n+1-p}}. \tag{10.22}$$

On equating the coefficients of $s^m t^n$ on both sides of this equation we obtain

$$\langle m|e^{-\beta z}|n\rangle = \sum_{p=0}^{\min(m,n)} \frac{(m+n-p)!}{(m-p)!(n-p)!p!} \frac{\alpha(\alpha-\beta)^p \beta^{m+n-2p}}{(\alpha+\beta)^{m+n+1-p}}. \tag{10.23}$$

In a similar fashion we obtain the additional results

$$\langle m|n\rangle = \delta_{mn}, \tag{10.24a}$$

$$\langle m|\delta(z)\frac{d}{dz}|n\rangle = -\alpha^2\left(n+\frac{1}{2}\right), \tag{10.24b}$$

$$\langle m|e^{-\beta z}\frac{d}{dz}|n\rangle = -\frac{2}{\alpha}a_{mn} - \frac{4}{\alpha}\sum_{p=0}^{n-1} a_{mp}, \tag{10.24c}$$

$$\langle m|\frac{d}{dz}|n\rangle = -\frac{\alpha}{2}\delta_{mn} - \alpha\theta(n-m-1), \tag{10.24d}$$

$$\langle m|\frac{d^2}{dz^2}|n\rangle = \frac{\alpha^2}{4}\delta_{mn} + \alpha^2(n-m)\theta(n-m-1), \tag{10.24e}$$

$$\langle m|e^{-\beta z}\frac{d^2}{dz^2}|n\rangle = a_{mn} + 4\sum_{p=0}^{n-1} a_{mp}(n-p), \tag{10.24f}$$

where $\theta(n) = 1$ for $n = 0, 1, 2, \ldots$, and $\theta(n) = 0$ for $n = -1, -2, -3, \ldots$, while

$$a_{mn} = \frac{\alpha^3}{4} \sum_{p=0}^{\min(m,n)} \frac{(m+n-p)!}{(m-p)!(n-p)!p!} \frac{(\alpha-\beta)^p \beta^{m+n-2p}}{(\alpha+\beta)^{m+n+1-p}}. \tag{10.25}$$

To illustrate the use of the preceding results, we choose

$$\hat{\rho}(z) = \hat{\rho}(\infty)(1+Ae^{-az}) \tag{10.26a}$$

$$\hat{C}_{44}(z) = \hat{C}_{44}(\infty)(1+Be^{-bz}). \tag{10.26b}$$

Table 10.1. The parameters defining the different speed profiles studied.

Case	A	a (μm^{-1})	B	b (μm^{-1})
1	1	1	0.5	1.5
2	1	1	0.5	1
3	1	1	0.5	2

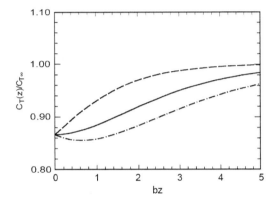

Fig. 10.1. The relative transverse speed of sound as a function of bz for the speed profiles generated by the parameters presented in Table 10.1. Cases 1, 2, and 3 correspond to the solid, dashed, and dash-dotted curves [10.6].

The different values of A, a, B, b used are given in Table 10.1. We define a depth dependent speed of transverse waves $c_T(z)$ by

$$\frac{c_T^2(z)}{c_T^2(\infty)} = \frac{1 + Be^{-bz}}{1 + Ae^{-az}}, \qquad (10.27)$$

where $c_T^2(\infty) = \hat{C}_{44}(\infty)/\hat{\rho}(\infty)$, and plot this speed in Fig. 10.1 for the values of the parameters given in Table 10.1. In each case the speed at the surface is smaller than the speed at infinite depth. This property of the speeds makes guided (Love-like) waves possible; therefore, at a given frequency ω several waves with different wave numbers k can exist.

The nonstandard eigenvalue problem posed by Eq. (10.16) can be solved, for example, by the EISPAK subroutine RGG [10.12].

The number of Laguerre functions kept in the expansion (10.10) in order to obtain a converged solution determines the size of the matrix eigenvalue equation (10.16). So, in general, at a given frequency more eigenvalues are obtained than the number of physically possible modes at that frequency. In all cases the unphysical modes have phase velocities greater than that of a wave with the transverse speed $c_T(\infty)$, the largest speed consistent with the infinite depth solution to the wave equation. Plots of the displacement field for these frequencies show increasing, rather than decreasing amplitudes as the depth increases.

The dispersion curves for the three lowest frequency Love-like modes corresponding to the Case 1 speed profile are plotted in Fig. 10.2. It should be emphasized that surface acoustic waves of this polarization do not exist in the absence of the depth dependence of $\hat{\rho}(z)$ and $\hat{C}_{44}(z)$. These modes are clearly dispersive, which is due to the presence of characteristic lengths, a^{-1} and b^{-1}, in the system being studied. In performing these calculations the parameter α entering Eq. (10.11) was chosen to equal the parameter b entering Eq. (10.26b). No effort was made to optimize its value to accelerate the convergence of the series (10.10). Except for small k/b and large k/b, where 20 to 30 Laguerre functions were needed in Eq. (10.10) to obtain four significant figure accuracy for the frequencies, only the first three Laguerre functions were needed in Eq. (10.10) to achieve

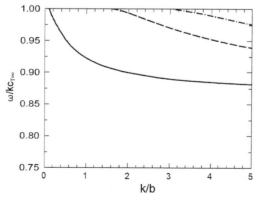

Fig. 10.2. The three lowest frequency branches of the dispersion curve for a shear horizontal surface wave corresponding to the Case 1 speed profile [10.6].

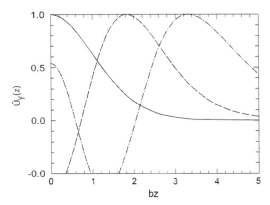

Fig. 10.3. The displacement amplitude $\hat{u}_y(k,\omega|z)$ as a function of bz for the modes at $k/b = 5$ shown in Fig. 10.2. The line types correspond to those in the latter figure [10.6].

the same kind of accuracy. In general, the frequency of a given mode converges more rapidly than the frequency of the next higher frequency mode. In Fig. 10.3 we show the displacement $\hat{u}_y(k,\omega|z)$ calculated for $k/b = 5$ as a function of depth for the three modes whose dispersion curves are shown in Fig. 10.2. The amplitudes of these physically acceptable solutions clearly decay with increasing z.

10.2. Surface Acoustic Waves on Curved Surfaces

We have seen in Sec. 7.2 that the curvature of a vacuum-metal interface along which a surface plasmon polariton propagates can produce surface electromagnetic waves that cannot exist at a planar interface. The same is true of the propagation of a surface acoustic wave on a curved stress-free surface of a solid. The curvature of such a surface can bind to it a surface acoustic wave that does not exist on a planar surface, such as a wave of shear horizontal polarization. That this can happen can be understood in a qualitative way by noting that if a coordinate transformation is carried out that maps the curved surface into a planar surface, the equation of motion in the new coordinate system contains additional terms that give it the form of a wave equation with a speed of transverse sound that increases with distance from the surface. As we have seen in Sec. 10.1 this is just the condition required for the binding of a shear horizontal surface acoustic wave to a planar surface. In addition, the curvature of a surface can make a Rayleigh surface acoustic wave, which is non-dispersive when it propagates on a planar surface, into a dispersive surface wave.

In this section we will investigate the propagation of surface acoustic waves on two different types of curved surfaces. We begin with a study of the propagation of a surface wave circumferentially around a portion of a cylindrical surface, a situation in which the radius of curvature of the surface is a constant. We then consider an example of the propagation of a surface wave circumferentially around a surface with a variable radius of curvature. These two forms of surface curvature can make Rayleigh waves dispersive, and can bind shear horizontal surface waves to the surface. However, as was discussed in Sec. 10, in this discussion of surface acoustic waves on curved surfaces we will confine our attention to the case of surface acoustic waves of shear horizontal polarization, because their existence requires the presence of the curvature, while surface acoustic waves of sagittal polarization already exist in its absence. Finally, we will study the propagation of a shear horizontal surface acoustic wave on a periodically corrugated surface and a randomly rough surface. We will find that these kinds of departures of a surface from planarity can also bind a wave of this polarization, which does not exist in the absence of the roughness.

Although in some of these investigations we will be using curvilinear coordinate systems in which the surface over which the wave propagates is a surface of constant coordinate, it is convenient to set the stage for these calculations with some general comments concerning the propagation of a wave on a curved surface.

The system we consider consists of an elastic medium in the region $y < \zeta(x)$ and vacuum in the region $y > \zeta(x)$. Thus the system is invariant in the z direction. The surface profile function $\zeta(x)$ is assumed to be a single-valued function of x that is differentiable.

The equations of motion of the elastic medium within the linear theory of elasticity, are [10.7]

$$\rho \ddot{u}_\alpha = \sum_\beta \frac{T_{\alpha\beta}}{\partial x_\beta}, \quad \alpha = x, y, z, \tag{10.28}$$

where ρ is the mass density of the medium, assumed to be constant, $u_\alpha(\mathbf{x}; t)$ is the α Cartesian component of the displacement of the medium at the point $\mathbf{x}$ and the time t, and $T_{\alpha\beta}$ is an element of the stress tensor. The latter is given by Hooke's law.

$$T_{\alpha\beta} = \sum_{\mu\nu} C_{\alpha\beta\mu\nu} \frac{\partial u_\mu}{\partial x_\nu}, \quad \alpha, \beta = x, y, z, \qquad (10.29)$$

where the $\{C_{\alpha\beta\mu\nu}\}$ are the elements of a fourth rank tensor called the *elastic modulus tensor*.

When we combine Eqs. (10.28) and (10.29) we obtain the equation of motion of the medium in the form

$$\rho \ddot{u}_\alpha = \sum_{\beta\mu\nu} C_{\alpha\beta\mu\nu} \frac{\partial^2 u_\mu}{\partial x_\beta \partial x_\nu}, \quad \alpha = x, y, z. \qquad (10.30)$$

If the elastic medium occupies a volume V bounded by a surface S that is assumed to be stress free, the boundary conditions on the displacement field that express this assumption can be written as

$$\sum_\beta T_{\alpha\beta} \hat{n}_\beta \bigg|_S = 0, \quad \alpha = x, y, z, \qquad (10.31)$$

where $\hat{\mathbf{n}}$ is the unit vector normal to the surface S at each point of it, directed away from the volume V.

For an isotropic elastic medium, which we assume here due to its simplicity, the elastic modulus tensor is given by [10.13]

$$C_{\alpha\beta\mu\nu} = \rho(c_\ell^2 - 2c_t^2)\delta_{\alpha\beta}\delta_{\mu\nu} + \rho c_t^2 (\delta_{\alpha\mu}\delta_{\beta\nu} + \delta_{\alpha\nu}\delta_{\beta\mu}), \qquad (10.32)$$

where c_ℓ and c_t are the speeds of longitudinal and transverse sound waves in the medium, respectively.

We apply the preceding results to the case of a shear horizontal wave whose sagittal plane is the xy plane. The single nonzero component of the displacement field can be written in the form $u_z(x, y|\omega) \exp(-i\omega t)$. With the use of Eq. (10.32) this component of the elastic displacement field is found to satisfy the equation of motion

$$\left(\frac{\partial^2}{\partial x^2} + \frac{\partial^2}{\partial y^2} + \frac{\omega^2}{c_t^2}\right) u_z(x,y|\omega) = 0, \quad (10.33)$$

in the region $y < \zeta(x)$.

The unit vector normal to the surface $y = \zeta(x)$ and directed from the elastic medium into the vacuum is

$$\hat{\mathbf{n}} = \frac{(-\zeta'(x), 1, 0)}{[1 + (\zeta'(x))^2]^{\frac{1}{2}}}, \quad (10.34)$$

where the prime denotes differentiation with respect to x. In the geometry assumed there are only two nonzero elements of the stress tensory. These are

$$T_{xz} = T_{zx} = \rho c_t^2 \frac{\partial u_z}{\partial x}, \quad (10.35a)$$

$$T_{zy} = T_{yz} = \rho c_t^2 \frac{\partial u_z}{\partial y}. \quad (10.35b)$$

On combining Eqs. (10.34) and (10.35), the boundary condition on the surface $z = \zeta(x)$ becomes

$$(T_{zx}\hat{n}_x + T_{zy}\hat{n}_y)|_{y=\zeta(x)} = \rho c_t^2 \left(\frac{\partial u_z}{\partial x}\hat{n}_x + \frac{\partial u_z}{\partial y}\hat{n}_y\right)_{y=\zeta(x)}$$

$$= \frac{\rho c_t^2}{[1 + (\zeta'(x))^2]^{\frac{1}{2}}}\left[-\zeta'(x)\frac{\partial u_z}{\partial x} + \frac{\partial u_z}{\partial y}\right]_{y=\zeta(x)}$$

$$= \rho c_t^2 \frac{\partial}{\partial n} u_z(x,z|\omega)\bigg|_{y=\zeta(x)} = 0, \quad (10.36)$$

where

$$\frac{\partial}{\partial n} = \frac{1}{[1 + (\zeta'(x))^2]^{\frac{1}{2}}}\left[-\zeta'(x)\frac{\partial}{\partial x} + \frac{\partial}{\partial y}\right], \quad (10.37)$$

is the derivative along the normal to the surface $y = \zeta(x)$ at each point of it, directed from the solid into the vacuum.

10.2.1. Surface acoustic waves on a cylindrical surface

The circumferential propagation of surface acoustic waves around a cylindrical boundary was studied first by Lord Rayleigh [10.14], who showed that in addition to the Rayleigh wave other types of waves bound to the curved surface can exist. These waves have come to be called "whispering gallery" waves.

The study of surface acoustic waves propagating circumferentially around a circular cylinder was subsequently studied theoretically [10.15–10.20] and experimentally [10.16] by several authors. In this section we follow primarily the treatment of Brekhovskikh [10.19].

The system we consider consists of a cylindrical surface of radius R in contact with vacuum. The cylindrical surface can be either convex toward the vacuum or concave toward it. We introduce a cylindrical coordinate system (r, θ, z), where the z axis is directed along the axis of the cylinder, while r and θ are polar coordinates in a plane perpendicular to the z axis. In this and the next section we consider only the plane problem in which the displacement field is independent of the coordinate z. We also assume a harmonic time dependence of the displacement field of the form $\exp(-i\omega t)$.

In cylindrical coordinates the elements of the stress tensor σ_{ij} for an isotropic elastic medium are given by [10.21]

$$\sigma_{ii} = \lambda \theta + 2\mu e_{ii}, \tag{10.38a}$$

$$\sigma_{ij} = 2\mu e_{ij} \quad (i \neq j), \tag{10.38b}$$

where λ and μ are the Lamé constants,

$$\theta = \nabla \cdot \mathbf{u} = e_{rr} + e_{\theta\theta} + e_{zz}, \tag{10.39}$$

and the elements of the strain tensor are

$$e_{rr} = \frac{\partial u_r}{\partial r}, \tag{10.40a}$$

$$e_{\theta\theta} = \frac{1}{r}\frac{\partial \mu_\theta}{\partial \theta} + \frac{1}{r}u_r, \tag{10.40b}$$

$$e_{zz} = \frac{\partial u_z}{\partial z}, \tag{10.40c}$$

$$e_{r\theta} = \frac{1}{2}\left(\frac{1}{r}\frac{\partial u_r}{\partial \theta} + \frac{\partial u_\theta}{\partial r} - \frac{u_\theta}{r}\right) = e_{\theta r}, \qquad (10.40\text{d})$$

$$e_{rz} = \frac{1}{2}\left(\frac{\partial u_z}{\partial r} + \frac{\partial u_r}{\partial z}\right) = e_{zr}, \qquad (10.40\text{e})$$

$$e_{\theta z} = \frac{1}{2}\left(\frac{\partial u_\theta}{\partial z} + \frac{1}{r}\frac{\partial u_z}{\partial \theta}\right) = e_{z\theta}. \qquad (10.40\text{f})$$

In these expressions $u_r(r,\theta,z;t)$, $u_\theta(r,\theta,z;t)$, and $u_z(r,\theta,z;t)$ are the components of the displacement vector.

The equations of motion of the medium are

$$\rho \ddot{u}_r = \frac{\partial \sigma_{rr}}{\partial r} + \frac{1}{r}\frac{\partial \sigma_{r\theta}}{\partial \theta} + \frac{\partial \sigma_{rz}}{\partial z} + \frac{\sigma_{rr} - \sigma_{\theta\theta}}{r}, \qquad (10.41\text{a})$$

$$\rho \ddot{u}_\theta = \frac{\partial \sigma_{r\theta}}{\partial r} + \frac{1}{r}\frac{\partial \sigma_{\theta\theta}}{\partial \theta} + \frac{\partial \sigma_{\theta z}}{\partial z} + \frac{2}{r}\sigma_{r\theta}, \qquad (10.41\text{b})$$

$$\rho \ddot{u}_z = \frac{\partial \sigma_{rz}}{\partial r} + \frac{1}{r}\frac{\partial \sigma_{\theta z}}{\partial \theta} + \frac{\partial \sigma_{zz}}{\partial \theta} + \frac{1}{r}\sigma_{rz}, \qquad (10.41\text{c})$$

where ρ is the mass density of the medium. These equations have to be supplemented by the boundary conditions on the surface $r = R$ of the cylinder, namely that the stresses acting on the surface vanish.

We now apply these results to the study of the propagation of surface acoustic waves of shear horizontal polarization around a portion of the cylinder. The displacement vector for such waves is parallel to the axis of the cylinder, and has the form

$$\mathbf{u}(r,\theta,z;t) = \hat{\mathbf{z}} u_z(r,\theta|\omega)\exp(-i\omega t). \qquad (10.42)$$

From Eqs. (10.40) we find that the nonzero elements of the strain tensor in this case are

$$e_{rz} = \frac{1}{2}\frac{\partial u_z}{\partial r}, \qquad e_{\theta z} = \frac{1}{2}\frac{\partial u_z}{\partial \theta}, \qquad (10.43)$$

so that the nonzero elements of the stress tensor are

$$\sigma_{rz} = \mu \frac{\partial u_z}{\partial r}, \qquad \sigma_{\theta z} = \frac{\mu}{r}\frac{\partial u_z}{\partial \theta}. \qquad (10.44)$$

Equations (10.41c) and (10.44) yield the equation satisfied by the amplitude $u_z(r,\theta|\omega)$,

$$\left(\frac{\partial^2}{\partial r^2} + \frac{1}{r}\frac{\partial}{\partial r} + \frac{1}{r^2}\frac{\partial^2}{\partial \theta^2} + \frac{\omega^2}{c_t^2}\right) u_z(r,\theta|\omega) = 0, \qquad (10.45)$$

where we have introduced the speed of transverse acoustic waves c_t through the relation $\mu/\rho = c_t^2$. This equation must be supplemented by the condition that the stresses acting on the surface $r = R$ of the cylinder must vanish. The only stress acting on the surface is σ_{rz}, so that from Eq. (10.44) the boundary condition can be written as

$$\sigma_{rz}|_{r=R} = \mu \frac{\partial u_z}{\partial r}\bigg|_{r=R} = 0. \qquad (10.46)$$

We solve Eq. (10.45) by separating the variables. We write

$$u_z(r,\theta|\omega) = R(r)\Theta(\theta), \qquad (10.47)$$

and find that $\Theta(\theta)$ and $R(r)$ satisfy the equations

$$\left(\frac{d^2}{d\theta^2} + \nu^2\right)\Theta = 0, \qquad (10.48a)$$

$$\left(\frac{d^2}{dr^2} + \frac{1}{r}\frac{d}{dr} + \frac{\omega^2}{c_t^2} - \frac{\nu^2}{r^2}\right)R = 0, \qquad (10.48b)$$

respectively, where ν^2 is the separation constant. The sign of ν^2 has been chosen in such a way that the solution of Eq. (10.48a) describes a wave propagating around the cylinder in a clockwise fashion, namely

$$\Theta(\theta) = e^{i\nu\theta}. \qquad (10.49)$$

Because we are considering propagation of the wave over only a portion of a circular boundary, it is not necessary to impose a single-valuedness requirement on the displacement component $u_z(r,\theta|\omega)$, and hence on $\Theta(\theta)$. Thus, ν need not be an integer. If we rewrite Eq. (10.49) in the form

$$\Theta(\theta) = e^{i\frac{\nu}{R}(R\theta)}, \qquad (10.50)$$

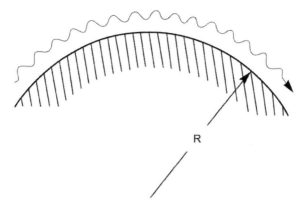

Fig. 10.4. A shear horizontal surface acoustic wave propagating circumferentially around a portion of a cylindrical surface on an isotropic elastic medium that is convex toward the vacuum.

and recall that $R\theta$ is the path length measured along the cylindrical surface, we see that $\nu/R \equiv k$ can be regarded as the wave number characterizing the propagation of this circumferential wave.

In solving Eq. (10.48b) there are two cases to consider: (i) the solid is convex toward the vacuum; and (ii) the solid is concave toward the vacuum. We consider these two cases in turn.

(i) Solid Convex Toward the Vacuum

In the case that the solid is convex to the vacuum (Fig. 10.4) we seek a solution of Eq. (10.48b) that increases with increasing r, as is required of a wave that is bound to the surface $r = R$. The solutions of Eq. (10.48b) are Bessel functions, and from among the several types of these functions we choose for the solution of Eq. (10.48b)

$$R(r) = J_\nu\left(\frac{\omega}{c_t}r\right), \qquad (10.51)$$

where $J_\nu(x)$ is a Bessel function of the first kind of order ν. The displacement component $u_z(r,\theta|\omega)$ thus takes the form

$$u_z(r,\theta|\omega) = \text{const.} J_\nu\left(\frac{\omega}{c_t}r\right)e^{i\nu\theta}. \qquad (10.52)$$

Substitution of Eq. (10.52) into the boundary condition (10.46) yields the equation that relates the wave number $k = \nu/R$ and the frequency ω of the circumferential wave,

$$J'_\nu\left(\frac{\omega}{c_t}R\right) = 0, \tag{10.53}$$

where the prime denotes differentiation with respect to argument.

Our choice of the Bessel function $J_\nu(\frac{\omega}{c_t}r)$ as the solution of Eq. (10.48b) was dictated by the following property of this function. For a fixed value of (real, nonzero) ν, $J_\nu(x)$ increases exponentially with increasing x, until a value $x \sim \nu$ is reached, at which it acquires an oscillatory dependence on x that continues for $x > \nu$. Since we are interested in a solution for r in the range $0 < r < R$ that is localized for r in the vicinity of R, i.e. tends to zero as $r \to 0$, $J_\nu((\omega/c_t)r)$ has this behavior provided that ν is of the order of $(\omega/c_t)R$. This requirement on ν is consistent with the boundary condition (10.53), which requires that $r = R$ be in the range of r where $J_\nu((\omega/c_t)r)$ has an oscillatory behavior rather than an exponentially increasing nature.

To solve Eq. (10.53) we therefore need a representation of $J'_\nu(x)$ in the transition region where the order ν is comparable to the argument x. For a fixed value of z, large $|\nu|$, and $|\arg \nu| < \pi/2$, we have the result that [10.22]

$$\begin{aligned}J'_\nu(\nu + z\nu^{\frac{1}{3}}) &= -\frac{2^{2/3}}{\nu^{2/3}}Ai'(-2^{\frac{1}{3}}z)\left\{1 - \frac{4}{5}\frac{z}{\nu^{2/3}} + \cdots\right\} \\ &+ \frac{2^{1/3}}{\nu^{4/3}}Ai(-2^{\frac{1}{3}}z)\left\{\left(\frac{3}{5}z^3 - \frac{1}{5}\right) + \cdots\right\} + O(\nu^{-2}),\end{aligned} \tag{10.54}$$

where $Ai(-x)$ is an Airy function. If we now set

$$\frac{\omega}{c_t}R = \nu + z\nu^{2/3}, \tag{10.55}$$

then

$$\frac{\omega}{c_t} = k\left[1 + \frac{z}{(kR)^{2/3}}\right]. \tag{10.56}$$

Thus, if we wish to obtain (ω/c_t) to the lowest nonzero order in $(kR)^{-1}$, we see from Eq. (10.56) that we need only the approximation to z that is independent of $\nu = kR$. From Eqs. (10.53) and (10.54) we see that this approximation is given by the solutions of

$$Ai'(-2^{1/3}z) = 0. \tag{10.57}$$

The only zeros of $Ai'(x)$ occur for negative values of x. If we denote these zeros by $-x_i$, $i = 1, 2, 3 \ldots$, the first few of them are [10.23]

$$x_1 = 1.019, \quad x_2 = 3.248, \quad x_3 = 4.820, \ldots \tag{10.58}$$

The dispersion relation (10.56) thus finally takes the form

$$\left(\frac{\omega}{c_t}\right)_i = k\left[1 + \frac{x_i}{2^{1/3}(kR)^{2/3}}\right], \quad i = 1, 2, 3, \ldots \tag{10.59}$$

We have shown that a homogeneous, isotropic elastic cylinder with a stress-free surface that is convex to vacuum can support an infinite number of shear horizontal surface acoustic waves that propagate circumferentially around its surface. These waves are dispersive, i.e. their phase velocities are functions of their wavelength. This feature is due to the presence of a characteristic length in the problem, namely the radius R of the cylinder. These phase velocities are also larger than the speed c_t of bulk shear waves in the medium. Such waves are not possible if the surface is planar. They are bound to the surface by its curvature. Indeed, from Eq. (10.59) we see that in the limit as $R \to \infty$ the phase velocities of all of the waves approach c_t, i.e. they become surface skimming bulk transverse waves.

The surface acoustic waves studied in this section are the whispering gallery waves that were first studied by Lord Rayleigh [10.14].

We noted in Sec. 10.2 that the binding of a shear horizontal wave to the surface of a circular cylinder arises because a coordinate transformation that maps the circular surface of the cylinder into a planar surface introduces terms into the transformed equation of motion that impart a depth dependence to the speed of transverse waves of the kind that gives rise to Love-like waves. To see this we carry out the coordinate transformation

$$z = R\ln\frac{R}{r}, \tag{10.60}$$

and define $R(r) = \hat{R}(z) = \hat{R}(R\ln(R/r))$. The region $z > 0$ corresponds to the interior of the cylinder ($0 < r < R$), while the region $z > 0$ corresponds to the vacuum outside the cylinder ($r > R$). Equation (10.48b) is then transformed into

$$\left[\frac{d^2}{dz^2} - k^2 + \frac{\omega^2}{c_t^2(z)}\right]\hat{R}(z) = 0, \qquad (10.61)$$

where $k = \nu/R$, and

$$c_t^2(z) = c_t^2 \exp\left(2\frac{z}{R}\right). \qquad (10.62)$$

The boundary condition satisfied by $\hat{R}(z)$ becomes

$$\left.\frac{d}{dz}\hat{R}(z)\right|_{z=0} = 0. \qquad (10.63)$$

Equation (10.61) resembles the equation of motion of shear horizontal waves on a planar surface of a medium whose speed of transverse sound waves increases exponentially with increasing distance into the medium. A comparison of Eq. (10.61) with Eq. (10.9), shows that this analogy is imperfect, however, because the latter equation of motion contains terms not present in Eq. (10.61) that are associated with a spatial derivative of the elastic modulus $\hat{C}_{44}(z)$. Nevertheless, the analogy is a good one for a slowly varying $\hat{C}_{44}(z)$. Since $c_t(z)$ is smaller at the surface of the medium than in the interior, this is just the kind of situation that gives rise to Love waves. Thus, we can say that shear horizontal surface acoustic waves propagating circumferentially around the stress-free surface of an isotropic elastic cylinder that is convex toward vacuum are analogous to Love waves on a planar stress-free surface in which c_t increases exponentially with increasing distance into the medium from the surface.

In the immediate vicinity of the surface, where z/R is small, Eq. (10.61) takes the form

$$\left[\frac{d^2}{dz^2} - k^2 + \frac{\omega^2}{c_t^2} + 2\frac{\omega^2}{c_t^2}\frac{z}{R}\right]\hat{R}(z) = 0, \qquad (10.64)$$

which has the solution

$$\hat{R}(z) = Ai\left[\left(\frac{Rc_t^2}{2\omega^2}\right)^{2/3}\left(2\frac{\omega^2}{c_t^2}\frac{z}{R} + k^2 - \frac{\omega^2}{c_t^2}\right)\right]. \quad (10.65)$$

The dispersion relation obtained from the boundary condition (10.63) is

$$Ai'\left(\left(\frac{Rc_t^2}{2\omega^2}\right)^{2/3}\left(k^2 - \frac{\omega^2}{c_t^2}\right)\right) = 0. \quad (10.66)$$

Thus, we have that

$$\left(\frac{Rc_t^2}{2\omega^2}\right)^{2/3}\left(k^2 - \frac{\omega^2}{c_t^2}\right) = -x_i, \quad (10.67)$$

from which we obtain

$$\left(\frac{\omega}{c_t}\right)_i = k\left[1 + \frac{x_i}{2^{1/3}(kR)^{2/3}}\right], \quad (10.68)$$

which is just the result obtained earlier (Eq. (10.59)).

These whispering gallery waves are often interpreted in terms of rays propagating along the boundary and undergoing successive reflections from it (Fig. 10.5). A whispering gallery wave whose phase velocity is $c_i = c_t[1 + x_i/2^{1/3}(kR)^{2/3}]$ impinges on the surface of the cylinder at an angle χ_i measured from the tangent to the surface at

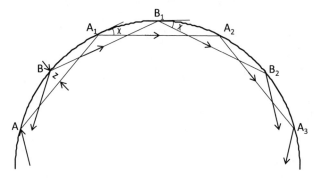

Fig. 10.5. Whispering gallery waves interpreted as rays propagating along the cylindrical boundary of a solid and undergoing successive reflections from it (rays $AA_1A_2\ldots,BB_1B_2\ldots$) [10.19].

the point of impact, and is reflected from the surface at the same angle from the tangent. This tangent is perpendicular to the radius vector to this point. The value of χ_i is obtained from

$$\operatorname{tg} \chi_i = \frac{[(\omega_i/c_t)^2 - k^2]^{\frac{1}{2}}}{k}, \tag{10.69}$$

where $k = (\omega_i/c_t)[1 + x_i/2^{1/3}(kR)^{2/3}]^{-1}$. In the limit that $x_i/2^{1/3}(kR)^{2/3} \ll 1$, Eq. (10.69) becomes

$$\operatorname{tg} \chi_i \cong \left(\frac{2}{kR}\right)^{1/3} x_i^{1/2}, \tag{10.70}$$

or

$$\chi_i \cong \left(\frac{2}{kR}\right)^{1/3} x_i^{1/2}. \tag{10.71}$$

As the wave travels around the boundary, due to the circularity of the latter and the fact that the tangent to it is perpendicular to the radius vector at each point of it, the wave strikes the surface and is scattered from it at equidistant points along it, with the angles of incidence and scattering given by χ_i measured from the tangent at each point of contact. It follows from simple geometrical considerations that the maximum departure of the ray from the boundary is $z = (R/2)\chi_i^2$.

(ii) Solid Concave to Vacuum

When the cylinder is concave to the vacuum (Fig. 10.6) we choose the solution of Eq. (10.48b) to be

$$R(r) = H_\nu^{(1)}\left(\frac{\omega}{c_t}r\right), \tag{10.72}$$

where $H_\nu^{(1)}(x) = J_\nu(x) + iY_\nu(x)$ is a Hankel function of the first kind and order ν. For a fixed value of ν and $|x| \to \infty$ $H_\nu^{(1)}(x)$ has the asymptotic form

$$H_\nu^{(1)}(x) \sim \left(\frac{2}{\pi x}\right)^{\frac{1}{2}} \exp\left[i\left(x - \frac{1}{2}\nu\pi - \frac{\pi}{4}\right)\right]. \tag{10.73}$$

This choice for $R(r)$ describes a wave that radiates energy into the interior of the medium from the surface. It is therefore attenuated as

Shear Horizontal Surface Acoustic Waves on Graded Index Media

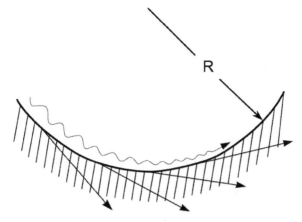

Fig. 10.6. A shear horizontal surface acoustic wave propagating circumferentially around a portion of a cylindrical surface on an isotropic elastic medium that is concave toward the vacuum.

it propagates around the cylinder. This means that the wave number $k = \nu/R$ that is the solution of the dispersion relation

$$H_\nu^{(1)}\left(\frac{\omega}{c_t}R\right)' = 0, \tag{10.74}$$

which follows from Eqs. (10.36), (10.37), and (10.59) is complex for real ω, $k = k_R + ik_I$, with both k_R and k_I positive.

Solutions to the equation $H_\nu^{(1)}(x)' = 0$ in the limit as $x \to \infty$ were obtained by Franz [10.24]. The first three of them are

$$\nu = x + e^{i\frac{\pi}{3}}x^{1/3}0.808617 - e^{-i\frac{\pi}{3}}x^{-1/3}0.145463 + \cdots \tag{10.75a}$$

$$\nu = x + e^{i\frac{\pi}{3}}x^{1/3}2.578096 - e^{-i\frac{\pi}{3}}x^{-1/3}0.260341 + \cdots \tag{10.75b}$$

$$\nu = x + e^{i\frac{\pi}{3}}x^{1/3}3.825715 - e^{-i\frac{\pi}{3}}x^{-1/3}0.514009 + \cdots \tag{10.75c}$$

Let us consider the solution given by Eq.(10.75a). With the replacements $\nu = kR$ and $x = \omega R/c_t$, it becomes

$$kR = \frac{\omega}{c_t}R + e^{i\frac{\pi}{3}}\left(\frac{\omega}{c_t}R\right)^{\frac{1}{3}}0.808617 - e^{-i\frac{\pi}{3}}\left(\frac{\omega}{c_t}R\right)^{-\frac{1}{3}}0.145463 + \cdots \tag{10.76}$$

On separating the right-hand side of this equation into its real and imaginary parts we obtain $k = k_R + ik_I$ where

$$k_R = \frac{\omega}{c_t}\left[1 + \frac{0.40431}{(\frac{\omega}{c_t}R)^{2/3}} - \frac{0.07273}{(\frac{\omega}{c_t}R)^{4/3}} + 0\left(\left(\frac{\omega}{c_t}R\right)^{-2}\right)\right], \quad (10.77a)$$

$$k_I = \frac{\omega}{c_t}\left[\frac{0.70028}{(\frac{\omega}{c_t}R)^{2/3}} + \frac{0.12597}{(\frac{\omega}{c_t}R)^{4/3}} + 0\left(\left(\frac{\omega}{c_t}R\right)^{-2}\right)\right]. \quad (10.77b)$$

Similar results are obtained for the remaining modes.

Thus, when the elastic medium is concave to the vacuum an infinite number of leaky shear horizontal, surface acoustic waves can propagate circumferentially around its cylindrical boundary. These waves are attenuated as they propagate because they radiate energy into the interior of the solid. The wave number k of each mode is greater than that of the surface skimming bulk shear acoustic wave on a planar surface, ω/c_t, and approaches the latter as the radius R of the cylinder approaches infinity. In this limit the attenuation of the wave vanishes. Surface acoustic waves of this nature do not exist on a planar surface.

10.2.2. A variable radius of curvature

The cylinder considered in the preceding section, around which a shear horizontally polarized surface acoustic wave propagated circumferentially, had a constant radius of curvature, namely its radius R. It was found that when the cylinder is convex to the vacuum an infinite number of surface acoustic waves exists, while when the cylinder is concave to the vacuum an infinite number of leaky surface acoustic waves exists. It is natural to ask what happens when the cylinder has a variable radius of curvature.

In this section we examine the propagation of a shear horizontal surface acoustic wave circumferentially around a portion of the surface of a solid bounded by a parabolic profile. In this case it is convenient to work in the parabolic cylinder coordinate system (ξ, η, z) defined by [10.25]

$$x = \xi\eta, \quad -\infty \leq \xi \leq \infty, \quad 0 \leq \eta \leq \infty \quad (10.78a)$$

$$y = \frac{1}{2}(\eta^2 - \xi^2), \qquad (10.78b)$$

$$z = z. \qquad (10.78c)$$

In this coordinate system Eq. (10.33) takes the form [10.26]

$$\left[\frac{1}{\xi^2 + \eta^2}\left(\frac{\partial^2}{\partial \xi^2} + \frac{\partial^2}{\partial \eta^2}\right) + \frac{\omega^2}{c_t^2}\right]\hat{u}_z(\xi, \eta \,|\, \omega) = 0, \qquad (10.79)$$

where we have introduced the definition

$$u_z(x, y|\omega) = \hat{u}_z(\xi, \eta \,|\, \omega). \qquad (10.80)$$

Equation (10.79) can be rewritten as

$$\left[\frac{\partial^2}{\partial \xi^2} + \frac{\partial^2}{\partial \eta^2} + \frac{\omega^2}{c_t^2}(\xi^2 + \eta^2)\right]\hat{u}_z(\xi, \eta \,|\, \omega) = 0. \qquad (10.81)$$

We assume that the elastic medium occupies the region $0 \leq \eta \leq \eta_0$, $-\infty \leq \xi \leq \infty$ (see Fig. 10.7). Thus, we are dealing with the case in which the solid is convex toward the vacuum.

From Eq. (10.78) it is straightforward to show that the surface $\eta = \eta_0$ in parabolic cylinder coordinates corresponds to the surface

$$y = \frac{1}{2}\left(\eta_0^2 - \frac{x^2}{\eta_0^2}\right), \qquad (10.82)$$

in Cartesian coordinates, and that the elastic medium occupies the region $y \leq \frac{1}{2}[\eta_0^2 - (x/\eta_0)^2]$.

Equation (10.81) must be supplemented by a boundary condition that expresses the requirement that the surface $\eta = \eta_0$ be free of stresses. Equation (10.36) in this case becomes

$$\left.\frac{\partial}{\partial \eta}\hat{u}_z(\xi, \eta \,|\, \omega)\right|_{\eta=\eta_0} = 0. \qquad (10.83)$$

We solve Eq. (10.81) by separation of variables. Thus we write $\hat{u}_z(\xi, \eta|\omega)$ as the product

$$\hat{u}_z(\xi, \eta|\omega) = F(\xi)G(\eta), \qquad (10.84)$$

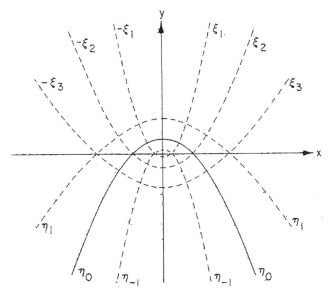

Fig. 10.7. Parabolic cylinder coordinates. This figure shows the cross sections of the surfaces of constant ξ and η. The z axis is perpendicular to the drawing. The boundary of the elastic medium is the surface $\eta = \eta_0$. The solid occupies the region $\eta < \eta_0$.

and find that $F(\xi)$ and $G(\eta)$ satisfy the equations

$$F''(\xi) + \left(\frac{\omega^2}{c_t^2}\xi^2 + \mu^2\right) F(\xi) = 0, \qquad (10.85a)$$

$$G''(\eta) + \left(\frac{\omega^2}{c_t^2}\eta^2 - \mu^2\right) G(\eta) = 0, \qquad (10.85b)$$

where μ^2 is the separation constant.

We begin by considering the solution of Eq. (10.85b). With the changes of variables

$$\eta = \left(\frac{c_t}{2\omega}\right)^{\frac{1}{2}} y, \quad G(\eta) = g(y) = g\left(\left(\frac{2\omega}{c_t}\right)^{\frac{1}{2}}\eta\right), \qquad (10.86)$$

we find that the function $g(y)$ satisfies the equation

$$\frac{d^2 g(y)}{dy^2} + \left(\frac{1}{4}y^2 - a\right) g(y) = 0, \qquad (10.87)$$

where

$$a = \frac{c_t}{2\omega}\mu^2. \tag{10.88}$$

The solutions of Eq. (10.87) are parabolic cylinder functions [10.27]. We seek a solution of this equation that increases exponentially with increasing y, as we require of a surface wave, and then becomes an oscillatory function of y so that the boundary condition (10.83) can be satisfied. The solutions of Eq. (10.87) are of exponential type for $-2\sqrt{a} < y < 2\sqrt{a}$, and are oscillatory functions of y for $|y| > 2\sqrt{a}$ [10.28]. The standard solutions of Eq. (10.87) are denoted by $W(a, \pm y)$, and are defined by [10.29]

$$W(a, \pm y) = \frac{1}{2^{3/4}}\left(\sqrt{\frac{G_1(a)}{G_3(a)}}y_1(a, y) \mp \sqrt{\frac{2G_3(a)}{G_1(a)}}y_2(a, y)\right), \tag{10.89}$$

where

$$G_1(a) = \left|\Gamma\left(\frac{1}{4} + \frac{1}{2}ia\right)\right|, \quad G_3(a) = \left|\Gamma\left(\frac{3}{4} + \frac{1}{2}ia\right)\right|, \tag{10.90}$$

and $\Gamma(x)$ is the gamma function. The functions $y_1(a, x)$ and $y_2(a, x)$ are

$$y_1(a, x) = 1 + a\frac{x^2}{2!} + \left(a^2 - \frac{1}{2}\right)\frac{x^4}{4!} + \left(a^3 - \frac{7}{2}a\right)\frac{x^6}{6!} + \cdots \tag{10.91a}$$

$$y_2(a, x) = x + a\frac{x^3}{3!} + \left(a^2 - \frac{3}{2}\right)\frac{x^5}{5!} + \left(a^3 - \frac{13}{2}a\right)\frac{x^7}{7!} + \cdots \tag{10.91b}$$

in which non-zero coefficients a_n of $x^n/n!$ are connected by

$$a_{n+2} = aa_n - \frac{1}{4}n(n-1)a_{n-2}, \tag{10.92}$$

with $a_0 = 1$ and $a_1 = 1$. Of the two standard solutions $W(a, \pm y)$ of Eq. (10.87) it is $W(a, -y)$ that increases exponentially with y until y

reaches a value of approximately $2\sqrt{a}$, at which it begins to oscillate as y increases beyond this value. The boundary condition satisfied by the function $g(y)$ then becomes

$$\frac{d}{dy}g(y)\bigg|_{y=y_0} = \frac{d}{dy}W(a,-y)\bigg|_{y=y_0} \qquad (10.93)$$

where

$$y_0 = \left(\frac{2\omega}{c_t}\right)^{\frac{1}{2}}\eta_0. \qquad (10.94)$$

This is the equation that connects the frequency ω of the wave to the parameter μ for a given value of η_0.

Equation (10.93) has to be solved numerically. However, for y in the vicinity of $2\sqrt{a}$ an approximate solution of Eq. (10.87) is

$$g(y) \cong Ai(-a^{\frac{1}{6}}(y - 2\sqrt{a})), \qquad (10.95)$$

where $Ai(z)$ is an Airy function, provided that $|y - 2\sqrt{a}| \ll 4\sqrt{a}$, a condition that can be readily satisfied. The boundary condition (10.93) now becomes

$$Ai'(-a^{\frac{1}{6}}(y_0 - 2\sqrt{a})) = 0, \qquad (10.96)$$

so that

$$a^{\frac{1}{6}}(y_0 - 2\sqrt{a}) = x_i, \qquad (10.97)$$

where we have denoted the zeros of $Ai'(x)$ by $-x_i$. The first few of them are given by Eq. (10.58). Equation (10.97) can be rearranged into

$$\frac{\omega}{c_t} = \frac{\mu}{\eta_0} + \frac{1}{2\eta_0}\left(\frac{2\omega}{c_t}\right)^{2/3}\frac{x_i}{\mu^{1/3}}, \qquad (10.98)$$

or

$$\left(\frac{\omega}{c_t}\right)_i = \frac{\mu}{\eta_0}\left[1 + \frac{x_i}{2^{1/3}(\eta_0\mu)^{2/3}}\right], \quad i = 1, 2, 3, \ldots, \qquad (10.99)$$

to lowest nonzero order in $(\eta_0\mu)^{-1}$. This result is valid provided that $(\eta_0\mu)^{2/3} \gg x_i/2^{1/3}$. It only remains to relate the parameter μ to an effective wavenumber for the wave.

To this end we now turn to the solution of Eq. (10.85a). We introduce the changes of variables

$$\xi = \left(\frac{c_t}{2\omega}\right)^{\frac{1}{2}} x, \quad F(\xi) = f(x) = f\left(\left(\frac{2\omega}{c_t}\right)^{\frac{1}{2}} \xi\right), \quad (10.100)$$

and find that the equation for $f(x)$ is

$$\frac{d^2 f(x)}{dx^2} + \left(\frac{1}{4}x^2 + a\right) f(x) = 0, \quad (10.101)$$

where a has been defined in Eq. (10.88). We seek a solution of this equation that describes a wave propagating in the positive x direction. As a is real and positive, the solutions of Eq. (10.101) are oscillatory for all x [10.28]. Two linearly independent solutions of Eq. (10.101) are $y_1(-a,x)$ and $y_2(-a,x)$, where $y_{1,2}(a,x)$ have been defined by Eqs. (10.91).

We seek the linear combination of $y_1(-a,x)$ and $y_2(-a,x)$ that has the form of a wave propagating in the positive x direction. It is found to be

$$f(x) = y_1(-a,x) + ia^{\frac{1}{2}} y_2(-a,x). \quad (10.102)$$

This is easily seen if we seek a solution of Eq.(10.101) in the form

$$f(x) = \exp[i(a_1 x + a_2 x^2 + a_3 x^3 + \cdots)], \quad (10.103)$$

and require that $a_1 = a^{\frac{1}{2}}$, since the solution of Eq. (10.101) that has the form of a wave traveling in the positive x direction when $a \gg \frac{1}{4}x^2$ is $f(x) \cong \text{const.} \exp(i\sqrt{a}x)$. On substituting Eq. (10.103) into Eq. (10.101) we obtain the equation

$$i(2a_2 + 6a_3 x + 12a_4 x^2 + 20a_5 x^3 + \cdots)$$
$$-(a_1 + 2a_2 x + 3a_3 x^2 + 4a_4 x^3 + \cdots)^2 + \frac{1}{4}x^2 + a = 0. \quad (10.104)$$

By equating to zero the coefficient of each power of x on the left-hand side of this equation we obtain for the first few coefficients a_n

$$\begin{aligned} a_1 &= a^{\frac{1}{2}}, \\ a_2 &= 0, \\ a_3 &= 0, \\ a_4 &= \frac{i}{48}, \\ a_5 &= \frac{a^{\frac{1}{2}}}{120}, \\ &\cdots \end{aligned} \qquad (10.105)$$

Thus Eq. (10.103) becomes

$$f(x) = \exp\left(-\frac{1}{48}x^4 + \frac{a}{360}x^6 + \cdots\right)$$
$$\times \exp\left[i\left(a^{\frac{1}{2}}x + \frac{a^{\frac{1}{2}}}{120}x^5 - \frac{a^{3/2}}{1260}x^7 + \cdots\right)\right]. \qquad (10.106)$$

This solution is of the form we seek. If we expand the product of exponentials in powers of x, we find that the result is that given by Eq. (10.102). With the use of Eqs. (10.100) we obtain $F(\xi)$ from the result given by Eq. (10.106) in the form

$$F(\xi) = \exp\left[-\frac{1}{12}\left(\frac{\omega}{c_t}\right)^2 \xi^4 \left(1 - \frac{1}{5}\mu^2 \xi^2 + \cdots\right)\right]$$
$$\times \exp\left\{i\mu\left[\xi + \frac{1}{30}\left(\frac{\omega}{c_t}\right)^2 \xi^5 - \frac{\mu^2}{315}\left(\frac{\omega}{c_t}\right)^2 \xi^7 + \cdots\right]\right\}. \qquad (10.107)$$

It is convenient to rewrite the solution for $F(\xi)$ in a form in which the arc length of the distance traveled by the wave appears in the exponent. The coefficient multiplying the arc length can then be identified as an effective wave number of the surface wave. Because the radius of curvature of the parabolic boundary is not a constant, we expect that this wavenumber will not be a constant as it is in the case of a shear horizontal surface wave propagating around a cylindrical boundary.

In parabolic cylindrical coordinates the fundamental metrical form is [10.30]

$$ds^2 = (\xi^2 + \eta^2)(d\xi^2 + d\eta^2). \tag{10.108}$$

If we specialize to the case that $\eta = \eta_0$, so that $d\eta = 0$, the arc length along this curve is

$$s = \int_0^\xi d\xi' (\xi'^2 + \eta_0^2)^{\frac{1}{2}}$$

$$= \frac{1}{2}\eta_0^2 \left\{ \sinh^{-1}\frac{\xi}{\eta_0} + \frac{\xi}{\eta_0}\left[1 + \left(\frac{\xi}{\eta_0}\right)^2\right]^{\frac{1}{2}} \right\}$$

$$= \eta_0 \xi \left[1 + \frac{1}{6}\left(\frac{\xi}{\eta_0}\right)^2 - \frac{1}{40}\left(\frac{\xi}{\eta_0}\right)^4 + 0\left(\frac{\xi}{\eta_0}\right)^6 \right]. \tag{10.109}$$

We next invert this result to obtain

$$\xi = \frac{s}{\eta_0} - \frac{1}{6}\frac{s^3}{\eta_0^5} + \frac{13}{120}\frac{s^5}{\eta_0^9} + O(s^6). \tag{10.110}$$

When this expression is substituted into Eq. (10.107), the result can be written as

$$F(\xi) = e^{i(k_R + ik_I)s}, \tag{10.111}$$

where

$$k_R = \frac{\mu}{\eta_0} \left\{ 1 - \frac{1}{6}\frac{s^2}{\eta_0^4} + \left[\frac{13}{120} + \frac{1}{30}\left(\frac{\omega}{c_t}\right)^2 \eta_0^4\right]\frac{s^4}{\eta_0^8} + \cdots \right\}, \tag{10.112a}$$

$$k_I = \frac{1}{2}\left(\frac{\omega}{c_t}\right)^2 \left[\frac{s^3}{\eta_0^4} - \left(\frac{2}{3} + \frac{\mu^2 \eta_0^2}{5}\right)\frac{s^5}{\eta_0^8} + \cdots \right]. \tag{10.112b}$$

The presence of an imaginary part in the effective wave number of the surface wave is due to the non-constant radius of curvature of the parabolic surface of the elastic medium. For, if we adopt the ray picture discussed at the end of Sec. 10.2.1 to describe the propagation of these surface waves around the parabolic boundary, we find that for a grazing angle of incidence and reflection χ_1 at some point on the boundary, the angle of incidence and reflection of the ray at the next point on the surface struck by it will no longer be χ_1, but will have a different value χ_2. As the ray continues to propagate along the boundary its angles of incidence and reflection will continue to change, and the points on the boundary at which it is struck by the ray are not equally spaced. Eventually a point is reached at which the angle of reflection is so large that the boundary is not struck again by the ray, which is then reflected into the interior of the elastic medium, i.e. the wave peels off from the boundary and radiates energy into the medium. The wave is attenuated thereby, not by any losses in the medium but by scattering out of the incident beam.

10.3. Surface Acoustic Waves on Rough Surfaces

We have noted earlier in this chapter that a Rayleigh surface acoustic wave is non-dispersive because there is no characteristic length in the system that supports it, namely a planar vacuum-solid interface, and that a shear horizontal surface acoustic wave does not exist in the same system. However, if the planar surface is roughened, either periodically or randomly, the Rayleigh wave becomes dispersive, and a shear horizontal surface acoustic wave can now be supported by the surface. These effects are examples of what can be termed roughness-induced dispersion.

The motivation for studies of surface acoustic waves on such rough surfaces arises in part from the technological applications of the propagation of Rayleigh waves on periodically corrugated surfaces in the acousto-electrical fields. It is characteristic of periodic structures, such as a grating ruled on a planar surface, that they cause wave slowing, and create band gaps, i.e. stop bands. The degree of this slowing, and the positions and widths of the bandgaps are useful characteristics in the design of Rayleigh wave delay lines, filters, and resonators. Moreover, periodic surface roughness can convert, in a controllable way, surface wave energy into bulk waves, and *vice versa*, and can thus serve as a surface-bulk or bulk-surface transducer..

The propagation of a Rayleigh wave on a one-dimensional periodically corrugated surface has been studied by several authors [10.31–10.35]. We are not aware of any studies of the propagation of a Rayleigh wave on a doubly periodic surface. Rayleigh waves on one-dimensional randomly rough surfaces were studied in Refs. [10.36, 10.37], and on two-dimensional randomly rough surfaces in Refs. [10.37–10.44]. These waves have their frequencies shifted (depressed) from their values for a Rayleigh wave on a planar surface. They are also attenuated as they propagate along the surface due to their roughness-induced scattering into bulk acoustic waves in the solid, and into other Rayleigh waves. The former is the dominant attenuation mechanism. In Refs. [10.38–10.40] only the attenuation of a Rayleigh wave was calculated. In Refs. [10.41–10.44] the roughness-induced shift in the frequency of a Rayleigh wave, as well as its attenuation, was calculated.

A primary reason for the study of Rayleigh waves on randomly rough surfaces is that surface roughness appears to be the dominant mechanism for the attenuation of Rayleigh waves [10.45]. However, large amplitude random roughness is difficult to treat theoretically, so that all studies of it until now have been perturbative in nature, which implies small-amplitude, small slope, roughness.

Surface acoustic waves of shear horizontal polarization on one-dimensional periodically corrugated surfaces have been studied by many authors [10.37, 10.46–10.49]. Their propagation on doubly periodic surfaces does not appear to have been studied until now.

The study of shear horizontal surface acoustic waves on periodically corrugated surfaces has been motivated in part because they constitute a new type of surface acoustic wave that does not exist on a planar surface, and in part because they display a significant wave slowing, which can be useful in technological applications.

The properties of shear horizontal surface acoustic waves on one-dimensional randomly rough surfaces have been studied theoretically in Refs. [10.36, 10.37] and [10.50]. These waves are dispersive, display wave slowing, and are attenuated due to their roughness-induced scattering into bulk waves in the solid. The frequency shift and attenuation rate of shear horizontal surface acoustic waves on a two-dimensional randomly rough surface were calculated in Refs. [10.37, 10.51]. Some errors in Ref. [10.51] were corrected in Ref. [10.37].

In this section we discuss the properties of surface acoustic waves on periodic and randomly rough surfaces. We restrict ourselves, as in the preceding sections of this chapter, to surface waves of shear horizontal polarization on surfaces defined by a one-dimensional surface profile function. The results we obtain display the general features found in the dispersion curves of surface acoustic waves of sagittal or shear horizontal polarization, on two-dimensional randomly rough surfaces, their determination is simpler than for two-dimensional rough surfaces, and since surface waves of this polarization do not exist on a planar surface, their existence is a particularly dramatic consequence of surface roughness for surface acoustic waves.

We begin by considering the general problem of the propagation of a shear horizontal surface acoustic wave on the surface of an isotropic elastic medium, characterized by a mass density ρ and a speed of transverse sound c_t that occupies the region $z > \zeta(x)$. Its sagittal plane is the xz plane. The region $z < \zeta(x)$ is vacuum. The surface $z = \zeta(x)$ is assumed to be stress-free. The surface profile function $\zeta(x)$ is assumed to be a single-valued function of x that is differentiable.

The elastic displacement field in the region $z > \zeta(x)$ in this case has the form

$$\mathbf{u}(\mathbf{x};t) = (0, u_y(x,z|\omega), 0) \exp(-i\omega t). \tag{10.113}$$

The equation of motion satisfied by $u_y(x,z|\omega)$ in this region is

$$-\omega^2 u_y = c_t^2 \left(\frac{\partial^2}{\partial x^2} + \frac{\partial^2}{\partial z^2} \right) u_y. \tag{10.114}$$

The stress-free boundary condition at the surface $z = \zeta(x)$ can be written as

$$\left[-\zeta'(x)\frac{\partial}{\partial x} + \frac{\partial}{\partial z} \right] u_y \bigg|_{z=\zeta(x)} = 0. \tag{10.115}$$

In addition, we require that $u_y(x,z|\omega)$ vanish as $z \to \infty$.

The solution of Eq. (10.114) in the region $z > \zeta(x)_{\max}$ that satisfies the boundary condition at infinity can be written as

$$u_y(x,z|\omega) = \int_{-\infty}^{\infty} \frac{dp}{2\pi} A(p,\omega) \exp[ipx - \beta(p,\omega)z], \tag{10.116}$$

where

$$\beta(p,\omega) = [p^2 - (\omega/c_t)^2]^{\frac{1}{2}}, \quad \mathrm{Re}\,\beta(p,\omega) > 0, \ \mathrm{Im}\,\beta(p,\omega) < 0. \tag{10.117}$$

We next invoke the Rayleigh hypothesis and use the representation (10.116) in satisfying the boundary condition (10.115). This yields a homogeneous integral equation satisfied by the amplitude function $A(p,\omega)$:

$$\int_{-\infty}^{\infty} \frac{dp}{2\pi}[-ip\zeta'(x) - \beta(p,\omega)]\exp[ipx - \beta(p,\omega)\zeta(x)]A(p,\omega) = 0. \tag{10.118}$$

We now introduce the representations

$$\exp[-\gamma\zeta(x)] = \int_{-\infty}^{\infty} \frac{dQ}{2\pi}\hat{I}(\gamma|Q)\exp(iQx), \tag{10.119}$$

$$\zeta'(x)\exp[-\gamma\zeta(x)] = -\frac{i}{\gamma}\int_{-\infty}^{\infty} \frac{dQ}{2\pi}Q\hat{I}(\gamma|Q)\exp(iQx), \tag{10.120}$$

where

$$\hat{I}(\gamma|Q) = \int_{-\infty}^{\infty} dx\,\exp[-\gamma\zeta(x)]\exp(-iQx). \tag{10.121}$$

On substituting Eqs. (10.119) and (10.120) into Eq. (10.118), the latter becomes

$$\int_{-\infty}^{\infty} \frac{dq}{2\pi}\exp(iqx)\int_{-\infty}^{\infty}\frac{dp}{2\pi}\frac{\hat{I}(\beta(p,\omega)|q-p)}{\beta(p,\omega)}[pq - (\omega/c_t)^2]A(p,\omega) = 0. \tag{10.122}$$

When we equate to zero the qth Fourier coefficient on the left-hand side of Eq. (10.122), and then interchange the roles of q and p, we obtain the integral equation satisfied by $A(p,\omega)$ in the form

$$\int_{-\infty}^{\infty} \frac{dq}{2\pi}\frac{\hat{I}(\beta(q,\omega)|p-q)}{\beta(q,\omega)}[pq - (\omega/c_t)^2]A(q,\omega) = 0. \tag{10.123}$$

This is the exact equation for $A(k,\omega)$ within the Rayleigh hypothesis.

It will be convenient for some purposes to remove the delta function from the function $\hat{I}(\gamma|Q)$. We do this by writing Eq. (10.121) as

$$\hat{I}(\gamma|Q) = \int_{-\infty}^{\infty} dx \exp(-iQx)\{1 + \exp[-\gamma\zeta(x)] - 1\}$$
$$= 2\pi\delta(Q) - \gamma\hat{J}(\gamma|Q), \qquad (10.124)$$

where

$$\hat{J}(\gamma|Q) = \int_{-\infty}^{\infty} dx \exp(-iQx)\frac{1 - \exp[-\gamma\zeta(x)]}{\gamma}. \qquad (10.125)$$

With the use of Eq. (10.124), we obtain the equation satisfied by $A(p,\omega)$ in the alternative form

$$\beta(k,\omega)A(k,\omega) = \int_{-\infty}^{\infty} \frac{dq}{2\pi} \hat{J}(\beta(q,\omega)|k-q)[kq - (\omega/c_t)^2]A(q,\omega). \qquad (10.126)$$

We now apply the preceding results to the cases where the surface profile function $\zeta(x)$ is a periodic function of x and where it is a random function of x.

10.3.1. *A periodic surface*

When the surface profile function is a periodic function of x with a period a, $\zeta(x+a) = \zeta(x)$, the function $\hat{I}(\gamma|Q)$, Eq. (10.121) becomes

$$\hat{I}(\gamma|Q) = \sum_{m=-\infty}^{\infty} 2\pi\delta(Q - (2\pi m/a))\hat{\mathcal{I}}_m(\gamma), \qquad (10.127)$$

where

$$\hat{\mathcal{I}}_m(\gamma) = \frac{1}{a}\int_{-\frac{1}{2}a}^{\frac{1}{2}a} dx \exp(-i(2\pi m/a)x - \gamma\zeta(x)). \qquad (10.128)$$

We also need to express the amplitude function $A(q,\omega)$ in the form

$$A(q,\omega) = \sum_{n=-\infty}^{\infty} 2\pi\delta(q - k - (2\pi n/a))A_n(k), \qquad (10.129)$$

so that the displacement field $u_y(x,z|\omega)$, Eq. (10.116), takes a form

$$u_y(x,z|\omega) = \sum_{n=-\infty}^{\infty} A_n(k) \exp[ik_n x - \beta(k_n,\omega)z], \quad (10.130)$$

that satisfies the Bloch-Floquet theorem,

$$u_y(x+a,z|\omega) = \exp(ika) u_y(x,z|\omega). \quad (10.131)$$

When Eqs. (10.127) and (10.129) are substituted into Eq. (10.123) we obtain as the equation for the coefficients $\{A_n(k)\}$

$$\sum_{n=-\infty}^{\infty} \frac{\hat{\mathcal{I}}_{m-n}(\beta(k_n,\omega))}{\beta(k_n,\omega)} [k_m k_n - (\omega/c_t)^2] A_n(k) = 0,$$

$$m = 0, \pm 1, \pm 2, \ldots \quad (10.132)$$

The dispersion relation for shear horizontal surface acoustic waves propagating normally to the grooves and ridges of a periodically corrugated surface is obtained by equating to zero the determinant of the matrix of coefficients in Eq. (10.132).

As in the case of surface plasmon polaritons propagating on a periodic corrugated vacuum-metal interface, discussed in Sec. 7.3.1, the solutions $\omega(k)$ of the dispersion relation obtained from Eq. (10.132) are periodic functions of k with a period $2\pi/a$, and are even functions of k. True surface waves are found only in the non-radiative region of the (ω,k) plane bounded from the left by the dispersion curve of bulk transverse waves, $\omega = c_t k$, and from the right by the boundary of the first Brillouin zone of the periodic surface, $k = \pi/a$.

Dispersion curves obtained by a numerical solution of the dispersion relation have been calculated for a sinusoidal surface profile function

$$\zeta(x) = \zeta_0 \cos(2\pi x/a). \quad (10.133)$$

The function $\hat{\mathcal{I}}_m(\gamma)$ defined by Eq. (10.128) is given by

$$\hat{\mathcal{I}}_m(\gamma) = (-1)^m I_m(\zeta_0 \gamma), \quad (10.134)$$

where $I_m(z)$ is a modified Bessel function of the first kind and order m.

By restricting m and n to run from $-N$ to N the infinite determinant formed from the coefficients in Eq. (10.132) was replaced by the determinant of a $(2N+1) \times (2N+1)$ matrix. A value of k was selected and the value of the determinant was calculated as ω was increased from 0 to $c_t k$ in small increments $\Delta \omega$. A change in sign of the determinant was the signal of a zero at that value of ω. A plot of the dispersion curve was generated in this way. The convergence of the solution was tested by increasing N and seeing if it approached a stable limiting value.

We recall that a surface acoustic wave of shear horizontal polarization does not exist on a planar surface. A planar surface supports a surface-skimming bulk transverse wave whose dispersion relation is $\omega = c_t k$, but this is not a true surface wave. The existence of a surface wave of this polarization on a periodically corrugated surface is due entirely to the corrugation of the surface.

In Fig. 10.8(a) we plot this dispersion curve for the case that $\zeta_0/a = 0.5$ and $c_t = 3 \times 10^3$ ms^{-1}. It consists of a single branch that is tangent to the dispersion curve of the surface skimming bulk transverse waves as $k \to 0$, and displays wave slowing as k increases. The entire range of frequencies $\omega > \omega(\pi/a)$ is a stop band for shear horizontal surface acoustic waves in this case.

As the ratio ζ_0/a increases, this branch is shifted to lower frequencies, and at a critical value $\zeta_0/a = 0.58$ a second, higher frequency, branch enters the non-radiative region of the (ω, k) plane at the point $[c_t(\pi/a), \pi/a]$. With a further increase of ζ_0/a both branches shift to lower frequencies. In Fig. 10.8(b) the two branches of the dispersion curve are plotted for the case $\zeta_0/a = 1.0$.

The results presented in Figs. 10.8(a) and 10.8(b) were obtained by the use of determinants no larger than 43×43.

These results are in agreement with those obtained by Baghai-Wadji and Maradudin [10.49], which were obtained by a different approach for surface acoustic waves of shear horizontal polarization propagating perpendicularly to the grooves of a lamellar grating ruled on the surface of a cubic elastic medium.

These results also indicate that the Rayleigh hypothesis can be used in numerical studies of the propagation of surface acoustic waves

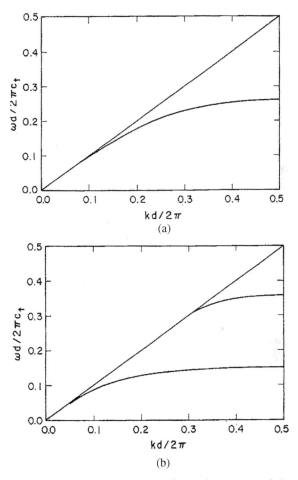

Fig. 10.8. Branches of the dispersion curve for surface waves of shear horizontal polarization propagating normally to the grooves and ridges of the grating defined by Eq. (10.133). An isotropic elastic medium characterized by a speed of transverse sound $c_t = 3 \times 10^3 \text{ms}^{-1}$ has been assumed. (a) $\zeta_0/a = 0.5$; (b) $\zeta_0/a = 1.0$. [10.35].

on periodically corrugated surfaces that are significantly rougher that those for which the Rayleigh hypothesis is rigorously valid. For the sinusoidal profile defined by Eq. (10.133) the Rayleigh hypothesis is expected to be valid only as long as $\zeta_0/a < 0.07126$ [10.35]. It is argued that the convergence of the results obtained for values of ζ_0/a as large as 1.0 is asymptotic in nature [10.52, 10.53]. By this is meant that as more terms are kept in the expansion (10.130), i.e. as N is increased, the frequencies of the surface acoustic wave approach limiting values, only to diverge from them as the number of plane waves increases beyond some critical value. Thus, in the calculations that produced Fig. 10.8 the number of plane waves used was increased until the difference in going from $\mathcal{N}$ to $\mathcal{N}+2$ plane was not smaller than it was in going from $\mathcal{N}-2$ to $\mathcal{N}$ plane waves. The results obtained with $\mathcal{N}$ plane waves were used in obtaining Fig. 10.8.

10.3.2. *A randomly rough surface*

In discussing the propagation of a surface acoustic wave of shear horizontal polarization on a one-dimensional randomly rough surface we assume that the surface profile function $\zeta(x)$ possesses the statistical properties described in Sec. 7.3. The starting point for this discussion is Eq. (10.126). In the small roughness limit, which is defined by the approximation $\hat{J}(\beta(q,\omega)|k-q) \cong \hat{\zeta}(k-q)$, this equation becomes

$$\beta(k,\omega)A(k,\omega) = \int_{-\infty}^{\infty} \frac{dq}{2\pi} \hat{\zeta}(k-q)[kq - (\omega/c_t)^2]A(q,\omega). \quad (10.135)$$

As in Sec. 7.3.2 we seek the equation satisfied by the mean wave propagating on the random surface. To this end we operate on both sides of Eq. (10.135) with the smoothing operator P introduced in Sec. 7.3.2 and obtain

$$\beta(k,\omega)PA(k,\omega)$$
$$= \int_{-\infty}^{\infty} \frac{dq}{2\pi} P\hat{\zeta}(k-q)[kq - (\omega/c_t)^2][PA(q,\omega) + QA(q,\omega)]$$
$$= \int_{-\infty}^{\infty} \frac{dq}{2\pi} P\hat{\zeta}(k-q)[kq - (\omega/c_t)^2]QA(q,\omega). \quad (10.136)$$

We next apply the complementary operator $Q = 1 - P$ to both sides of Eq. (10.135), with the result that

$$\beta(q,\omega)QA(q,\omega)$$
$$= \int_{-\infty}^{\infty} \frac{dr}{2\pi} Q\hat{\zeta}(q-r)[qr - (\omega/c_t)^2][PA(r,\omega) + QA(r,\omega)].$$
$$(10.137)$$

We wish to obtain the right-hand side of Eq. (10.136) only to $0(\zeta^2)$. Since from Eq. (10.137) $QA(q,\omega)$ is of $O(\zeta)$, we can write the solution of Eq. (10.137) as

$$QA(q,\omega) = \frac{1}{\beta(q,\omega)} \int_{-\infty}^{\infty} \frac{dr}{2\pi} \hat{\zeta}(q-r)[qr - (\omega/c_t)^2] PA(r,\omega). \tag{10.138}$$

When Eq. (10.138) is substituted into Eq. (10.136), we obtain the equation satisfied by $PA(k,\omega)$:

$$\beta(k,\omega) PA(k,\omega) = \int_{-\infty}^{\infty} \frac{dq}{2\pi} \int_{-\infty}^{\infty} \frac{dr}{2\pi} P\hat{\zeta}(k-q)\hat{\zeta}(q-r)$$

$$\times \frac{[kq - (\omega/c_t)^2][qr - (\omega/c_t)^2]}{\beta(q,\omega)} PA(r,\omega)$$

$$= \delta^2 \int_{-\infty}^{\infty} \frac{dq}{2\pi} g(k-q) \left[\frac{[kq - (\omega/c_t)^2]^2}{\beta(q,\omega)} PA(k,\omega) \right]. \tag{10.139}$$

The dispersion relation for a shear horizontal surface acoustic wave is finally obtained in the form

$$\beta(k,\omega) = \delta^2 \int_{-\infty}^{\infty} \frac{dq}{2\pi} g(k-q) \frac{[kq - (\omega/c_t)^2]^2}{\beta(q,\omega)}. \tag{10.140}$$

To solve Eq. (10.140) to obtain ω as a function of k we write

$$\frac{\omega^2}{c_t^2} = k^2 - \delta^4 \Delta^2(k), \tag{10.141}$$

so that

$$\omega(k) = c_t k \left[1 - \delta^4 \frac{\Delta^2(k)}{2k^2} \right], \tag{10.142}$$

to lowest nonzero order in δ. If we note that the departure of $(\omega/c_t)^2$ from k is of $O(\delta^4)$, the expression for $\Delta(k)$ that follows from Eq. (10.140) becomes

$$\Delta(k) = \Delta_1(k) + i\Delta_2(k), \tag{10.143}$$

where

$$\Delta_1(k) = \int_{-\infty}^{-k} \frac{dq}{2\pi} g(k-q) \frac{(kq-k^2)^2}{(q^2-k^2)^{\frac{1}{2}}}$$
$$+ \int_{k}^{\infty} \frac{dq}{2\pi} g(k-q) \frac{(kq-k^2)^2}{(q^2-k^2)^{\frac{1}{2}}}, \qquad (10.144a)$$

$$\Delta_2(k) = \int_{-k}^{k} \frac{dq}{2\pi} g(k-q) \frac{(kq-k^2)^2}{(k^2-q^2)^{\frac{1}{2}}}. \qquad (10.144b)$$

If we make the changes of variable $k - q = 2ku$ and $k - q = -2ku$ in the first and second integrals on the right hand side of Eq. (10.144a), respectively, and the change of variable $q - k = 2ku$ in the integral on the right-hand side of Eq. (10.144b), we obtain the simpler expressions

$$\Delta_1(k) = \frac{2k^4}{\pi} \left\{ \int_1^\infty du \frac{u^{3/2} g(2ku)}{(u-1)^{\frac{1}{2}}} + \int_0^\infty du \frac{u^{3/2} g(2ku)}{(u+1)^{\frac{1}{2}}} \right\},$$
$$(10.145a)$$

$$\Delta_2(k) = \frac{2k^4}{\pi} \int_0^1 du \frac{u^{3/2} g(2ku)}{(1-u)^{\frac{1}{2}}}. \qquad (10.145b)$$

For the Gaussian power spectrum $g(Q)$ given by Eq. (7.105), the expressions for $\Delta_1(k)$ and $\Delta_2(k)$ can be transformed into

$$\Delta_{1,2}(k) = \frac{2x^4}{\sqrt{\pi} a^3} d_{1,2}(x), \qquad (10.146)$$

where

$$d_1(x) = 2 \int_0^\infty d\theta \cosh^4 \theta \exp(-x^2 \cosh^4 \theta)$$
$$+ 2 \int_0^\infty d\theta \sinh^4 \theta \exp(-x^2 \sinh^4 \theta), \qquad (10.147a)$$

$$d_2(x) = 2 \int_0^{\pi/2} d\theta \sin^4 \theta \exp(-x^2 \sin^4 \theta), \qquad (10.147b)$$

and we have defined $ka = x$. The behaviors of these two functions for small x are

$$d_1(x) \sim \frac{1}{x^2} - \frac{3}{4}\ln x + O(1), \tag{10.148a}$$

$$d_2(x) \sim \frac{3\pi}{8} - \frac{35\pi}{128}x^2 + O(x^4). \tag{10.148b}$$

The frequency of the surface wave can now be obtained from Eqs. (10.142) and (10.146) in the form

$$\omega(k) = c_t k \left[1 + \frac{\delta^4}{a^4}w_1(x) - i\frac{\delta^4}{a^4}w_2(x)\right], \tag{10.149}$$

where the universal functions $w_1(x)$ and $w_2(x)$ are defined by

$$w_1(x) = -\frac{2}{\pi}x^6[d_1^2(x) - d_2^2(x)], \tag{10.150a}$$

$$w_2(x) = \frac{4}{\pi}x^6 d_1(x)d_2(x). \tag{10.150b}$$

In the long wavelength limit Eq. (10.149) becomes

$$\omega(k) = c_t k \left(1 - \frac{2}{\pi}\frac{\delta^4}{a^4}x^2\right) - ic_t k \frac{3}{2}\frac{\delta^4}{a^4}x^4. \tag{10.151}$$

Finally, the inverse decay length of the displacement field with increasing distance into the solid from the surface is obtained from Eqs. (10.117), (10.141), and (10.146) in the form

$$\beta(k,\omega(k)) = \delta^2 \frac{2x^4}{\sqrt{\pi}a^3}[d_1(x) + id_2(x)]. \tag{10.152}$$

In the long wavelength limit this expression becomes

$$\beta(k,\omega(k)) = \frac{2}{\sqrt{\pi}}\frac{\delta^2}{a^3}x^2 + i\frac{3\sqrt{\pi}}{4}\frac{\delta^2}{a^3}x^4. \tag{10.153}$$

In Fig. 10.9 we have plotted $d_1(x)$ and $d_2(x)$ as functions of x. The functions $w_1(x)$ and $w_2(x)$ are plotted in Fig. 10.10.

The functions $d_1(x)$ and $d_2(x)$ are seen to be positive for all values of x. One of the consequences of these results is that the inverse decay length of the displacement field into the solid, $\text{Re}\beta(k,\omega(k))$, is always positive. Thus, the wave is bound to the surface for all

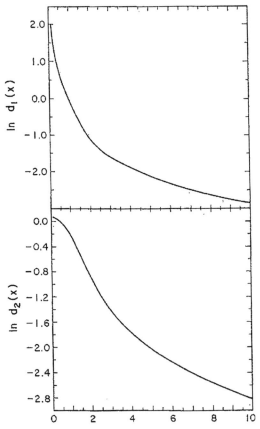

Fig. 10.9. The functions $d_1(x)$ and $d_2(x)$ defined by Eqs. (10.147) of the text [10.36].

values of $x = ka$. In the long wavelength limit $\text{Re}\beta(k,\omega(k))$ is proportional to k^2, Eq. (10.153). A second consequence of the positivity of $d_1(x)$ and $d_2(x)$ is that $\text{Im}\omega(k) = -c_t k(\delta/a)^4(4x^6/\pi)d_1(x)d_2(x)$ is negative for all x. This means that the surface wave is attenuated as it propagates along the randomly rough surface for all values of x. In the long wavelength limit $\text{Im}\omega(k)$ is proportional to k^5 which, in view of Eq. (10.151) means that it is proportional to the fifth power of its frequency. The explanation for this dependence lies in the fact that the frequency dependence of Rayleigh scattering is ω^{d+1}, where d is the dimensionality of the scatterer. The ridges and grooves responsible for the scattering of a shear horizontal surface wave in the

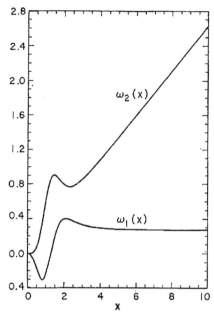

Fig. 10.10. The functions $\omega_1(x)$ and $\omega_2(x)$ defined by Eqs. (10.149) and (10.150) of the text, for shear horizontal surface acoustic waves on a random grating [10.36].

present case are two dimensional, since they are defined by the equation $x = \zeta(x)$. Thus, the Rayleigh scattering law in the present case gives us an ω^3 frequency of the scattering rate in the low-frequency, long-wavelength limit. The remaining factor of ω^2 arises because the penetration depth of this wave into the solid is proportional to the square of its wavelength parallel to the surface (see Eq. (10.153)), increasing the interaction volume thereby.

Since in the absence of surface roughness a shear horizontal surface acoustic wave cannot exist, the attenuation of the roughness-induced shear horizontal surface acoustic waves studied here is due entirely to their scattering into bulk elastic waves. These waves also display the phenomenon of wave slowing ($\omega_1(x) < 0$) for small x. However, $\omega_1(x)$ becomes positive for values of x greater than some critical value $x_c \simeq 1.5$. Thus, although in this range of x values the wave is still bound to the surface, $\mathrm{Re}\beta(k,\omega) > 0$, its phase velocity is greater than the speed of bulk transverse waves c_t, so that the surface wave turns out to be in the radiative region of the (ω, k) plane [10.37].

Experimental confirmation of the results obtained in this section has yet to be achieved.

Comments and Conclusions to Chapter 10

An approach to the calculation of properties of acoustic waves on planar surfaces of inhomogeneous semi-infinite elastic materials, where the inhomogeneities can be represented as functions of the distance z from the surface, has been presented in Sec. 10.1. This approach, which is based on the use of a coordinate-dependent mass density and coordinate-dependent elastic modulus tensor, and on an expansion of the elastic displacement field in a series of orthonormal functions, can be used to study surface acoustic waves of sagittal or shear horizontal polarization. We have illustrated it here by determining the dispersion relation and displacement field of a shear horizontal surface wave on the surface of such a medium, a wave that does not exist on the planar surface of a homogeneous medium.

Although this approach is computational in nature, the calculations required are not difficult, and it has the attractive feature that it is not tied to a particular dependence of the mass density and the elastic modulus tensor of the semi-infinite medium on the coordinate z. It is readily extended to the situation in which the surface wave propagates in an arbitrary direction on a surface of low symmetry and hence its displacement field is not perpendicular to the sagittal plane.

The method is probably best suited for problems where only the first few lowest frequency modes are of interest, because the number of terms in the Laguerre series increases as the mode number increases. If only a few modes are required greater computational efficiency will be achieved by the use of eigen-methods designed to produce just those few. The method used here is not designed to do this.

The discovery of a new type of surface wave is interesting for basic science reasons and for applications of such waves. An addition to the taxonomy of surface acoustic waves enlarges the range of possible applications of these waves, but also provides new insights into mechanisms that bind waves to surfaces.

In Sec. 10.2.1 we have shown that shear horizontal acoustic waves can propagate without attenuation circumferentially on a portion of an elastic cylinder of circular cross section, when the elastic medium is convex to the surrounding vacuum. In fact, these waves have the nature of the shear horizontal waveguide modes propagating on the planar surface of an elastic medium whose shear elastic modulus increases with increasing distance into the medium from its surface.

When the elastic medium is concave to the vacuum it supports an infinite number of leaky shear horizontal guided acoustic waves. They are attenuated as they propagate because they radiate energy into the interior of the solid, i.e. due to scattering out of the beam.

The experimental observation of these modes, and their application in devices has yet to be realized.

An isotropic elastic medium bounded by a parabolic boundary that is convex to the vacuum surrounding it is shown in Sec. 10.2.2 to support dispersive leaky surface acoustic waves of shear horizontal polarization propagating circumferentially on it. Again, we emphasize that surface acoustic waves of this polarization do not exist on the planar surface of an isotropic elastic medium. In the present case they are trapped by the curvature of the surface. This medium, by a coordinate transformation, can be shown to be equivalent to an elastic medium bounded by a planar surface, whose shear elastic modulus is a function of the coordinate z perpendicular to the planar surface that increases with increasing distance into the solid. This kind of graded medium has been shown in Sec. 10.1 to support guided waves analogous to Love waves. However, due to the variable radius of curvature of the parabolic boundary, the shear modulus of the graded medium is also a function of the coordinate x parallel to the planar surface. A surface or guided wave impinging on this kind of elastic inhomogeneity is scattered by it into the interior of the solid. This gives rise to the attenuation of the surface wave by scattering out of the incident beam, not by dissipation.

We conjecture that such waves will also exist on other curved surfaces with a variable radius of curvature. Their existence under the conditions assumed in Sec. 10.2.2 is an example of geometrical dispersion.

The result that the Rayleigh method can be used to calculate dispersion curves of surface acoustic waves on periodically corrugated surfaces defined by the surface profile function (10.133) for values of ζ_0/a significantly larger than that for which the method is rigorously convergent, in particular for values for which additional branches occur, should be tested on surfaces defined by other profile functions, to determine the range of its validity. If it is found to be valid for surface profiles other than the one given by Eq. (10.133) this will be a very useful result due to its simplicity in calculations of dispersion curves of surface acoustic waves.

The binding of a shear horizontal surface acoustic wave to a randomly rough surface is additional evidence that the surface skimming bulk transverse acoustic wave on a planar surface is unstable. A slight change in the boundary condition is enough to convert it into a surface wave.

Bibliography

[10.1] Lord Rayleigh, On waves propagated along the plane surface of an elastic solid, *Proc. London Math. Soc.* **17**, 4–11 (1887).

[10.2] A. E. H. Love, *Some Problems of Geodynamics* (Cambridge University Press, London, 1911).

[10.3] A. A. Oliner, Introduction, in *Acoustic Surface Waves*, ed. A. A. Oliner (Springer-Verlag, New York, 1978), pp. 1–12.

[10.4] L. M. Brekhovskikh, *Waves in Layered Media*, 2nd ed. (Academic Press, New York, 1980).

[10.5] W. M. Ewing, W. S. Jardetzky and F. Press, *Elastic Waves in Layered Media* (McGraw-Hill, New York, 1957).

[10.6] J. E. Gubernatis and A. A. Maradudin, A Laguerre series approach to the calculation of wave properties for surfaces of inhomogeneous elastic materials, *Wave Motion* **9**, 111–121 (1987).

[10.7] L. D. Landau and E. M. Lifshitz, *Theory of Elasticity*, 3rd ed. (Pergamon Press, New York, 1986), Secs. 10 and 22.

[10.8] D. Royer, J. M. Bonnet and E. Dieulesaint, in *The Mechanical Behavior of Electromagntic Solid Continua* ed. G. A. Maugin (North-Holland, Amsterdam, 1984), p. 23.

[10.9] M. Abramowitz and I. A. Stegun, eds., *Handbook of Mathematical Functions* (Dover, New York, 1964), Chapter 22.

[10.10] Ref. [10.9], p. 782, entry 22.7.12.

[10.11] Ref. [10.9], p. 784, entry 22.9.15.

[10.12] B. S. Garbow, J. M. Boyle, J. J. Dongarra and C. B. Moles, *Matrix Eigensystem Routines-EISPAK Guide Extension* (Springer-Verlag, New York, 1977).

[10.13] See, for example, W. Prager, *Introduction to Mechanics of Continua* (Ginn and Co., Boston, 1961), p. 91.
[10.14] Lord Rayleigh, *The Theory of Sound*, vol. II (Dover, New York, 1945), Sec. 287.
[10.15] K. Sezawa, Dispersion of elastic waves propagated on the surface of stratified bodies and on curved surfaces, *Bull. Earthquake Res. Inst. (Tokyo)* **3**, 1–18 (1927).
[10.16] J. Oliver, Rayleigh waves on a cylindrical curved surface, *Earthquake Notes* **26**, 24–25 (1955).
[10.17] I. A. Viktorov, Rayleigh-type waves on a cylindrical surface, *Soviet Physics-Acoustics* **4**, 131–136 (1958).
[10.18] O. D. Grase and R. R. Goodman, Circumferential waves on solid cylinders, *J. Acoust. Soc. Am.* **39**, 173–174 (1966).
[10.19] L. M. Brekhovskikh, Surface waves confined to the curvature of the boundary in solids, *Soviet-Physics-Acoustics* **13**, 462–472 (1968).
[10.20] B. Rulf, Rayleigh Waves on Curved Surfaces, *J. Acoust. Soc. Am.* **45**, 493–499 (1969).
[10.21] I. S. Sokolnikoff, *Mathematical Theory of Elasticity* (McGraw-Hill, New York, 1946), p. 202.
[10.22] Ref. [10.9] p. 367, entry 9.3.27.
[10.23] Ref. [10.9], p. 478, Table 10.13.
[10.24] W. Franz, Über die greensche Funktionen des Zylinders und der Kugel, *Z. fur Naturforschung* **9a**, 705–716 (1954).
[10.25] H. Margenau and G. M. Murphy, *The Mathematics of Physics and Chemistry*, 2nd ed. (D. van Norstrand, New York, 1956), Sec. 5.13.
[10.26] Ref. [10.25], Secs. 5.1, 5.2 and 5.13.
[10.27] Ref. [10.9], Chapter 19, Sec. 19.16.
[10.28] J. P. Miller, On the choice of standard solutions to Weber's equation, *Proc. Camb. Philos. Soc.* **48**, 428–435 (1952).
[10.29] Ref. [10.9], Secs. 19.16.1, 19.16.2, 19.16.3 and Secs. 19.17.1, 19.17.2 and 19.17.3.
[10.30] Ref. [10.25], Sec. 5.1.
[10.31] L. M. Brekhovskikh, Propagation of surface Rayleigh waves along the uneven boundary of a elastic body, *Soviet Physics-Acoustics* **5**, 288–295 (1960).
[10.32] P. V. H. Sabine, Rayleigh wave propagation on a periodically roughened surface, *Electron. Lett.* **6**, 149–151 (1970).
[10.33] Yu. V. Gulyaev, T. N. Kurach and V. P. Plesskii, Reflection of surface acoustic waves from a finite system of periodic perturbations, *Soviet Technical Physics Letters* **5**, 111–112 (1979).
[10.34] N. E. Glass, R. Loudon and A. A. Maradudin, Propagation of Rayleigh surface waves across a large amplitude grating, *Phys. Rev. B* **24**, 6843–6861 (1981).
[10.35] A. A. Maradudin and W. Zierau, Surface acoustic waves of sagittal and shear-horizontal polarizations on large-amplitude gratings, *Geophys. J. Int.* **118**, 325–332 (1994).

[10.36] X. Huang and A. A. Maradudin, Propagation of surface acoustic waves across random gratings, *Phys. Rev. B* **36**, 7827–7839 (1987).

[10.37] V. V. Kosachev and A. V. Shchegrov, Dispersion and attenuation of surface acoustic waves of various polarizations on a stress-free randomly rough surface of solid, *Ann. Phys. (N.Y.)* **240**, 225–265 (1995).

[10.38] A. A. Maradudin and D. L. Mills, Attenuation of Rayleigh surface waves by surface roughness, *Ann. Phys. (N.Y.)* **100**, 262–309 (1976).

[10.39] V. G. Polevoi, Scattering of surface acoustic waves by three-dimensional boundary discontinuities, *Soviet Physics — Acoustics* **29**, 52–55 (1983).

[10.40] V. V. Kosachev, Yu. N. Lokhov and V. N. Chukov, Theory of attenuation of Rayleigh surface acoustic waves on a free randomly rough surface of a solid, *Soviet Physics — JETP* **67**, 1825–1830 (1988).

[10.41] E. I. Urazakov and L. A. Fal'kovskii, Propagation of a Rayleigh wave along a rough surface, *Soviet Physics — JETP* **36**, 1214–1216 (1973).

[10.42] A. G. Eguiluz and A. A. Maradudin, Effective boundary conditions for a semi-infinite elastic medium bounded by a rough planar stress-free surface, *Phys. Rev. B* **28**, 711–727 (1983).

[10.43] A. G. Eguiluz and A. A. Maradudin, Frequency shift and attenuation length of a Rayleigh wave due to surface roughness, *Phys. Rev. B* **28**, 728–747 (1983).

[10.44] V. V. Krylov and V. E. Lyamov, Dispersion and damping of a Rayleigh wave propagating along a rough surface, *Soviet Physics — Technical Physics* **24**, 1424–1425 (1979).

[10.45] R. F. Wallis, D. L. Mills and A. A. Maradudin, Attenuation of Rayleigh waves by point defects, *Phys. Rev. B* **19**, 3981–3995 (1979).

[10.46] B. A. Auld, J. J. Gagnepain and M. Tan, Horizontal shear surface waves on corrugated surfaces, *Electron. Lett.* **12**, 650–652 (1976).

[10.47] Yu. V. Gulyaev and V. Plesskii, Slow shear surface acoustic waves in a slow wave structure on a solid surface, *Soviet Physics — Technical Physics* **23**, 266–269 (1975).

[10.48] N. E. Glass and A. A. Maradudin, Shear surface elastic waves on large amplitude gratings, *Electron. Lett.* **17**, 773–774 (1981).

[10.49] A. R. Baghai-Wadji and A. A. Maradudin, Shear horizontal surface acoustic waves on large amplitude gratings, *Appl. Phys. Lett.* **59**, 1841–1843 (1991).

[10.50] A. A. Bulgakov and S. I. Khankina, Quasi-steady-state surface acoustic waves at a rough solid surface, *Solid State Commun.* **44**, 55–57 (1982).

[10.51] O. Hardouin Duparc and A. A. Maradudin, Roughness trapped shear horizontal surface acoustic waves, *J. Elect. Spect. and Rel. Phenom.* **30**, 145–150 (1983).

[10.52] F. O. Goodman, Scattering of atoms by a stationary sinusoidal hardwall: rigorous treatment in $(n+1)$ dimensions and comparison with the Rayleigh method, *J. Chem. Phys.* **66**, 976–982 (1977).

[10.53] N. R. Hill and V. Celli, Limits of convergence of the Rayleigh method for surface scattering, *Phys. Rev. B* **17**, 2478–2481 (1978).

APPENDIX

FABRICATION OF GRADED-INDEX FILMS

Gradient-index optics has a long history that dates back to Maxwell's development of the fish-eye lens, which has a spherically symmetric index gradient that sharply images every point of a region of space [A.1], and continues with Wood's creation of a lens having two plane surfaces and an index of refraction that varies radially from a symmetry axis that is perpendicular to the surface [A.2], and Luneberg's design of a lens with a spherically symmetric index gradient that focuses every bundle of parallel rays into a point [A.3]. Despite this long history it is only since about the 1970's that methods for fabricating and characterizing gradient-index materials have been developed. It is still the case that theoretical developments in gradient-index optics outpace the development of techniques for producing good quality gradient materials.

The major applications of gradient-index media in optics at the present time are to telecommunications in the form of gradient-index optical fibers and gradient-index waveguides, and to imaging systems in the form of gradient-index lenses.

There are three types of index of refraction gradients. The first is the axial gradient, in which the index of refraction varies in a continuous fashion along the optical axis of a lens system. The surfaces of constant index of refraction are planes perpendicular to the optical axis. Gradient index anti-reflection coatings are examples of this type of gradient. Rugate optical filters [A.4] are usually defined as optical coatings with a continuous variation of refractive index in the direction perpendicular to the plane of the film, i.e. with an axial index gradient. Some authors reserve this name for filters with a sinusoidal or more generally periodic variation of the refractive index in the direction normal to the plane of the film.

The second type of gradient is a cylindrical gradient, in which the index of refraction varies continuously with distance from the optical axis. The surfaces of constant refractive index in this case are cylinders whose axis is the optical axis. An example of an optical system that is based on this type of gradient is the gradient-index fiber, in which the index of refraction varies radially from its center, so that it is larger along the center of the fiber than it is at its surface.

The third type of gradient is the spherical gradient, in which the index of refraction is symmetric about a point. The surfaces of constant refractive index in this case are spheres. The center of symmetry of the gradient need not coincide with the center of curvature of the surface. The Maxwell fish-eye lens [A.1] and the Luneberg lens [A.3] are examples of gradient-index optical systems in which the centers of symmetry of the gradient and of the curvature of the surface coincide.

These three types of refractive index gradients are discussed in detail in the book by Marchand [A.5], which is devoted primarily to the application of gradient-index media to the creation of gradient-index lenses. Axial gradient media in the form of thin films and coatings are treated in the books by Willey [A.6] and Baumeister [A.7]. The topics covered in these books include how to design a gradient-index optical system that, for example, acts as an anti-reflection coating over some range of wavelengths of the incident light; ray-tracing and other approaches to the determination of the paths light rays follow in propagating through an inhomogeneous medium; how to determine the index profile experimentally given a gradient-index medium; and methods for producing gradient-index materials. However, there is little discussion in these books and in the periodical literature about the aspect of gradient-index optics of greatest interest in the context of this book, namely how to produce a gradient-index system with a *specified* continuous spatial variation of the index of refraction.

Of the three types of index of refraction gradients described above, it is the axial gradient that is most relevant to the wave effects discussed in this book. In this section we present brief descriptions of several approaches that offer the possibility of fabricating samples with specified axial index gradients together with examples of films prepared by them.

A.1. Co-Evaporation

A continuous change of the chemical composition of a film with position in it changes its electrical, mechanical, and optical properties. A smooth graded refractive index change is required for the fabrication of rugate filters for use in optical applications. It can be achieved by producing a film with a prescribed variation of composition within it.

Such a film can be fabricated by the coevaporation of materials of low and high refractive indices by the use of two thermal sources [A.8]. Thus, Boivin and St.-Germain [A.9] produced inhomogeneous films with a graded index of refraction designed to serve as broadband or narrowband filters, i.e. to possess a given spectral reflectivity. The approach to producing such a film was based on the result that under some simple assumptions the logarithm of the index profile with respect to the optical thickness is the Fourier transform of the reflection amplitude of the film [A.9]. The optical thickness γ is defined by

$$\gamma = 2 \int_0^z n(z')dz', \qquad (A.1)$$

where $n(z)$ is the refractive index profile as a function of the mechanical thickness of the film measured from the substrate. These films were prepared by the coevaporation of lead chloride ($n = 2.18$) and cryolite ($n = 1.35$) inside a vacuum chamber where the pressure was lower than 10^{-8} Torr. The rate of deposition of the low-index material was kept constant, while the rate of deposition of the high-index material was controlled by an automated shutter that delivered a rate of deposition proportional to a preset reference function. The reference function corresponding to a desired index profile was stored in a memory, and as the deposition proceeded the rate controller delivered the appropriate amount of material in the mixture to produce the desired refractive index. The reference function was determined in the following way.

The index of refraction of the mixture was calculated as a function of the volume fraction C of the dispersed material by the use of the Maxwell Garnett formula extended for ellipsoidal particles [A.10, A.11]

$$\frac{\epsilon - \epsilon_H}{L\epsilon + (1-L)\epsilon_H} = C \frac{\epsilon_d - \epsilon_H}{L\epsilon_d + (1-L)\epsilon_H}. \quad (A.2)$$

Here ϵ is the dielectric constant of the mixture, ϵ_D is the dielectric constant of the dispersed material, ϵ_H is the dielectric constant of the host material, L is the shape factor of an ellipsoid, and C is the volume fraction.

When the cryolite was deposited at a rate of 0.5 nm/s, the shape factor for lead chloride was found to be $L_p = 0.78$ for $0 < C < 1$.

Since the refractive index is known as a function of the mechanical thickness z, it can be re-expressed as a function of time. Then Eq. (A.2) yields the volume fraction C as a function of time. If r_c and r_p are the rates of deposition of the cryolite and the lead chloride, respectively, one obtains

$$r_p = \frac{Cr_c}{1-C}, \quad (A.3)$$

for the rate of deposition of lead chloride as a function of time. It is this function that was stored in memory.

Coevaporation has been used to produce gradient index profiles of a variety of inhomogeneous films, for example, mixtures of Ge with MaF_2, CeF_3, ZnS, and CdTe [A.12], mixtures of Na_3AlF_6 and $PbCl_2$ and of MgF_2 and $PbCl_2$ [A.11], mixtures of SiO_2 and TiO_2 [A.13], and mixtures of Ge and ThF_2 [A.14].

A positive feature of this approach to the fabrication of graded index films is that large surface area films can be produced by its use.

A.2. Physical and Chemical Vapor Deposition

Several vapor deposition methods have been developed over the years for the creation of films on substrates. They are described briefly in this section and in the next.

In physical vapor deposition (PVD) [A.15], the plate to be coated is placed at the top of a bell jar that is then sealed and evacuated. An evaporation source is also present in the bell jar below the plate. It is electrically heated. The vapor emitted by the evaporation source rises and impinges on the plate, where it sublimes to form a solid film with a composition that has essentially the same stoichiometry as the evaporation source.

An advantage of using physical vapor deposition to produce a film coating a substrate is that relatively small amounts of material may be evaporated. This is important when expensive materials such as gold or rhodium are to be deposited.

Physical vapor deposition has its drawbacks, however. The substrates are usually heated to make the deposited film mechanically hard. Substrates such as plastics cannot survive under such treatment. Moreover, it is difficult and expensive to evaporate downward or sideways. The upward evaporation used requires that expensive tooling must be created to hold the plates to be coated.

A second deposition technique that is used for coating a substrate is chemical vapor deposition (CVD) [A.16–A.18]. In this method the substrate that is to be coated is exposed to one or several vaporized compounds or reagent gases, some or all of which contain constituents of the material to be deposited. A chemical reaction is then initiated, often by the application of heat. This reaction preferably occurs near or on the substrate surface, and produces the material to be deposited as a solid-phase reaction product that condenses on the substrate. The reaction often produces gases, e.g. HCl, CO, and H_2, which are then removed from the deposition chamber. By adjusting the deposition conditions so that the reaction takes place near or on the substrate surface (a heterogeneous reaction) the formation of a powdery deposit, which results if the reaction takes place in the gas phase (a homogeneous reaction), is avoided.

Chemical vapor deposition is similar to physical vapor deposition in that in both methods the deposit is formed from a vapor phase. The main difference between these two deposition methods is that in chemical vapor deposition the formation of the deposit occurs due to a chemical reaction near or on the substrate surface, and does not involve a mean free path of the gas molecules that is larger than or comparable with the dimensions of the deposition chamber as a major necessity for the functioning of the deposition process [A.19]. Moreover, chemical vapor deposition may be carried out at low pressures or in a high vacuum, depending on the structure of the deposit one desires or on effectively transporting the reactant species to or from the substrate.

Advantages of chemical vapor deposition are that it produces a coating of uniform thickness, if the temperature in the deposition chamber is relatively uniform, and a greater packing density than is achieved with physical vapor deposition. The greater packing density results in a layer of higher refractive index than is produced by physical vapor deposition.

Disadvantages of chemical vapor disposition are that usually higher substrate temperatures are required than for physical vapor deposition; the layers produced are usually mechanically stressed, which can limit the thickness of the coating; the reactive gases used in the deposition process and their reaction products are often highly toxic, explosive, or corrosive; and the uniformity of the layer is often hard to control.

Physical and chemical vapor deposition are generally not used to produce films with a graded index of refraction. We have described them in this section as an introduction to another vacuum deposition approach that has been, and is being used for the fabrication of compositionally varying thin films, and hence films with a spatially varying refractive index. This method is plasma-enhanced chemical vapor deposition, and we now turn to a description of it.

A.3. Plasma-Enhanced Chemical Vapor Deposition (PECVD)

An approach to the fabrication of a film with a prescribed variation of composition within it that produces a given variation of its refractive index is provided by plasma-enhanced chemical vapor deposition (PECVD) [A.20–A.23]. In this approach a substrate is situated in a vacuum chamber. A gas mixture is introduced into the chamber. Electrical energy is then used to transform the gas mixture into reactive radicals, ions, neutral atoms and molecules, and other highly excited species. In earlier times this was accomplished by the use of a DC glow discharge, but in recent times RF or microwave pulses are used for this purpose. Any gas in which a significant fraction of the atoms or molecules are ionized is called a plasma. With each microwave pulse the gases in the chamber decompose and react chemically to form a solid layer on the substrate whose composition is determined by which gases have been introduced into the chamber. Again the residual gases created in this reaction are then removed from the chamber.

In plasmas with low fractional ionization the electrons are so light compared with atoms and molecules that energy exchange between the electrons and the neutral gas is very inefficient. Consequently, the electrons can be kept at a very high equivalent temperature — tens of thousands of kelvins — while the neutral atoms or molecules remain at the ambient temperature. The energetic electrons can induce many processes, such as dissociation of molecules and creation of large quantities of free radicals, that are improbable at low temperatures. Since the formation of the reactive and energetic species in the gas phase occurs by collisions in the gas phase, the substrate can be kept at a relatively low temperature, of the order of 300°C. This film formation can occur on substrates at lower temperatures than is possible by the conventional chemical vapor deposition method. The thickness of the film produced by this technique is governed by the number of rf or microwave pulses.

In an early application of PECVD to the creation of compositionally inhomogeneous dielectric films, Lim et al. [A.23] used it to realize inhomogeneous silicon oxynitride (SiON) layers. The process gases used for the growth of these layers were silane (SiH_4) diluted in helium (2.01% SiH_4 in He), nitrogen (N_2), and nitrous oxide (N_2O) diluted in helium (5% N_2O in He). Nitrous oxide is very reactive compared with nitrogen. Consequently small variations of the N_2O flow rate lead to large variations in the refractive index of the film. For this reason the flow rates of SiH_4 and N_2, as well as the RF power, chamber pressure, and substrate temperature, were kept constant, only the N_2O flow rate was subject to real time control, by the use of a programmable microprocessor. The realization of a given compositional profile was achieved by the use of a calibration chart in which the refractive index and deposition of uniform layers of SiON are plotted as functions of the N_2O/N_2 flow rate. Linearly graded SiON layers with refractive indices varying from 1.46 to 2.05 were designed and fabricated in this manner on silicon substrates. The compositional profile was analyzed by Auger electron spectroscopy sputter profiling.

In the work of Lim et al. [A.23] as well as in that of Greenham et al. [A.24], rugate filters of silicon oxynitride were fabricated on substrates of silicon and silica, respectively, by PECVD.

An extensive review of the deposition of transparent dielectric optical films and coatings by PECVD has been written by Martinu and Poitras [A.25]. Included in this review is a discussion of the fabrication of graded index films.

The benefits of using PECVD in producing gradient-index films include lower deposition temperatures than are required in co-evaporation, a rapid deposition rate, and the production of mechanically hard and dense films.

There are drawbacks to the use of PECVD for the preparation of films with a graded index of refraction. The number of parameters involved in this technique is large, which makes the deposition difficult to control. Moreover, the substrate temperature can be high enough to be incompatible with important optical materials, such as the polycarbonates, which decompose at relatively moderate temperatures. In some versions of a plasma enhanced chemical vapor deposition reactor, the microwave field must penetrate the substrate. Such an apparatus could not be used to coat a metallic substrate.

A variant of PECVD is electron cyclotron resonances plasma-enhanced chemical vapor deposition (ECRPECVD) [A.26]. In this approach the ionized plasma is produced by superimposing a static magnetic field, which causes the electrons to move in circular orbits at an angular frequency called the cyclotron frequency, and a high-frequency electromagnetic field at the electron cyclotron frequency. The energy added to the electrons through their motion in cyclotron orbits increases their effective temperature. The collisions of these energetic electrons with the atoms or molecules in the gas mixture in the vacuum chamber lead to more ionized atoms or molecules in the plasma than is the case in the absence of the magnetic field. A higher plasma density enables a low temperature deposition process with deposition rates comparable to those achievable by chemical vapor deposition. This method has been used to fabricate graded index oxynitride films with specific linear and parabolic index profiles [A.27].

A.4. Pulsed Laser Deposition (PLD)

An approach to preparing a film with a graded index of refraction that is free from one of the drawbacks of PECVD, namely the high temperature of the substrate, is provided by pulsed laser deposition (PLD) [A.28]. In one application of this method [A.29] a high intensity excimer laser beam is incident on a sintered Si_3N_4 target, typically oriented at a 45° angle with respect to the beam, in the presence of oxygen gas. A silicon (100) wafer, positioned near and directly opposite to the target, serves as the substrate on which the laser ablated silicon nitride and the oxygen are deposited to produce a silicon oxynitride (SiO_xN_y) film. Because the plume from the ablated target strikes a region of the substrate approximately 5 millimeters in length, this substrate needs to be rotated during the deposition to produce uniformly deposited layers [A.30]. These layers have a diameter of approximately 1 cm, and their thickness decreases in the radial direction outside this region. The ablation process is controlled by means of the number of laser pulses, the output power, the kind of gas, and the gas pressure. In the work reported in Ref. [A.29] the laser fluence, number of pulses, and repetition rate were kept fixed in all depositions. To obtain different stoichiometries the partial pressure of the oxygen gas was smoothly varied. The growth of the film was monitored by a phase modulated ellipsometer at a fixed photon energy. An effective medium approximation [A.31] was used to analyze the ellipsometric equation.

In this way Machorro et al. [A.29] were able to produce inhomogeneous films of SiO_xN_y on a silicon substrate.

Pulsed laser deposition is a low temperature technique for producing such films. It does not require that the substrate be at a high temperature. However, the small sample sizes produced by this method reduces its applicability to the fabrication of practical optical filters. In addition, software for controlling the film growth process so that it produces a specified spatial variation of the film's refractive index is lacking at this time.

A.5. Graded Porosity

A porous air-glass interface reduces the index of refraction of a glass film by creating a mixture of air and glass at the interface that in turn reduces the reflection from the glass caused by the index mismatch there. Fraunhofer created porous antireflection coatings by etching a glass surface with acid [A.31].

When the porosity of a layer is structured suitably, a continuous gradient can be formed in it that can reduce the reflection even more. More generally, a continuously graded porosity can produce a smooth variation of the refractive index with distance from a surface of the kind needed for the observation of many of the effects considered in this book.

Thin films deposited at non-normal angles of incidence grow with densities lower than that of the bulk material, and at sufficiently large angles of incidence self-shadowing becomes the dominant growth mechanism, resulting in extremely porous films [A.33, A.34].

This result underlies the technique of glancing angle deposition (GLAD) for the fabrication of films with graded refractive indices (GRIN). Kennedy and Brett [A.35] exploited this property of oblique deposition, and used the angle of incidence as a means of controlling the porosity and therefore the refractive index of a dielectric coating. The density ρ of an obliquely deposited film was modeled as a function of the angle of incidence α by Tait et al. [A.36], and a similar method was used by Robbie et al. [A.37] to produce rugate filters with a sinusoidal index variation. When the density as a function of

α is inverted to obtain α as a function of the density, the following result is obtained for the angle of incidence as a function of film thickness:

$$\alpha(z) = \cos^{-1}\{[2\rho_r(z) - 1]^{-1}\}, \qquad (A.4)$$

where z is the film thickness. In their work Kennedy and Brett [A.35] chose a Gaussian profile for $\rho_r(z)$,

$$\rho_r(z) = \frac{\rho(z)}{\rho_0} = \exp\left[-m\left(\frac{z_0 - z}{z_0}\right)^2\right], \qquad (A.5)$$

where z_0 is the total thickness of the film, $\rho_0 = \rho(z_0)$, and m is a parameter that can be adjusted to obtain the profile that produces the minimum amount of reflection from the antireflection coating they were studying. Ideally, one would produce a GRIN GLAD antireflection coating by depositing the same material as the substrate to eliminate an index mismatch at the substrate interface. In fact, in their experimental work Kennedy and Brett deposited SiO_2 on a barium borosilicate glass substrate.

In fabricating the graded index film the substrate was rotated rapidly during the deposition around an axis normal to it. This produced a film with a vertical, columnar, structure of its surface profile, rather than the slanted posts that would result in the absence of the rotation. The software controlling the motion of the substrate was programmed with Eq. (A.5) for the angle of incidence, so that the porosity as a function of thickness was accurately controlled during the deposition. During the deposition the angle of incidence was varied from normal incidence to highly oblique angles according to Eq. (A.4). The deposition rate was monitored by a crystal thickness monitor, and the film thickness was obtained by integrating the rate. As the crystal thickness monitor measured the deposition rate only for deposition normal to the substrate, to obtain the film's thickness due to oblique growth at non-normal incidence, an empirical formula was used for the ratio of the deposition rate at the substrate to that at the crystal thickness monitor as a function of the angle of incidence in degrees, namely

$$\frac{R_{film}}{R_{ctm}} = 0.98 + 0.0033\alpha - 0.00014\alpha^2. \qquad (A.6)$$

This relation is specific to the system studied by Kennedy and Brett. A different dependence of the density $\rho(z)$ on z than the one given by Eq. (A.5) will lead to a different form for the right hand side of Eq. (A.6).

The transmissivity of the graded index films fabricated by the method described was measured as a function of the wavelength of the incident light in the interval from 400 nm to 1000 nm. The experimental results were compared with calculated results obtained by the use of effective medium theories applied to the known porosity of the film. Two such theories were used: the simple Drude model that is based on the assumption that the electromagnetic field in the effective medium is not affected by the polarization of the matrix medium, and the Maxwell Garnett theory [A.38, A.39] supplemented by a λ^{-4} Rayleigh factor that accounts for the increased scattering from the rough porous medium that decreases the transmissivity as the wavelength approaches the size of the porous projections of the film. Good agreement between the modified Maxwell Garnett model results and experimental data was found, suggesting that the method of fabricating the porous graded index film is able to reproduce the desired index profile.

A drawback to porous films is that the structures of the films created are delicate. Scraping the surface can break off the columnar projections that are crucial for the performance of such films as antireflection coatings. Moreover, porosity implies a rapid ambient degradation. Water, mainly from the atmosphere, enters the pores and reacts with the layer or creates a fungus. Under controlled laboratory conditions, however, where optics can be protected against rough treatment, GRIN GLAD films can provide performance that is superior to that of other types of antireflection coatings. It is also the case that such layers display notable loss in transmission when the wavelength of the incident light in the blue part of the visible spectrum, due to scattering from surface structures and finite-sized microstructures [A.35]. This loss is not significant for wavelengths greater than approximately 500 nm.

A.6. Ion-Assisted Deposition (IAD)

Situations can arise in which a graded index film with a columnar surface structure is undesirable. Several deposition techniques now exist that eliminate the formation of columnar surface structures. One such technique is ion-assisted deposition (IAD) [A.40, A.41].

In this approach a separate ion source directs a beam of ions at the growing film during the deposition process. The fact that the ion source is a separate source allows the ion energy, current density, angle of incidence, and species to be controlled independently from the material deposition process. The thin films deposited in this fashion can display an increased packing density, improved stability, and improved stoichiometry [A.40, A.42–A.45].

Although this technique has been extensively used to produce films with a good stoichiometry [A.45] or to change their index of refraction [A.41, A.43] for example, the films produced have mostly been homogeneous in their composition and hence in the dependence of their refractive indices on film thickness. Comparatively little use of this technique has been made in producing films with a graded index profile, especially films with a specified index profile. An example of the fabrication of such a film by IAD is provided by the preparation of a graded refractive film of silicon oxynitride (SiO_xN_y) by Snyder et al. [A.46]. In this work the film was deposited on a 50 nm diameter Si(100) substrate in a Balzers 760 system [A.47]. The substrate was heated to about 300°C by a quartz halogen heater. The vacuum in the deposition system was kept at a pressure of 1.50×10^{-6} Torr. High purity silicon was evaporated by an electron evaporation source. The substrate was simultaneously bombarded by nitrogen ions from a Kaufmann ion source [A.48], with beam energy and current set at 500 eV and 100 mA, respectively. The relative arrival ratio of nitrogen ions to silicon atoms was about 1:1. The composition of the film as a function of its thickness was controlled by linearly varying the ratio of nitrogen flowing into the ion source and oxygen in the backfill.

The optical properties of the film were characterized by spectroscopic ellipsometry (SE) [A.49]. In ellipsometry linearly polarized light is incident non-normally on a surface. The reflected light is generally elliptically polarized. Its polarization state is determined by the use of a polarizer prism. The ellipsometrically measured data are $\tan\psi$ and $\cos\Delta$, defined by $R_p/R_s = \tan\psi \exp(i\Delta)$, where R_p and R_s are the complex reflection coefficients for p-polarized and s-polarized light, respectively. The ellipsometrically measured spectra were analyzed with several filling models that were constructed on the basis of the variation of the film's composition profile measured by sputter depth profiling by Auger electron spectroscopy. These filling models were produced by the use of the Bruggeman effective medium approximation [A.50], and the assumption that SiO_xN_y is a physical mixture of two distinct phases, silicon dioxide (SiO_2) and silicon nitride (Si_3N_4). To take into account that the film produced in the manner described was heterogeneous, two fitting models were constructed to analyze the SE data. In the first a simple linear decrease of the relative volume fraction of Si_3N_4 from 100% at the film/substrate interface to an unknown value at the surface of the film was assumed. In the second model another linearly graded interfacial layer was added near the substrate, which was less steeply graded than the layer above it. Each of the two graded layers was subdivided into many (~ 10) sublayers of equal thickness. Each sublayer had a homogeneous composition, and the composition from one sublayer to the next was varied in a linear staircase manner.

The experimentally determined values of the ellipsometric parameters ψ and Δ were numerically fitted by the results of calculations of the reflectivities of p- and s-polarized light from each of the two models of the inhomogeneous deposited layer. The fitting parameters were the thickness of each linearly graded layer, and its initial and final layer volume fractions. The refractive indices of SiO_2, Si_3, N_4, and the Si substrate were taken to be the bulk values given in the literature [A.52]. The quality of a fit was judged quantitatively by the biased estimator σ defined by

$$\sigma = \frac{1}{N} \sum_{i=1}^{N} \left[\frac{(\psi_i^m - \psi_i^c)^2}{\delta\psi_i^2} + \frac{(\Delta_i^m - \Delta_i^c)^2}{\delta\Delta_i^2} \right], \qquad (A.7)$$

where N is the total number of measurements, and the superscripts m and c denote the measured and calculated data. The experimental errors in ψ_i and Δ_i are given by $\delta\psi_i$ and $\delta\Delta_i$, and are estimated values.

The best fit to the experimental data was obtained from the second fitting model. The results showed that the refractive index profile of the SiO_xN_y film was graded nonuniformly, and its dependence on the distance from the substrate was determined, together with the thickness of the film.

We note that the ellipsometric technique used to characterize thin films produced by ion-assisted deposition has also been used to characterize thin films fabricated by the method of pulsed laser deposition described in Sec. A.4.

Ion assisted deposition is a low temperature method. Thus it is suitable for the preparation of graded index films from materials that decompose at elevated temperatures, such as polycarbonates.

For this method to be useful for fabricating films with the kinds of graded index profiles that are investigated in this book, a method must be devised to control the ion beam and the material deposition process so that the film produced has a specified graded refractive index profile. This has not been done yet.

A.7. Sputtering

The oldest vacuum process for producing thin films is sputtering. In this method high energy (10–5000 eV) positive ions bombard a solid surface, called the target or the cathode, knocking out atoms from the surface. The ejected atoms then travel until they collide with a solid surface, the substrate, to which they give up their energy and condense, producing the film sought.

The ejection of atoms from the surface of the target by highly energetic ions is called sputtering. It is the cause of the erosion of the cathode in glow discharge. Pulker [A.52] indicates that sputtering was discovered by Grove [A.53] in 1852 and Plücker [A.54] in 1858, in gas discharge experiments. Soon after, in 1877, metal sputtering was used to deposit mirror coatings by Wright [A.55]. However, the widespread use of sputtering for the production of optical coatings did not begin until approximately the 1960s. This was likely due to the slow rates of deposition achievable by this method in the early days of its use, the difficulty of depositing insulating materials, and the rapid growth in the use of evaporation and condensation in a high vacuum for the deposition of optical films.

The technology of sputtering has evolved significantly in the past five decades, which has made it attractive for a variety of applications, including the production of optical coatings. It is beyond the scope of this Appendix to cover the subject of sputtering in any detail. We will instead describe briefly two versions of this technique for producing optical coatings, namely ion-beam sputtering and magnetron sputtering. Together with ion-assisted deposition these are the most important energetic deposition processes used in producing optical coatings today.

In ion-beam sputtering [A.56, A.57] an ion source is used to bombard a target with high energy ions. The sputtered atoms are then deposited on a substrate to produce a filter. This technique was developed at about the same time as ion-assisted deposition (Sec. A.6). However, the deposition rate is much slower than that of ion-assisted deposition. As a consequence for many applications it is not an economical process.

In magnetron sputtering [A.58–A.60] a magnetic field is used to confine a plasma above a target, which is kept at a negative voltage. This causes ions from the plasma to bombard the target, and the atoms ejected from the target are deposited on the substrate to be coated. When it is used in a suitable low pressure environment magnetron sputtering produces high quality coatings. In comparison with ion-beam sputtering it possesses a higher deposition rate and is therefore a less expensive method.

An important feature of planar magnetron sputtering methods is that the energy of the sputtered atoms arriving at the substrate is at least an order of magnitude greater than that of an atom produced by vacuum evaporation. This produces films that are physically more dense and adhere more strongly to the substrate. It is also comparatively simple to sputter mixtures of coating materials either by bombarding a target composed of a mixture or by co-bombarding two different targets.

Comments and Conclusions to the Appendix

The various wave effects discussed in the preceding chapters would be of limited interest if samples with the required continuous spatial variation of their index of refraction could not be fabricated. In this section we have described several approaches to the fabrication of films with axial index gradients. The development of these approaches has been stimulated in recent years by a renewal of interest in gradient-index thin films due to their use as optical filters, as broadband and narrow band antireflection coatings, as interfaces between two media with different refractive indices, as waveguides, etc. The approaches described do not exhaust the methods available, but serve to indicate that methods exist that offer the possibility of being used for fabricating the kinds of gradient-index media considered in the preceding chapters of this book, even if they have not been used for this purpose up to now. It is hoped that the theoretical predictions made in this book will stimulate experimental efforts to do so.

Bibliography

[A.1] J. C. Maxwell, Solutions of problems, Cambridge and Dublin Mathematical Journal VIII, 188–193 (1854); also *Scientific Papers of James Clerk Maxwell* Vol. 1, ed. W. D. Niven (Cambridge University Press. Cambridge, 1890) pp. 74–79 [reprinted by Dover, New York, 1965].
[A.2] R. W. Wood, *Physical Optics* (Macmillan, New York, 1905), p. 71.
[A.3] R. K. Luneberg, *Mathematical Theory of Optics* (University of California Press, Berkeley, 1964), Sec. 29.
[A.4] W. E. Johnson and R. L. Crane, An overview of rugate filter technology, in Technical Digest, Topical Meeting on Optical Interference Coatings (Optical Society of America, Washington, DC, 1988), pp. 118–121.
[A.5] E. W. Marchand, *Gradient Index Optics* (Academic Press, New York, 1978).
[A.6] R. R. Willey, *Practical Design and Production of Optical Thin Films*, 2nd ed. (Marcel Dekker, New York, 2002).
[A.7] P. W. Baumeister, *Optical Coating Technology* (SPIE Press, Bellingham, WA, 2004).

[A.8] R. Jacobsson, Inhomogeneous and coevaporated homogeneous films for optical applications, in *Physics of Thin Films*, Vol. 8, eds. G. Haas, M. H. Francombe, and R. W. Hoffman (Academic Press, New York, 1975), pp. 51–98.

[A.9] G. Boivin and D. St.-Germain, Synthesis of gradient-index profiles corresponding to spectral reflectance derived by inverse Fourier transform, *Appl. Opt.* **26**, 4209–4213 (1987).

[A.10] W. Southwell, Coating design using very thin high- and low-index layers, *Appl. Opt.* **24**, 457–460 (1985).

[A.11] D. St-Germain and G. Boivin, Indices de réfraction des mélanges Mg F_2-$PbCl_2$ et Na_3AlF_6-$PbCl_2$, *Can. J. Phys.* **64**, 316–319 (1986).

[A.12] H. Sankur, W. J. Gunning and J. F. DeNatale, Intrinsic stress and structural properties of mixed composition thin films, *Appl. Opt.* **27**, 1564–1567 (1988).

[A.13] W. J. Gunning, R. L. Hall, F. J. Woodberry, W. H. Southwell and N. S. Gluck, Codeposition of continuous composition rugate filters, *Appl. Opt.* **28**, 2945–2948 (1989).

[A.14] R. Bertram, M. F. Ouellette and P. Y. Tse, Inhomogeneous optical coatings: an experimental study of a new approach, *Appl. Opt.* **28**, 2935–2939 (1989).

[A.15] Ref. [A.7], p. 3–3.

[A.16] W. M. Feist, S. R. Steele and D. W. Readey, The preparation of films by chemical vapor deposition, in *Physics of Thin Films*, Vol. 5, eds. G. Haas and R. E. Thun (Academic Press, New York, 1969), pp. 237–322.

[A.17] W. Kern and V. S. Ban, Chemical vapor deposition of inorganic thin films, in *Thin Film Processes*, eds. J. L. Vossen and W. Kern (Academic Press, New York, 1978), pp. 257–331.

[A.18] J.-O. Carlsson, Chemical vapor deposition, in *Handbook of Deposition Technologies for Films and Coatings*, 2nd ed., ed. R. F. Bunshah (Noyes Publications, Park Ridge, NJ, 1994), pp. 374–433.

[A.19] Ref. [A.16], p. 238.

[A.20] Ref. [A.7], pp. 3–7.

[A.21] H. K. Pulker, *Coatings on Glass* (Elsevier, Amsterdam, 1984), p. 135.

[A.22] A. Sherman, Plasma-enhanced chemical vapor deposition, in *Handbook of Deposition Technologies for Films and Coatings*, 2nd ed., ed. R. F. Bunshah (Noyes Publications, Park Ridge, NJ, 1994), pp. 459–479.

[A.23] S. Lim, J. H. Ryu, J. F. Wager and L. M. Caras, Inhomogeneous dielectrics grown by plasma-enhanced chemical vapor deposition, *Thin Solid Films* **236**, 64–66 (1993).

[A.24] A. C. Greenham, B. A. Nichols, R. M. Wood, N. Nourshargh and K. L. Lewis, Optical interference filters with continuous refractive index modulations by microwave plasma-assisted chemical vapor deposition, *Opt. Eng.* **32**, 1018–1024 (1993).

[A.25] L. Martinu and D. Poitras, Plasma deposition of optical films and coatings: A review, *J. Vac. Sci. Technol. A* **18**, 2619–2645 (2000).

[A.26] M. Heming, J. Segner and J. Otto, Plasma impulse chemical vapor deposition, in *Optical Society of America 1992 Technical Digest*, See Vol. **15**, 296–298 (1992).

[A.27] S. Callard, A. Gagnaire and J. Joseph, Fabrication and characterization of graded refractive index silicon oxynitride thin film, *J. Vac. Sci. Technol. A* **15**, 2088–2094 (1997).

[A.28] E. Fogarassy, C. Fuchs, A. Stacui, S. de Unamuno, J. P. Stoquert, W. Marine and B. Lang, Low-temperature synthesis of silicon oxide, oxynitride and nitride films by pulsed excimer laser ablation, *J. Appl. Phys.* **76**, 2612–2620 (1994).

[A.29] R. Machorro, E. C. Samano, G. Soto, F. Villa and L. Cota-Araiza, Modification of refractive index in silicon oxynitride films during deposition, *Materials Lett.* **45**, 47–50 (2000).

[A.30] E. C. Samano, R. Machorro, G. Soto and L. Cote-Araiza, In situ ellipsometric characterization of SiN_x films grown by laser ablation, *J. Appl. Phys.* **84**, 5296–5305 (1998).

[A.31] D. E. Aspnes, Optical properties of thin films, *Thin Solid Films* **84** 249–262 (1982).

[A.32] J. Fraunhofer, Versuch über die Ursachen des anlaufens und Mattverdens des Glases und die Mittel densleben zuvorzukommenen (1817, Appendix 1819), *Gesammelte Schriften* (F. Hommel, Munich, 1888).

[A.33] K. Robbie, L. J. Friedrich, S. K. Dew, T. Smy and M. J. Brett, Fabrication of thin films with highly porous microstructures, *J. Vac. Sci. Technol. A* **13**, 1032–1035 (1995).

[A.34] K. Robbie and M. J. Brett, Sculptured thin films and glancing angles deposition: growth mechanisms and applications, *J. Vac. Sc. Technol. A* **15**, 1460–1465 (1997).

[A.35] S. R. Kennedy and M. J. Brett, Porous broadband antireflection coating by glancing angle deposition, *Appl. Opt.* **42**, 4573–4579 (2003).

[A.36] R. N. Tait, T. Smy and M. J. Brett, Modelling and characterization of columnar growth in evaporated films, *Thin Solid Films* **226**, 196–201 (1993).

[A.37] K. Robbie, A. J. P. Hnatiw, M. J. Brett, R. I. MacDonald and J. N. McMullin, Inhomogeneous thin film optical filters fabricated using glancing angle deposition, *Electron. Lett.* **33**, 1213–1214 (1997).

[A.38] J. C. Maxwell Garnett, Colours in metal glasses and in metallic films, *Philos. Trans. Roy. Soc. London* **203**, 385–420 (1904).

[A.39] J. C. Maxwell Garnett, Colours in metallic galsses, in metallic films and in metallic solutions, II. *Philos. Trans. Roy. Soc. London* **205**, 237–288 (1906).

[A.40] G. A. Al-Jumaily, J. J. McNally, J. R. McNeil and W. C. Hermann, Jr., Effect of ion assisted deposition on optical scatter and surface microstructure of thin films, *J. Vac. Sci. Technol. A* **3**, 651–655 (1985).

[A.41] J. J. McNally, K. C. Jungling, F. L. Williams and J. R. McNeil, Optical coatings deposited using ion assisted deposition, *J. Vac. Sci. Technol. A* **5**, 2145–2149 (1987).

[A.42] P. J. Martin, H. A. Macleod, R. P. Netterfield and C. G. Sainty, Ion-beamassisted deposition of thin films, *Appl. Opt.* **22**, 178–184 (1983).

[A.43] T. H. Allen, Properties of ion assisted deposited silica and titania films, *Proc. SPIE* **325**, 93–100 (1982).

[A.44] R. D. Bland, G. J. Kominiak and D. M. Mattox, Effect of ion bombardment during deposition on thick metal and ceramic deposits, *J. Vac. Sci. Technol.* **11**, 671–674 (1974).
[A.45] J. R. McNeil, A. C. Barron, S. R. Wilson and W. C. Hermann, Jr., Ion-assisted deposition of optical thin films: low energy vs high energy bombardment, *Appl. Opt.* **23**, 552–559 (1984).
[A.46] P. G. Snyder, Yi-M. Xiong, J. A. Woolam, G. A. Al-Jumaily and F. J. Gagliardi, Graded refractive index silicon oxynitride thin film characterized by spectroscopic ellipsometry, *J. Vac. Sci. Technol. A* **10**, 1462–1466 (1992).
[A.47] Balzers Group, 8 Sagamore Park Road, Hudson, NH 03051-4914.
[A.48] H. R. Kaufmann, Technology of ion beam sources, *J. Vac. Sci. Technol.* **15**, 272–276 (1978).
[A.49] P. G. Snyder, M. C. Rost, G. H. Bu-Abbud, J. A. Woollam and S. A. Alterovitz, Variable angle of incidence spectroscopic ellipsometry: Application to GaAs-Al_xGa_{1-x}As multiple heterostructures, *J. Appl. Phys.* **60**, 3293–3302 (1986).
[A.50] D. A. G. Bruggeman, Berechnung verschiedener physikalischer Konstanten von heterogenen substanzen, *Ann. Phys. (Leipzig)* **24**, 636–679 (1935).
[A.51] E. D. Palik, *Handbook of Optical Constants of Solids* (Academic, New York, 1985).
[A.52] Ref. [A.21], p. 213.
[A.53] W. R. Grove, On the electro-chemical polarity of gases, *Phil. Trans. Roy. Soc. (London)* **142**, 87–101 (1852).
[A.54] J. Plücker, Ueber die Einwirkung des magneten auf die elektrischen Entladungen in verdünnten Gasen, *Annalen der Physik und Chemie (Poggendorff)* **103**, 88–106 (1858).
[A.55] A. W. Wright, On the production of transparent metallic films by the electrical discharge in exhausted tubes, *American Journal of Science and Arts* **13**, 49–55 (1877).
[A.56] J. M. E. Harper, Ion beam deposition, in *Thin Film Processes*, eds. J. L. Vossen and W. Kern (Academic Press, New York, 1978), pp. 175–206, Sec. III.
[A.57] J. A. Thornton and J. E. Greene, Sputter deposition processes, in *Handbook of Deposition Technologies for Films and Coatings*, 2nd ed. R. F. Bunshah (Noyes Publications, Park Ridge, NJ, 1994), pp. 249–319, Sec. 3.5.
[A.58] R. K. Waits, Planar magnetron sputtering, in *Thin Film Processes* eds. J. L. Vossen and W. Kern (Academic Press, New York, 1978) 131–173.
[A.59] R. Parsons, Sputter deposition processes, in *Thin Film Processes II*, eds. J. L. Vossen and W. Kern (Academic Press, New York, 1991), pp. 177–208, Sec. II.B.
[A.60] J. A. Thornton and J. E. Greene, Sputter deposition processes, in Ref. [A.58], Secs. 3.3–3.4.

INDEX

acoustic wave, 251
adjacent films, 29
Airy function, 270, 280
anisotropic elastic medium, 254
anomalous dispersion, 21
antievanescent waves, 90
antireflection coating, 124, 304, 313
asymmetric graded index waveguides, 152
asymmetric Photonic Barriers, 47
asymmetric planar waveguides, 151
attenuation length, 192, 193
Auger electron spectroscopy, 316
autocorrelation function, 188, 189
auxiliary barrier, 210
axial index gradients, 303, 319
azimuthal surface electromagnetic waves, 167

backward waves, 22
band spectrum, 237
Bessel equation, 52, 169, 170
Bessel function, 172, 269, 270
Bloch-Floquet theorem, 183, 289
boundary condition, 158, 171, 173, 179, 238, 239, 264, 265, 272, 273, 280
Brewster effect, 130
Brillouin zone boundary, 186
broadband, 305
broadband antireflection nanocoatings, 128

Bruggeman effective medium approximation, 316
bulk-surface transducer, 284
Cauchy distribution, 62
characteristic frequency, 21
Chemical Vapor Deposition (CVD), 307, 309
circularly cylindrical interface, 153
circumferential wave, 270
co-evaporation, 305
complex reflection coefficient, 22, 316
concave profile, 20
conduction electron number density, 154
continuity condition, 22
continuous gradient, 312
convex profile, 21
corrugated conducting cylinders, 167
corrugated structure, 184
cubic crystal, 255
cubic elastic medium, 290
curved surfaces, 262
cut-off frequency, 21
cyclotron frequency, 310
cyclotron orbits, 311
cylindrical boundary, 282
cylindrical coordinates, 266
cylindrical gradient, 304
cylindrical surface, 263, 266
cylindrical vacuum-metal interface, 167, 175, 198

de Broglie waves, 67
decay, 67

degenerate perturbation theory, 186
density profile, 204
depth dependent speed, 260
dielectric-clad conducting cylinders, 167
dielectric constant, 154, 160
dielectric function, 153, 173
dielectric function with the free electron form, 195
dielectric-metal interface, 171
dielectric permittivity, 15
dielectric structure, 151
Dirac delta function, 255
discontinuities of curvature, 31
discontinuities of the gradient, 28
dispersion curve, 154, 159, 165, 174, 175, 185, 253
dispersion curve for surface waves of shear horizontal polarization, 291
dispersion relation, 152, 154, 158, 159, 163, 164, 173, 177, 183, 184, 187, 189, 191, 198, 254, 257, 271, 273, 275
dispersion relation for surface plasmon polaritons, 182
displacement field, 254, 261, 289
displacement vector, 267
dissipation, 192
double acoustic barriers, 217
Drude dielectric function, 153, 197
duality principle, 117

eddy current, 240
effective medium approximation, 311
effective refractive index, 22
effective wave number, 283
Eikhenwald, 67
elastic displacement, 254
elastic displacement field, 286, 298
elastic media, 205
elastic modulus tensor, 254, 264, 272, 298
elastic rod, 236
electric field, 154
electric field amplitude, 163, 165

electric field profiles, 154
electromagnetic acoustic transducer (EMAT), 240
electron cyclotron resonance plasma-enhanced chemical vapor deposition (ECRPECVD), 310
electronic signal processing, 253
electronic Wannier–Stark ladder, 235
ellipsometric parameters, 316
energy mean free path, 194, 196, 198
energy transport velocity, 193
ensemble of realizations, 187
equation of motion, 256, 264, 267
evanescent modes, 73

Faraday rotation, 117
Faraday's law, 241
field profiles, 152
first Brillouin zone, 184, 186
fish-eye lens, 4, 303
Fourier inversion theorem, 188
frustrated total internal reflection, 67
function $K_s(z)$, 156
fundamental metrical form, 283

gamma function, 156, 279
Gamow, 67
gap, 185, 186
Gaussian power spectrum, 194, 294
Gaussian random process, 187
generalized Fresnel formulae, 21
generating function, 17, 257, 258
geometrical dispersion, 165, 299
glancing angle deposition (GLAD), 312
Goos–Hänchen Effect, 140
graded index dielectric medium, 159, 162–164
graded index film, 310, 313, 315
graded index of refraction, 199, 311
graded index profile, 315
graded porosity, 312
graded refractive indices (GRIN), 312
graded-index optical waveguides, 165
gradient acoustic barriers, 204

gradient dispersiveless barriers, 38
gradient elastic medium, 252
gradient index profiles, 306, 315
gradient photonic barriers, 18
gradient-index lenses, 303
gradient-index media, 198, 319
gradient-index optics, 303
gradient-index waveguides, 303
group velocity, 80, 177, 186, 193, 198, 251
guided acoustic waves, 253
guided electromagnetic waves, 159
guided wave, 151, 152, 163, 165

Hankel function, 125, 170, 274
Heaviside unit step function, 254
Helmholtz equation, 68
heterogeneity-induced dispersion, 18, 153
heterogeneous reaction, 307
Hooke's law, 264

implicit functions, 57
integrated optics, 152
interface, 165
inverse decay length, 194, 196, 295
Ion-Assisted Deposition (IAD), 315, 317
ion-beam sputtering, 318
isotropic elastic medium, 266, 269, 275, 286, 291
isotropic plate, 253

$K_s(z)$, 162
$K_{is_1}(z)$, 156, 157, 164
Kronig–Penny model, 96

Laguerre functions, 261
Laguerre polynomial, 256
Laguerre series, 298
Lamé constants, 266
layered media, 253
leaky shear horizontal guided acoustic waves, 276, 299
leaky surface acoustic waves, 276
Li_2O, 152

lifetime of the energy of the wave, 184
$LiNbO_3$, 152
linearly chirped Moiré grating, 235
linearly polarized wave, 15
Lorentz force, 240
Lorentz-type oscillator, 107
Love-like modes, 261, 271
Love waves, 253, 256, 272, 299
Luneberg lens, 303, 304

magnetooptical effects, 117
magnetron sputtering, 318
mass density, 234, 237, 254, 264, 267, 286
Mathieu functions, 84
Maxwell fish-eye lens, 304
Maxwell Garnett formula, 306
Maxwell Garnett theory, 314
Maxwell's equations, 151, 152, 154
mechanical thickness, 306
modified Bessel function, 103, 156, 170, 173, 184, 289

nanostructures, 28
narrowband filters, 305
narrowed waveguide, 83
natural dispersion, 94
near infrared, 161, 165
negative magnetic permeability, 108
Neumann functions, 129
non-Fresnel reflectance, 28
non-local dispersion, 15
non-radiative region, 184, 186, 199
nonstandard eigenvalue equation, 257, 260
normal dispersion, 32
n-type semiconductor, 154

oblique propagation, 124
optical fiber, 235
optical filters, 319
optical lattice, 235
optical spectrum, 161, 165
optical thickness, 305
optical Wannier–Stark ladder, 235
oscillator strength, 106

parabolic boundary, 282, 299
parabolic cylinder coordinate system, 276, 277, 283
parabolic cylinder functions, 279
parabolic profile, 276
parabolic surface, 283
parametric representations, 55
periodic interface, 178
periodic surface, 288
periodic transmittance spectra, 78
periodically corrugated interface, 178
periodically corrugated surface, 182, 263, 284
phase, 177, 186
phase modulated ellipsometer, 311
phase shifts, 31
phase time, 81
phase velocity, 199, 251, 297
photonic crystals, 235
Physical Vapor Deposition (PVD), 307
planar asymmetric graded-index waveguide, 174
planar interface between vacuum and a homogeneous metal, 177
planar surface, 299
planar vacuum-metal interface, 186, 187, 192, 193
planar vacuum-solid interface, 284
plasma, 309
Plasma-Enhanced Chemical Vapor Deposition (PECVD), 308, 310
plasma frequency, 153, 173, 177, 178, 185, 194, 195
Poisson's ratio, 204
polariton resonances, 95
polariton gap, 106
polarization, 121
polarization — dependent tunneling, 133
power spectrum, 188, 189
Poynting vector, 71
p-polarized electromagnetic wave, 167
Pulsed Laser Deposition (PLD), 311, 312, 317

radiative mode, 186
radiative region, 184, 186
radius of curvature, 263
randomly corrugated interface, 153, 178
randomly rough surface, 182, 199, 263, 292
randomly rough vacuum-metal interface, 187, 191
ray-tracing, 304
Rayleigh factor, 314
Rayleigh hypothesis, 287, 290
Rayleigh method, 300
Rayleigh profile, 40
Rayleigh scattering, 296
Rayleigh scattering law, 297
Rayleigh waves, 251, 263, 266, 284, 285
recurrence relation, 256
reduced zone scheme, 186
reflectance spectrum, 32
reflectionless tunneling, 78
refractive index, 152
refractive index gradients, 304
refractive index profile, 305
rms height of a random surface, 188
rough surfaces, 284
roughness-induced damping constant, 195
roughness-induced dispersion, 178, 284
rugate filters, 303, 305, 310, 312

sagittal plane, 178, 251, 255, 264, 286, 298
sagittal polarization, 252
sandwich structures, 28
Schrodinger equation, 68
semiconductor superlattices, 235
separation constant, 168, 268, 278
separation of variables, 168, 172, 173, 268, 277
shear acoustic waves, 204
shear elastic modulus, 234, 237
shear horizontal polarization, 251, 298

shear horizontal surface acoustic
  wave, 264, 269, 271, 275, 282, 284,
  296
shear modulus, 299
simple free electron dielectric
  function, 173, 177, 185
single-mode waveguide, 82
single-negative metamaterial, 108
sinusoidal surface profile function,
  184, 289, 291
skin layer, 103
slab optical waveguides, 165
slab waveguides, 152
slope of the profile, 44
small roughness limit, 292
smooth transition layer, 34
smoothing operator, 190, 292
solvability condition, 171
sound phase shifter, 245
spatially varying refractive index, 308
spectral filtration, 31
spectral range, 161
spectroscopic ellipsometry (SE), 316
speed of transverse sound, 262
spherical gradient, 304
split-ring resonators, 95
S polarization, 151
S-polarized azimuthal surface
  electromagnetic waves, 167
S-polarized guided electromagnetic
  wave, 160, 161, 164
S-polarized guided plasmon
  polaritons, 171, 175
S-polarized guided waves, 153, 154
S-polarized surface electromagnetic
  wave, 165, 170
sputter depth profiling, 316
sputtering, 317
standing wave solutions, 152
standing waves, 253
stochastic function, 190
stochastic integral equation, 190
stop band, 95
strain tensor, 266
stratified elastic media, 236

stratified piezoelectric media, 236
stress tensor, 205, 254, 266
stress-free boundary condition, 253,
  255
strings, 229
substrate, 25
subwavelength slit, 82
superlattices, 109
surface acoustic wave of shear
  horizontal polarization, 290
surface acoustic waves, 251, 252, 255,
  262, 266, 276
surface electromagnetic wave, 151,
  153
surface-localized guided waves, 158
surface plasmon polariton, 177, 183,
  184, 185, 192–195, 198, 262, 289
surface plasmon polariton dispersion
  curve, 178
surface plasmon polariton on a planar
  surface, 186
surface profile function, 178, 182, 184,
  187–190, 263, 292
surface roughness, 192
surface skimming bulk shear acoustic
  wave, 276
surface skimming bulk transverse
  waves, 271, 290

torsional elastic waves, 239, 240
torsional normal modes, 240
torsional oscillations, 234
torsional vibrations, 234
torsional waves of special rods, 236
total internal reflection, 67
transfer matrix, 239
transfer-matrix method, 238
transmission coefficient, 25
transmission line, 247
transmissivity, 314
transmittance spectra, 27
transparency windows, 73
transverse and longitudinal vibration
  modes, 95
transverse correlation length, 189

trapezoidal barrier, 17
truncated profiles of gradient barriers, 49
tunneling of light, 67
tunneling of sound, 243

unharmonic periodical structure, 30

vacuum-metal interface, 153, 262
variable radius of curvature, 299
variable-density string, 231
vector potential, 18
Verdet constant, 117
visible region, 161, 165

Wannier–Stark ladder, 236, 242
wave barriers, 15
wave slowing, 186, 199, 285, 290, 297
waveguide modes, 165, 198, 199
Wentzel-Kramers-Brillouin (WKB), 38, 152
whispering gallery waves, 266, 271, 273

Young's modulus, 204

$\zeta(x)$, 187